FPGAs

World Class Designs

Newnes World Class Designs Series

Analog Circuits: World Class Designs
Robert A. Pease
ISBN: 978-0-7506-8627-3

Embedded Systems: World Class Designs
Jack Ganssle
ISBN: 978-0-7506-8625-9

Power Sources and Supplies: World Class Designs
Marty Brown
ISBN: 978-0-7506-8626-6

FPGAs: World Class Designs
Clive "Max" Maxfield
ISBN: 978-1-85617-621-7

Digital Signal Processing: World Class Designs
Kenton Williston
ISBN: 978-1-85617-623-1

Portable Electronics: World Class Designs
John Donovan
ISBN: 978-1-85617-624-8

RF Front-End: World Class Designs
Janine Sullivan Love
ISBN: 978-1-85617-622-4

For more information on these and other Newnes titles visit: www.newnespress.com

FPGAs
World Class Designs

Clive "Max" Maxfield

with

W Bolton
R.C. Cofer
André DeHon
Rick Gentile
Ben Harding
Scott Hauck
David Katz
Richard Munden
Richard Tinder
Peter R. Wilson
Bob Zeidman

AMSTERDAM · BOSTON · HEIDELBERG · LONDON
NEW YORK · OXFORD · PARIS · SAN DIEGO
SAN FRANCISCO · SINGAPORE · SYDNEY · TOKYO
Newnes is an imprint of Elsevier

ELSEVIER

Newnes

Newnes is an imprint of Elsevier
30 Corporate Drive, Suite 400, Burlington, MA 01803, USA
Linacre House, Jordan Hill, Oxford OX2 8DP, UK

Library of Congress Cataloging-in-Publication Data
Application submitted

British Library Cataloguing-in-Publication Data
A catalogue record for this book is available from the British Library.

ISBN: 978-1-85617-621-7

For information on all Newnes publications
visit our Web site at www.elsevierdirect.com

Transferred to Digital Printing, 2010

Printed and bound in the United Kingdom

Contents

Preface

I know that I'm enamored with just about everything to do with electronics and computing, but whichever way you look at them, today's Field Programmable Gate Arrays (FPGAs) *"stand proud in the crowd"* as amazingly cool devices.

I remember being a young lad at middle school in the late 1960s and early 1970s in Sheffield, Yorkshire, England. At that time, I was an avid subscriber to the electronics hobbyist magazine *Practical Electronics*. At the beginning of each month I would make a detour to the newspaper shop on the way to (and from) school every day to pester the old man behind the counter: *"Has it arrived yet?"* The day the latest issue was firmly clutched in my hot and sticky hands was a happy day indeed.

Of particular interest to me at that time was the monthly article called *Take Twenty*, which involved a project that required 20 or fewer components and cost 20 or fewer shillings (see sidebar), where this latter point was a decided advantage to a young man on a limited budget.

Pounds, Shillings, and Pence

In my younger years, the British currency system was based on the concept of *pounds, shillings*, and *pence*. This is also written as *LSD*, where the "L" comes from the Latin word *libra* and the "D" comes from the Latin word *denarius*, which was a type of coin in Roman times.

The way in which this worked was that there were 12 pennies in a shilling—also called a *bob*—and 20 shillings (240 pennies) in a pound. Then there was the *threepenny piece* that was referred to as a *threpney bit* (or *threpeni*, *thripeni*, or *thrupeni*—the actual pronunciation depended on where in the UK one lived) and the *sixpence* (worth six pennies), which was also known as a *tanner*. There was also the *florin, half-crown, double-florin, crown, half-sovereign, sovereign*, and *guinea* ... and then things started to get complicated.

It's easy to see the problems involved with such a scheme; for example, let's assume that something was priced at 1 pound, 8 shillings, and 9 pence ... take a moment to calculate how much it would cost for three of the little rascals (it's not easy, is it?). Thus, in February 1971, Great Britain retired the concept of pounds, shillings, and pence and officially adopted a decimal system in which a pound equaled 100 pennies (they were called "new pennies" at the time), much like the American dollar equals 100 cents. Strange as it may seem, however, the majority of British citizens fought this move toward decimalization tooth-and-nail claiming that the new scheme was far too complicated and would never catch on!

Now, be honest, when was the last time you discovered an interesting tidbit of trivia like this in a book on FPGAs?

It didn't matter what the project was—I was going to build it! As soon as the final bell released us from school, I'd leap on the first bus heading toward the nearest electronics hobbyist store—*Bardwells*—purchase the requisite components, and race home to start happily soldering away. [As a point of interest, although almost 40 years have passed since those far-off days, Bardwells is still around (**www.bardwells.co.uk**), although they long ago moved from their backstreet corner-shop location behind the old Abbeydale Cinema to a more prominent location on Abbeydale Road. I still drop in for "old-time's-sake" whenever I return to England to visit my mom and younger brother.]

But we digress... All of my early projects were constructed using discrete (individually packaged) components—transistors, resistors, capacitors, inductors. Later I moved to using simple "jelly-bean" integrated circuits (ICs). In the case of digital projects, these were almost invariably SN74xx-series transistor-transistor logic (TTL) parts from Texas Instruments presented in 14-pin or 16-pin dual-in-line (DIL) packages. A typical chip might contain four 2-input AND gates. We all thought these devices were amazingly clever at the time.

One of my big passions in my younger years was robots. I longed to create a machine that could "walk and talk," but this seemed like an impossible dream at the time. In the early days of digital design, control systems were essentially "hard-wired"—initially using discrete transistors and later using simple ICs. In addition to being largely fixed in function, this style of design severely limited the complexity that could be achieved in a timely and cost-effective manner. And then the microprocessor appeared on the scene...

The world's first commercially available microprocessor—the 4004 from Intel—made its debut in 1971. This was followed in November 1972 by the 8008, which was essentially an 8-bit version of the 4004. In turn, the 8008 was followed by the 4040, which extended the 4004's capabilities by adding logical and compare instructions and by supporting subroutine nesting using a small internal stack.

In reality, the 4004, 4040, and 8008 were all designed for specific applications; it was not until April 1974 that Intel presented the first true general-purpose microprocessor, the 8080. This 8-bit device, which contained around 4,500 transistors and could perform 200,000 operations per second, was destined for fame as the central processor of many of the early home computers.

Following the 8080, the microprocessor field exploded with devices such as the 6800 from Motorola in August 1974, the 6502 from MOS Technology in 1975, and the Z80 from Zilog in 1976 (to name but a few).

As you might imagine, the mid-to-late-1970s was a very exciting time for the designers of digital systems. The combination of a microprocessor with a few support chips allowed one to construct incredibly sophisticated control systems. And the great thing was that—if you discovered an error or conceived a cunning enhancement—all you had to do was to "tweak" the program that was running on the processor. But as amazing as all of this was, we had no idea of the wonders to come...

Around the beginning of the 1980s, I was designing application-specific integrated circuits (ASICs). By today's standards these were incredibly simple—as I recall, they were implemented at the 5 micron technology node and contained only around 2,000 equivalent logic gates, but this was still pretty amazing stuff at the time.

In 1984, Xilinx announced the first FPGAs (these devices didn't actually start shipping until 1985). Although they were significantly more sophisticated than the simple programmable logic devices (PLDs) available at the time, these components were seen by the majority of digital design engineers as being of interest only for implementing glue-logic, rather simple state machines, and relatively limited data processing tasks. How things have changed...

Today, FPGAs are one of the most exciting electronic device categories around. In addition to their programmable fabric—which can be used to implement anything from soft microprocessor cores to hardware accelerators—they can contain megabits of memory and hundreds of hard-macro multipliers, adders, and digital signal processing (DSP) functions. Also, in addition to thousands of programmable general purpose input/output (GPIO) pins, they can support multiple high-speed serial interconnect channels and ... the list goes on.

Different types of FPGAs are finding their way into an incredible range of applications, from portable, battery-powered, handheld devices, to automotive control and entertainment systems, to teraflop computing engines searching for extraterrestrial life as part of the SETI project.

This truly is an exhilarating time to be involved with FPGAs. Almost every day I see some announcement regarding a new family of devices or a new type of design tool or a new technique to increase processing performance or I/O bandwidth. For example, off the top of my head, I can recall the following items of interest that have crossed my desk over the last few weeks:

- Align Engineering (**www.aligneng.com**) introduced the concept of Align Locked Loops (ALLs) which allow every FPGA LVDS pair to be used as a complete SERDES (high-speed serial interconnect) solution.

- MagnaLynx (**www.magnalynx.com**) announced an extreme-performance, low pin-count interconnect technology for FPGAs and other devices used in high-speed embedded applications.

- Actel (**www.actel.com**) launched a variety of different FPGAs containing the equivalent of 15,000 logic gates for only 99 cents.

- Altera (**www.altera.com**) announced a monster member of its 65 nm Stratix III FPGA family boasting 340,000 logic elements.

- Lattice Semiconductor (**www.latticesemi.com**), which specializes in low-cost non-volatile FPGAs with high-end features, augmented its product portfolio with a selection of automotive-grade, AEC-Q100 certified devices.

- QuickLogic (**www.quicklogic.com**) launched ever-more sophisticated versions of its ultra-low-power Customer-Specific Standard Products (CSSPs) in which special selections of hard-macro functions are combined with FPGA-style programmable fabric.

- Xilinx (**www.xilinx.com**) and Synfora (**www.synfora.com**) announced an advanced algorithmic synthesis technology—PICO Express FPGA—that can accept an untimed ANSI C representation of the required algorithms and generate corresponding RTL that can be handed over to conventional synthesis and place-and-route engines (see sidebar).

Pico Express FPGA Algorithmic Synthesis Technology

In order to demonstrate the power of PICO Express FPGA, the folks at Synfora created a rather cool demo involving real-time color conversion and Sobel edge detection. As illustrated in the image below, the original untimed C-based algorithm first converts the incoming RGB video stream to YUV.

Next, a Sorbel algorithm is used to detect edges. This involves a compute-intensive convolution-based algorithm to calculate the gradients in the image and to identify edges based on the gradient difference between adjacent pixels.

Finally, RGB video is generated from the processed video stream. As seen in the illustration above, half of the image is displayed as edges, while the other half is displayed as normal video so as to provide a comparison. All of this is being performed on HD Video in real time using a 133 MHz "out-of-the-box" Xilinx Spartan 3E development board. This is very cool—I want to play with one of these little scamps myself!

And as to the future? Well, the current state-of-the-art as I pen these words[1] is for FPGAs to be created at the 65 nm technology node. By the summer of 2008, however, at least one FPGA vendor—Altera—will have announced a new family of devices implemented using a 45 nm process. These new FPGAs will set new standards for capacity and performance. For example, my understanding is that a single device will be able to implement logic equivalent to a 12-million gate ASIC, which is absolutely incredible.

Each new generation of FPGAs "opens doors" to applications that did not previously fall into the programmable logic arena. We can only imagine the types of designs that these next-generation devices will enable—and we can only imagine what new surprises and delights are in store over the coming years. I for one cannot wait! In the meantime...

... I selected the various topics in this book in order to provide a wide-ranging overview as to the types of design projects that are being implemented using FPGAs today, and also to cover associated design tools and methodologies. I really hope you enjoy reading about these subjects and—maybe,—one day, you'll be contributing some of your *own* designs to a future edition of this tome.

Clive "Max" Maxfield

---

[1] It's the 1st March 2008, which means I have to say *"White Rabbits"* because that's what everyone says to each other in England on this day — don't ask me why.

About the Editor

Clive "Max" Maxfield is six feet tall, outrageously handsome, English and proud of it. In addition to being a hero, a trendsetter, and a leader of fashion, he is widely regarded as an expert in all aspects of electronics and computing (at least by his mother).

After receiving his B.S. in Control Engineering in 1980 from Sheffield Polytechnic (now Sheffield Hallam University), England, Max began his career as a designer of central processing units for mainframe computers. During his career, he has designed everything from ASICs to PCBs and has meandered his way through most aspects of Electronics Design Automation (EDA). To cut a long story short, Max now finds himself President of TechBites Interactive (www.techbites.com). A marketing consultancy, TechBites specializes in communicating the value of technical products and services to non-technical audiences through a variety of media, including websites, advertising, technical documents, brochures, collaterals, books, and multimedia.

In addition to numerous technical articles and papers appearing in magazines and at conferences around the world, Max is also the author and coauthor of a number of books, including *Bebop to the Boolean Boogie (An Unconventional Guide to Electronics)*, *Designus Maximus Unleashed (Banned in Alabama)*, *Bebop BYTES Back (An Unconventional Guide to Computers)*, *EDA: Where Electronics Begins*, *The Design Warrior's Guide to FPGAs*, and *How Computers Do Math* (www.diycalculator.com).

In his spare time (Ha!), Max is coeditor and copublisher of the web-delivered electronics and computing hobbyist magazine *EPE Online* (www.epemag.com). Max also acts as editor for the Programmable Logic DesignLine website (www.pldesignline.com) and for the iDESIGN section of the Chip Design Magazine website (www.chipdesignmag.com).

On the off-chance that you're still not impressed, Max was once referred to as an *"industry notable"* and a *"semiconductor design expert"* by someone famous who wasn't prompted, coerced, or remunerated in any way!

About the Contributors

Bill Bolton (Chapters 11 and 12 available on line at www.elsevierdirect.com/companions/9781856176217) is the author of *Programmable Logic Controllers*. Formerly he was Head of Research, Development and Monitoring at the Business and Technician Education council (BTEC) in England, having also been a UNESCO consultant in Argentina and Thailand, a member of the British Technical Aid project to Brazil on technician education, a member of the Advanced Physics Nuffield Project for school physics, and a senior lecturer in a college. He is also the author of many successful engineering textbooks.

Robert W. Brodersen (Chapter 10) received a B.S. in both Electrical Engineering and Mathematics from California State Polytechnic University, Pomona, in 1966, and his M.S. and Ph.D. degrees in Electrical Engineering from MIT in 1968 and 1972, respectively. After spending three years with Texas Instruments in Dallas, he joined the faculty of the EECS Department at Berkeley in 1976, where he has pursued research in the areas of RF and digital wireless communications design, signal processing applications, and design methodologies. In 1994 he was the first holder of the John R. Whinnery Chair in Electrical Engineering and Computer Sciences. In 1998 he was instrumental in founding the Berkeley Wireless Research Center (BWRC), a consortium involving university researchers, industrial partners, and governmental agencies that is involved in all aspects of the design of highly integrated CMOS wireless systems. He retired in 2006 as Professor Emeritus, but remains active at BWRC, where he is Co-Scientific Director, and at the Donald O. Pederson Center for Electronics Systems Design.

R.C. Cofer (Chapter 7) is co-author of *Rapid System Prototyping with FPGAs*.

Chen Chang (Chapter 10) is a contributor to *Reconfigurable Computing*.

Rick Gentile (Chapters 8 and 9) joined Analog Devices Inc. (ADI) in 2000 as a Senior DSP Applications Engineer. He currently leads the Applications Engineering Group, where he is responsible for applications engineering work on the Blackfin, SHARC and TigerSHARC processors. Prior to joining ADI, Rick was a Member of the Technical Staff at MIT Lincoln Laboratory, where he designed several signal processors used in a wide range of radar sensors. He received a B.S. in 1987 from the University of Massachusetts at Amherst and an M.S. in 1994 from Northeastern University, both in Electrical and Computer Engineering.

Ben Harding (Chapter 7) is co-author of *Rapid System Prototyping with FPGAs*.

Scott Hauck (Chapter 10) is a Professor at the University of Washington's Department of Electrical Engineering, and an Adjunct in the Department of Computer Science and Engineering. His work is focused around FPGAs, chips that can be programmed and reprogrammed to implement complex digital logic. His interests are the application of FPGA technology to situations that can make use of their novel features: high-performance computing, reconfigurable subsystems for system-on-a-chip, computer architecture education, hyperspectral image compression, and other areas. His work encompasses VLSI, CAD, and applications of these devices.

David Katz (Chapters 8 and 9) has over 15 years of experience in circuit and system design. Currently, he is Blackfin Applications Manager at Analog Devices, Inc., where he focuses on specifying new processors. He has published over 100 embedded processing articles domestically and internationally, and he has presented several conference papers in the field. Additionally, he is co-author of *Embedded Media Processing* (Newnes 2005). Previously, he worked at Motorola, Inc., as a senior design engineer in cable modem and automation groups. David holds both a B.S. and M. Eng. in Electrical Engineering from Cornell University.

Richard Munden (Chapter 4) is a founder and CEO of Free Model Foundry and began using and managing Computer Aided Engineering (CAE) systems in 1987. He has been concerned with simulation and modeling issues since then. Richard authored *ASIC & FPGA Verification: A Guide to Component Modeling* published by Elsevier. He co-founded Free Model Foundry in 1995 and is its President and CEO. He was CAE/PCB manager at Siemens Ultrasound (previously Acuson Corp) in Mountain View, CA until 2006 when he took on full-time responsibilities at FMF. Prior to joining Acuson, he was a CAE manager at TRW in Redondo Beach, CA. He is a frequent and well-known contributor to several EDA users groups and industry conferences. Richard's primary focus over the years has been verification of board-level designs. You can read his blog at blog.FreeModelFoundry.com.

Brian C. Richards (Chapter 10) is a contributor to *Reconfigurable Computing.*

Richard Tinder (Chapter 5) holds a B.S., M.S. and Ph.D. all from the University of California, Berkeley. In the early '70s he spent one year as a visiting faculty member at the University of California, Davis, in what was then the Department of Mechanical Engineering and Materials Science. There he continued teaching materials science including solid state thermodynamics and advanced reaction kinetics. Following his return to WSU, he taught logic design and conducted research in that area until retirement in 2004. Currently, he is Professor Emeritus of the School of Electrical Engineering and Computer Science at WSU.

Peter R. Wilson (Chapters 3 and 6) is Senior Lecturer in Electronics at the University of Southampton. He holds degrees from Heriot-Watt University, an MBA from Edinburgh Business School and a PhD from the University of Southampton. He worked in the Avionics and Electronics Design Automation Industries for many years at Ferranti, GEC-Marconi and Analogy prior to rejoining academia. He has published widely in the areas of FPGA design, modeling and simulation, VHDL, VHDL-AMS, magnetics and power electronics. He is a Senior Member of the IEEE, member of the IET, and a Chartered Engineer. He is the author of *Design Recipes for FPGAs* and has contributed to *Circuit Design: Know It All* and *FPGAs: World Class Designs.*

John Wawrzynek (Chapter 10) is a contributor to *Reconfigurable Computing.*

Bob Zeidman (Chapter 2) author of *Designing with FPGAs and CPLDs* is the president of Zeidman Consulting (www.ZeidmanConsulting.com), a premiere contract research and development firm in Silicon Valley and president of Software Analysis and Forensic Engineering Corporation (www.SAFE-corp.biz), the leading provider of software intellectual property analysis tools. Bob is considered a pioneer in the fields of analyzing and synthesizing software source code, having created the SynthOS™ program for synthesizing operating systems and the CodeSuite® program for detecting software intellectual property theft. Bob is also one of the leading experts in the Verilog hardware description language as well as ASIC and FPGA design. He has written three engineering texts—*Verilog Designer's Library, Introduction to Verilog,* and *Designing with FPGAs and CPLDs*—in addition to numerous articles and papers. He teaches engineering and business courses at conferences throughout the world. Bob is an experienced and well-regarded expert in intellectual property disputes related to software source code, microprocessor design, networking, and other technologies. Bob holds four patents and earned two bachelor's degrees, in physics and electrical engineering, from Cornell University and a master's degree in electrical engineering from Stanford University.

CHAPTER 1

Alternative FPGA Architectures

Clive Maxfield

This chapter is taken from my own book, The Design Warrior's Guide to FPGAs (Devices, Tools, and Flows), *ISBN: 9780750676045. Actually, despite its title, I think it's only fair to warn you that the* Design Warrior's Guide *doesn't actually teach you how to implement designs using FPGAs; instead, it provides a detailed introduction to a wide range of generic concepts, such as what FPGAs are, how they work, how you program (configure) them, different design flows for various styles of design (schematic-based, HDL-based, C/ C++-based, DSP-based, embedded processor-based, etc.).*

Now, it may seem a little immodest for me to select a chapter from one of my own works as the opening topic in this book, but there are a plethora of different FPGA architectures available and it can be very tricky to wrap one's brain around all of the different possibilities. The situation is further confused by the fact that each FPGA vendor's website and supporting materials focuses only on their own technology and they tend to give the impression that other approaches are of no account.

In order to address this issue, this chapter ranges across the entire spectrum of potential architectures. First, we discover the differences between SRAM-based, antifuse-based, Flash-based, and hybrid (Flash and SRAM-based) technologies. Next, we consider MUX- versus LUT-based logic blocks. Then there is the use of LUTs as distributed RAMS or shift registers, the provision of fast carry chains, embedded block RAMs, embedded multipliers, DSP elements, soft and hard microprocessor cores, clock generators, high-speed serial transceivers, and . . . the list goes on.

Also tackled is the thorny question as to what is meant by the term "system gates" versus real logic gates. The answer may (or may not) surprise you . . .

—Clive "Max" Maxfield

1.1   A Word of Warning

In this chapter, we introduce a plethora of architectural features. Certain options—such as using antifuse versus SRAM configuration cells—are mutually exclusive. Some FPGA vendors specialize in one or the other; others may offer multiple device families based on these different technologies. (Unless otherwise noted, the majority of these discussions assume SRAM-based devices.)

In the case of embedded blocks such as multipliers, adders, memory, and microprocessor cores, different vendors offer alternative "flavors" of these blocks with different "recipes" of ingredients. (Much like different brands of chocolate chip cookies featuring larger or smaller chocolate chips, for example, some FPGA families will have bigger/better/badder embedded RAM blocks, while others might feature more multipliers, or support more I/O standards, or . . .)

The problem is that the features supported by each vendor and each family change on an almost daily basis. This means that once you've decided what features you need, you then need to do a little research to see which vendor's offerings currently come closest to satisfying your requirements.

1.2   A Little Background Information

Before hurling ourselves into the body of this chapter, we need to define a couple of concepts to ensure that we're all marching to the same drumbeat. For example, you're going to see the term *fabric* used throughout this book. In the context of a silicon chip, this refers to the underlying structure of the device (sort of like the phrase "the fabric of civilized society").

> The word "fabric" comes from the Middle English *fabryke*, meaning, "something constructed."

When you first hear someone using "fabric" in this way, it might sound a little snooty or pretentious (in fact, some engineers regard it as nothing more than yet another marketing term promoted by ASIC and FPGA vendors to make their devices sound more sophisticated than they really are). Truth to tell, however, once you get used to it, this is really quite a useful word.

When we talk about the *geometry* of an IC, we are referring to the size of the individual structures constructed on the chip—such as the portion of a *field-effect transistor* (FET) known as its *channel*. These structures are incredibly small. In the early to mid-1980s, devices were based on 3 µm geometries, which means that their smallest structures were 3 millionths of a meter in size. (In conversation, we would say, "This IC is based on a three-micron technology.")

> The "µ" symbol stands for "micro" from the Greek *micros*, meaning "small" (hence the use of "µP" as an abbreviation for microprocessor).

Each new geometry is referred to as a *technology node*. By the 1990s, devices based on 1 µm geometries were starting to appear, and feature sizes continued to plummet throughout the course of the decade. As we moved into the twenty-first century, high-performance ICs had geometries as small as 0.18 µm. By 2002, this had shrunk to 0.13 µm, and by 2003, devices at 0.09 µm were starting to appear.

> In the metric system, "µ" stands for "one millionth part of," so 1 µm represents "one millionth of a meter."

Any geometry smaller than around 0.5 µm is referred to as *deep submicron* (DSM). At some point that is not well defined (or that has multiple definitions depending on whom one is talking to), we move into the *ultradeep submicron* (UDSM) realm.

> DSM is pronounced by spelling it out as "D-S-M."
> UDSM is pronounced by spelling it out as "U-D-S-M."

Things started to become a little awkward once geometries dropped below 1 µm, not the least because it's a pain to keep having to say things like "zero point one three microns." For this reason, when conversing it's becoming common to talk in terms of *nano*, where one nano (short for nanometer) equates to a thousandth of a micron—that is, one thousandth of a millionth of a meter. Thus, instead of mumbling, "point zero nine microns" (0.09 µm), one can simply proclaim, "ninety nano" (90 nano) and have done with it. Of course, these both mean exactly the same thing, but if you feel moved to regale your friends on these topics, it's best to use the vernacular of the day and present yourself as hip and trendy rather than as an old fuddy-duddy from the last millennium.

1.3 Antifuse versus SRAM versus . . .

1.3.1 SRAM-Based Devices

The majority of FPGAs are based on the use of SRAM configuration cells, which means that they can be configured over and over again. The main advantages of this technique are that new design ideas can be quickly implemented and tested, while evolving standards and protocols can be accommodated relatively easily. Furthermore, when the system is first powered up, the FPGA can initially be programmed to perform one function such as a self-test or board/system test, and it can then be reprogrammed to perform its main task.

Another big advantage of the SRAM-based approach is that these devices are at the forefront of technology. FPGA vendors can leverage the fact that many other companies specializing in memory devices expend tremendous resources on *research and development (R&D)* in this area. Furthermore, the SRAM cells are created using exactly the same CMOS technologies as the rest of the device, so no special processing steps are required in order to create these components.

R&D is pronounced by spelling it out as "R-and-D."

In the past, memory devices were often used to qualify the manufacturing processes associated with a new technology node. More recently, the mixture of size, complexity, and regularity associated with the latest FPGA generations has resulted in these devices being used for this task. One advantage of using FPGAs over memory devices to qualify the manufacturing process is that, if there's a defect, the structure of FPGAs is such that it's easier to identify and locate the problem (that is, figure out what and where it is). For example, when IBM and UMC were rolling out their 0.09 μm (90 nano) processes, FPGAs from Xilinx were the first devices to race out of the starting gate.

Unfortunately, there's no such thing as a free lunch. One downside of SRAM-based devices is that they have to be reconfigured every time the system is powered up. This either requires the use of a special external memory device (which has an associated cost and consumes real estate on the board), or of an on-board microprocessor (or some variation of these techniques).

1.3.2 Security Issues and Solutions with SRAM-Based Devices

Another consideration with regard to SRAM-based devices is that it can be difficult to protect your intellectual property, or IP, in the form of your design. This is because the configuration file used to program the device is stored in some form of external memory.

IP is pronounced by spelling it out as "I-P."

Currently, there are no commercially available tools that will read the contents of a configuration file and generate a corresponding schematic or netlist representation. Having said this, understanding and extracting the logic from the configuration file, while not a trivial task, would not be beyond the bounds of possibility given the combination of clever folks and computing horsepower available today.

Let's not forget that there are reverse-engineering companies all over the world specializing in the recovery of "design IP." And there are also a number of countries whose governments turn a blind eye to IP theft so long as the money keeps rolling in (you know who you are). So if a design is a high-profit item, you can bet that there are folks out there who are ready and eager to replicate it while you're not looking.

In reality, the real issue here is not related to someone stealing your IP by reverse-engineering the contents of the configuration file, but rather their ability to clone your design, irrespective of whether they understand how it performs its magic. Using readily available technology, it is relatively easy for someone to take a circuit board, put it on a "bed of nails" tester, and quickly extract a complete netlist for the board. This netlist can subsequently be used to reproduce the board. Now the only task remaining for the nefarious scoundrels is to copy your FPGA configuration file from its boot PROM (or EPROM, E^2PROM, or whatever), and they have a duplicate of the entire design.

On the bright side, some of today's SRAM-based FPGAs support the concept of *bitstream encryption*. In this case, the final configuration data is encrypted before being stored in the external memory device. The encryption key itself is loaded into a special SRAM-based register in the FPGA via its JTAG port. In conjunction with some associated logic, this key allows the incoming encrypted configuration bitstream to be decrypted as it's being loaded into the device.

> JTAG is pronounced "J-TAG"; that is, by spelling out the 'J' followed by "tag" to rhyme with "bag."

The command/process of loading an encrypted bitstream automatically disables the FPGA's read-back capability. This means that you will typically use unencrypted configuration data during development (where you need to use read-back) and then start to use encrypted data when you move into production. (You can load an unencrypted bitstream at any time, so you can easily load a test configuration and then reload the encrypted version.)

The main downside to this scheme is that you require a battery backup on the circuit board to maintain the contents of the encryption key register in the FPGA when power is removed from the system. This battery will have a lifetime of years or decades because it need only maintain a single register in the device, but it does add to the size, weight, complexity, and cost of the board.

1.3.3 Antifuse-Based Devices

Unlike SRAM-based devices, which are programmed while resident in the system, antifuse-based devices are programmed off-line using a special device programmer.

The proponents of antifuse-based FPGAs are proud to point to an assortment of (not-insignificant) advantages. First of all, these devices are nonvolatile (their configuration data remains when the system is powered down), which means that they are immediately available as soon as power is applied to the system. Following from their nonvolatility, these devices don't require an external memory chip to store their configuration data, which saves the cost of an additional component and also saves real estate on the board.

One noteworthy advantage of antifuse-based FPGAs is the fact that their interconnect structure is naturally "rad hard," which means they are relatively immune to the effects of radiation. This is of particular interest in the case of military and aerospace applications because the state of a configuration cell in an SRAM-based component can be "flipped" if that cell is hit by radiation (of which there is a lot in space). By comparison, once an antifuse has been programmed, it cannot be altered in this way. Having said this, it should also be noted that any flip-flops in these devices remain sensitive to radiation, so chips intended for radiation-intensive environments must have their flip-flops protected by *triple redundancy design*. This refers to having three copies of each register and taking a majority vote (ideally all three registers will contain identical values, but if one has been "flipped" such that two registers say 0 and the third says 1, then the 0s have it, or vice versa if two registers say 1 and the third says 0).

> Radiation can come in the form of gamma rays (very high-energy photons), beta particles (high-energy electrons), and alpha particles.
>
> It should be noted that rad-hard devices are not limited to antifuse technologies. Other components, such as those based on SRAM architectures, are available with special rad-hard packaging and triple redundancy design.

But perhaps the most significant advantage of antifuse-based FPGAs is that their configuration data is buried deep inside them. By default, it is possible for the device programmer to read this data out because this is actually how the programmer works. As each antifuse is being processed, the device programmer keeps on testing it to determine when that element has been fully programmed; then it moves onto the next antifuse. Furthermore, the device programmer can be used to automatically verify that the configuration was performed successfully (this is well worth doing when you're talking about devices containing 50 million plus programmable elements). In order to do this, the device programmer requires the ability to read the actual states of the antifuses and compare them to the required states defined in the configuration file.

Once the device has been programmed, however, it is possible to set (grow) a special security antifuse that subsequently prevents any programming data (in the form of the presence or absence of antifuses) from being read out of the device. Even if the device is decapped (its top is removed), programmed and unprogrammed antifuses appear to be identical, and the fact that all of the antifuses are buried in the internal metallization layers makes it almost impossible to reverse-engineer the design.

Vendors of antifuse-based FPGAs may also tout a couple of other advantages relating to power consumption and speed, but if you aren't careful this can be a case of the quickness of the hand deceiving the eye. For example, they might tease you with the fact that an antifuse-based device consumes only 20% (approximately) of the standby power of an equivalent SRAM-based component, that their operational power consumption is also significantly lower, and that their interconnect-related delays are smaller. Also, they might casually mention that an antifuse is much smaller and thus occupies much less real estate on the chip than an equivalent SRAM cell (although they may neglect to mention that antifuse devices also require extra programming circuitry, including a large, hairy programming transistor for each antifuse). They will follow this by noting that when you have a device containing tens of millions of configuration elements, using antifuses means that the rest of the logic can be much closer together. This serves to reduce the interconnect delays, thereby making these devices faster than their SRAM cousins.

> It's worth noting that when the MRAM technologies come to fruition, these may well change the FPGA landscape.
>
> This is because MRAM fuses would be much smaller than SRAM cells (thereby increasing component density and reducing track delays), and they would also consume much less power.
>
> Furthermore, MRAM-based devices could be preprogrammed like antifuse-based devices (great for security) and reprogrammed like SRAM-based components (good for prototyping).

And both of the above points would be true . . . if one were comparing two devices implemented at the same technology node. But therein lies the rub, because antifuse technology requires the use of around three additional process steps after the main manufacturing process has been qualified. For this (and related) reasons, antifuse devices are always at least one—and usually several—generations (technology nodes) behind SRAM-based components, which effectively wipes out any speed or power consumption advantages that might otherwise be of interest.

> **1821:** England. Michael Faraday invents the first electric motor.

Of course, the main disadvantage associated with antifuse-based devices is that they are OTP, so once you've programmed one of these little scallywags, its function is set in stone. This makes these components a poor choice for use in a development or prototyping environment.

1.3.4 EPROM-Based Devices

This section is short and sweet because no one currently makes—or has plans to make—EPROM-based FPGAs.

1.3.5 E²PROM/Flash-Based Devices

E²PROM- or Flash-based FPGAs are similar to their SRAM counterparts in that their configuration cells are connected together in a long shift-register-style chain. These devices can be configured off-line using a device programmer. Alternatively, some versions are in-system programmable, or ISP, but their programming time is about three times that of an SRAM-based component.

Once programmed, the data they contain is nonvolatile, so these devices would be "instant on" when power is first applied to the system. With regard to protection, some of these devices use the concept of a multibit key, which can range from around 50 bits to several hundred bits in size. Once you've programmed the device, you can load your user defined key (bit-pattern) to secure its configuration data. After the key has been loaded, the only way to read data out of the device, or to write new data into it, is to load a copy of your key via the JTAG port (this port is discussed later in this chapter). The fact that the JTAG port in today's devices runs at around 20 MHz means that it would take billions of years to crack the key by exhaustively trying every possible value.

Two-transistor E²PROM and Flash cells are approximately 2.5 times the size of their one-transistor EPROM cousins, but they are still way smaller than their SRAM counterparts. This means that the rest of the logic can be much closer together, thereby reducing interconnect delays.

1821: England. Michael Faraday plots the magnetic field around a conductor.

On the downside, these devices require around five additional process steps on top of standard CMOS technology, which results in their lagging behind SRAM-based devices by one or more generations (technology nodes). Last but not least, these devices tend to have relatively high static power consumption due to their containing vast numbers of internal pull-up resistors.

1.3.6 Hybrid Flash-SRAM Devices

Last but not least, there's always someone who wants to add yet one more ingredient into the cooking pot. In the case of FPGAs, some vendors offer esoteric combinations of programming technologies. For example, consider a device where each configuration element is formed from the combination of a Flash (or E²PROM) cell and an associated SRAM cell.

In this case, the Flash elements can be preprogrammed. Then, when the system is powered up, the contents of the Flash cells are copied in a massively parallel fashion into their corresponding SRAM cells. This technique gives you the nonvolatility associated with antifuse devices, which means the device is immediately available when power is first applied to the system. But unlike an antifuse-based component, you can subsequently use the SRAM cells to reconfigure the device while it remains resident in the system. Alternatively, you can reconfigure the device using its Flash cells either while it remains in the system or off-line by means of a device programmer.

1.3.7   Summary

Table 1.1 briefly summarizes the key points associated with the various programming technologies described above:

Table 1.1: Summary of programming technologies

Feature	SRAM	Antifuse	E^{2}PROM/Flash
Technology node	State-of-the-art	One or more generations behind	One or more generations behind
Reprogrammable	Yes (in system)	No	Yes (in-system or offline)
Reprogramming speed (inc. erasing)	Fast	–	3x slower than SRAM
Volatile (must be programmed on power-up)	Yes	No	No (but can be if required)
Requires external configuration file	Yes	No	No
Good for prototyping	Yes (very good)	No	Yes (reasonable)
Instant-on	No	Yes	Yes
IP Security	Acceptable (especially when using bitstream encryption)	Very Good	Very Good
Size of configuration cell	Large (six transistors)	Very small	Medium-small (two transistors)
Power consumption	Medium	Low	Medium
Rad Hard	No	Yes	Not really

1.4 Fine-, Medium-, and Coarse-Grained Architectures

It is common to categorize FPGA offerings as being either *fine grained* or *coarse grained*. In order to understand what this means, we first need to remind ourselves that the main feature that distinguishes FPGAs from other devices is that their underlying fabric predominantly consists of large numbers of relatively simple programmable logic block "islands" embedded in a "sea" of programmable interconnect. (Figure 1.1).

In reality, the vast majority of the configuration cells in an FPGA are associated with its interconnect (as opposed to its configurable logic blocks). For this reason, engineers joke that FPGA vendors actually sell only the interconnect, and they throw in the rest of the logic for free!

In the case of a fine-grained architecture, each logic block can be used to implement only a very simple function. For example, it might be possible to configure the block to act as any 3-input function, such as a primitive logic gate (AND, OR, NAND, etc.) or a storage element (D-type flip-flop, D-type latch, etc.).

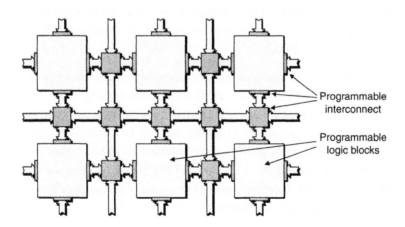

Programmable
interconnect

Programmable
logic blocks

Figure 1.1: Underlying FPGA fabric

1821: England. Sir Charles Wheatstone reproduces sound.

In addition to implementing glue logic and irregular structures like state machines, fine-grained architectures are said to be particularly efficient when executing *systolic algorithms* (functions that benefit from massively parallel implementations). These architectures are also said to offer some advantages with regard to traditional logic synthesis technology, which is geared toward fine-grained ASIC architectures.

The mid-1990s saw a lot of interest in fine-grained FPGA architectures, but over time the vast majority faded away into the sunset, leaving only their coarse-grained cousins. In the case of a coarse-grained architecture, each logic block contains a relatively large amount of logic compared to their fine-grained counterparts. For example, a logic block might contain four 4-input LUTs, four multiplexers, four D-type flip-flops, and some fast carry logic (see the following topics in this chapter for more details).

An important consideration with regard to architectural granularity is that fine-grained implementations require a relatively large number of connections into and out of each block compared to the amount of functionality that can be supported by those blocks. As the granularity of the blocks increases to medium-grained and higher, the amount of connections into the blocks decreases compared to the amount of functionality they can support. This is important because the programmable interblock interconnect accounts for the vast majority of the delays associated with signals as they propagate through an FPGA.

One slight fly in the soup is that a number of companies have recently started developing really coarse-grained device architectures comprising arrays of nodes, where each node is a highly complex processing element ranging from an algorithmic function such as a *fast Fourier transform (FFT)* all the way up to a complete general-purpose microprocessor. Although these devices aren't classed as FPGAs, they do serve to muddy the waters. For this reason, LUT-based FPGA architectures are now often classed as medium-grained, thereby leaving the coarse-grained appellation free to be applied to these new node-based devices.

1.5   MUX- versus LUT-Based Logic Blocks

There are two fundamental incarnations of the programmable logic blocks used to form the medium-grained architectures referenced in the previous section: MUX (multiplexer) based and LUT (lookup table) based.

MUX is pronounced to rhyme with "flux."

LUT is pronounced to rhyme with "nut."

1.5.1 MUX-Based

As an example of a MUX-based approach, consider one way in which the 3-input function $y = (a \& b) \mid c$ could be implemented using a block containing only multiplexers (Figure 1.2).

The device can be programmed such that each input to the block is presented with a logic 0, a logic 1, or the true or inverse version of a signal (a, b, or c in this case) coming from another block or from a primary input to the device. This allows each block to be configured in myriad ways to implement a plethora of possible functions. (The x shown on the input to the central multiplexer in Figure 1.2 indicates that we don't care whether this input is connected to a 0 or a 1.)

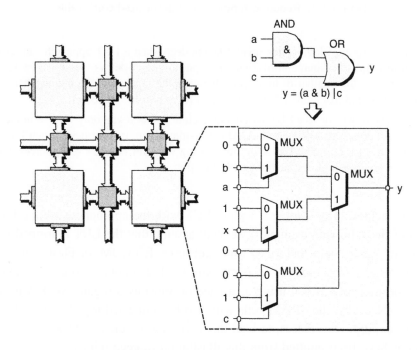

Figure 1.2: MUX-based logic block

1.5.2   LUT-Based

The underlying concept behind a LUT is relatively simple. A group of input signals is used as an index (pointer) to a lookup table. The contents of this table are arranged such that the cell pointed to by each input combination contains the desired value. For example, let's assume that we wish to implement the function:

$$y = (a \ \& \ b)|c$$

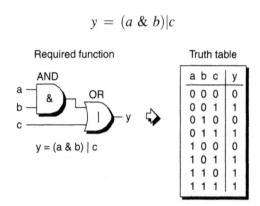

Figure 1.3: Required function and associated truth table

If you take a group of logic gates several layers deep, then a LUT approach can be very efficient in terms of resource utilization and input-to-output delays. (In this context, "deep" refers to the number of logic gates between the inputs and the outputs. Thus, the function illustrated in Figure 1.3 would be said to be two layers deep.)

However, one downside to a LUT-based architecture is that if you only want to implement a small function—such as a 2-input AND gate—somewhere in your design, you'll end up using an entire LUT to do so. In addition to being wasteful in terms of resources, the resulting delays are high for such a simple function.

This can be achieved by loading a 3-input LUT with the appropriate values. For the purposes of the following examples, we shall assume that the LUT is formed from SRAM cells (but it could be formed using antifuses, E^2PROM, or Flash cells, as discussed earlier in this chapter). A commonly used technique is to use the inputs to select the desired SRAM cell using a cascade of transmission gates as shown in Figure 1.4. (Note that the SRAM cells will also be connected together in a chain for configuration purposes—that is, to load them with the required values—but these connections have been omitted from this illustration to keep things simple.)

By comparison, in the case of mux-based architectures containing a mixture of muxes and logic gates, it's often possible to gain access to intermediate values from the signals linking the logic gates and the muxes. In this case, each logic block can be broken down into smaller fragments, each of which can be used to implement a simple function. Thus, these architectures may offer advantages in terms of performance and silicon utilization for designs containing large numbers of independent simple logic functions.

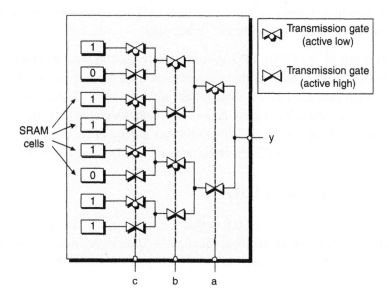

Figure 1.4: A transmission gate-based LUT (programming chain omitted for purposes of clarity)

If a transmission gate is enabled (active), it passes the signal seen on its input through to its output. If the gate is disabled, its output is electrically disconnected from the wire it is driving.

The transmission gate symbols shown with a small circle (called a *bobble* or a *bubble*) indicate that these gates will be activated by a logic 0 on their control input. By comparison, symbols without bobbles indicate that these gates will be activated by a logic 1. Based on this understanding, it's easy to see how different input combinations can be used to select the contents of the various SRAM cells.

1.5.3   MUX-Based versus LUT-Based?

Once upon a time—when engineers handcrafted their circuits prior to the advent of today's sophisticated CAD tools—some folks say that it was possible to achieve the best results using MUX-based architectures. (Sad to relate, they usually don't explain exactly how these results were better, so this is largely left to our imaginations.) It is also said that MUX-based architectures have an advantage when it comes to implementing control logic along the lines of "if this input is *true* and this input is *false*, then make that output *true*..."[1] However, some of these architectures don't provide high-speed carry logic chains, in which case their LUT-based counterparts are left as the leaders in anything to do with arithmetic processing.

Throughout much of the 1990s, FPGAs were widely used in the telecommunications and networking markets. Both of these application areas involve pushing lots of data around, in which case LUT-based architectures hold the high ground. Furthermore, as designs (and device capacities) grew larger and synthesis technology increased in sophistication, handcrafting circuits largely became a thing of the past. The end result is that the majority of today's FPGA architectures are LUT-based, as discussed below.

1.5.4   3-, 4-, 5-, or 6-Input LUTs?

The great thing about an *n*-input LUT is that it can implement any possible n-input combinational (or combinatorial) logic function. Adding more inputs allows you to represent more complex functions, but every time you add an input, you double the number of SRAM cells.

> Some folks prefer to say "combinational logic," while others favor "combinatorial logic."
>
> **1822:** England. Charles Babbage starts to build a mechanical calculating machine called the Difference Engine.

The first FPGAs were based on 3-input LUTs. FPGA vendors and university students subsequently researched the relative merits of 3-, 4-, 5-, and even 6-input LUTs into

---

[1] Some MUX-based architectures—such as those fielded by QuickLogic (www.quicklogic.com)—feature logic blocks containing multiple layers of MUXes preceded by primitive logic gates like ANDs. This provides them with a large fan-in capability, which gives them an advantage for address decoding and state machine decoding applications.

the ground (whatever you do, don't get trapped in conversation with a bunch of FPGA architects at a party). The current consensus is that 4-input LUTs offer the optimal balance of pros and cons.

In the past, some devices were created using a mixture of different LUT sizes, such as 3-input and 4-input LUTs, because this offered the promise of optimal device utilization. However, one of the main tools in the design engineer's treasure chest is logic synthesis, and uniformity and regularity are what a synthesis tool likes best. Thus, all of the really successful architectures are currently based only on the use of 4-input LUTs. (This is not to say that mixed-size LUT architectures won't reemerge in the future as design software continues to increase in sophistication.)

1.5.5   LUT versus Distributed RAM versus SR

The fact that the core of a LUT in an SRAM-based device comprises a number of SRAM cells offers a number of interesting possibilities. In addition to its primary role as a lookup table, some vendors allow the cells forming the LUT to be used as a small block of RAM (the 16 cells forming a 4-input LUT, for example, could be cast in the role of a 16×1 RAM). This is referred to as *distributed* RAM because (a) the LUTs are strewn (distributed) across the surface of the chip, and (b) this differentiates it from the larger chunks of *block RAM* (introduced later in this chapter).

Yet another possibility devolves from the fact that all of the FPGA's configuration cells—including those forming the LUT—are effectively strung together in a long chain (Figure 1.5).

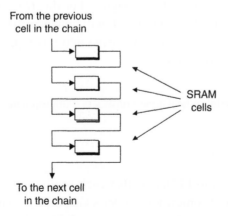

Figure 1.5: Configuration cells linked in a chain

The point here is that, once the device has been programmed, some vendors allow the SRAM cells forming a LUT to be treated independently of the main body of the chain and to be used in the form of a shift register. Thus, each LUT may be considered to be multifaceted (Figure 1.6).

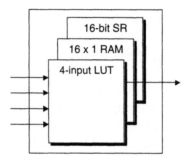

Figure 1.6: A multifaceted LUT

1.6   CLBs versus LABs versus Slices

"Man cannot live by LUTs alone," as the Bard would surely say if he were to be reincarnated accidentally as an FPGA designer. For this reason, in addition to one or more LUTs, a programmable logic block will contain other elements, such as multiplexers and registers. But before we delve into this topic, we first need to wrap our brains around some terminology.

1.6.1   A Xilinx Logic Cell

One niggle when it comes to FPGAs is that each vendor has its own names for things. But we have to start somewhere, so let's kick off by saying that the core building block in a modern FPGA from Xilinx is called a *logic cell (LC)*. Among other things, an LC comprises a 4-input LUT (which can also act as a 16 × 1 RAM or a 16-bit shift register), a multiplexer, and a register (Figure 1.7).

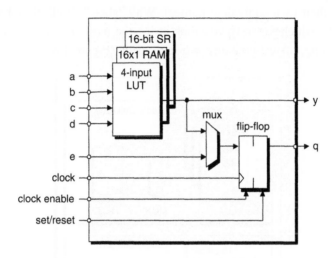

Figure 1.7: A simplified view of a Xilinx LC

It must be noted that the illustration presented in Figure 1.7 is a gross simplification, but it serves our purposes here. The register can be configured to act as a flip-flop, as shown in Figure 1.7, or as a latch. The polarity of the *clock* (rising-edge triggered or falling-edge triggered) can be configured, as can the polarity of the *clock enable* and *set/reset* signals (active-high or active-low).

In addition to the LUT, MUX, and register, the LC also contains a smattering of other elements, including some special fast carry logic for use in arithmetic operations (this is discussed in more detail a little later).

1827: Germany. Georg Ohm investigates electrical resistance and defines Ohm's Law.

1.6.2 An Altera Logic Element

Just for reference, the equivalent core building block in an FPGA from Altera is called a *logic element (LE)*. There are a number of differences between a Xilinx LC and an Altera LE, but the overall concepts are very similar.

1.6.3   Slicing and Dicing

The next step up the hierarchy is what Xilinx calls a *slice* (Altera and the other vendors doubtless have their own equivalent names). Why "slice"? Well, they had to call it something, and—whichever way you look at it—the term *slice* is "something." At the time of this writing, a slice contains two logic cells (Figure 1.8).

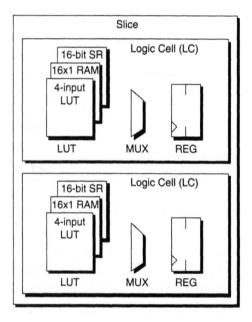

Figure 1.8: A slice containing two logic cells

The reason for the "at the time of this writing" qualifier is that these definitions can—and do—change with the seasons. The internal wires have been omitted from this illustration to keep things simple; it should be noted, however, that although each logic cell's LUT, MUX, and register have their own data inputs and outputs, the slice has one set of *clock*, *clock enable*, and *set/reset* signals common to both logic cells.

The definition of what forms a CLB varies from year to year. In the early days, a CLB consisted of two 3-input LUTs and one register. Later versions sported two 4-input LUTs and two registers.

Now, each CLB can contain two or four slices, where each slice contains two 4-input LUTS and two registers. And as for the morrow . . . well, it would take a braver man than I even to dream of speculating.

1.6.4 CLBs and LABs

And moving one more level up the hierarchy, we come to what Xilinx calls a *configurable logic block (CLB)* and what Altera refers to as a *logic array block (LAB)*. (Other FPGA vendors doubtless have their own equivalent names for each of these entities, but these are of interest only if you are actually working with their devices.)

Using CLBs as an example, some Xilinx FPGAs have two slices in each CLB, while others have four. At the time of this writing, a CLB equates to a single logic block in our original visualization of "islands" of programmable logic in a "sea" of programmable interconnect (Figure 1.9).

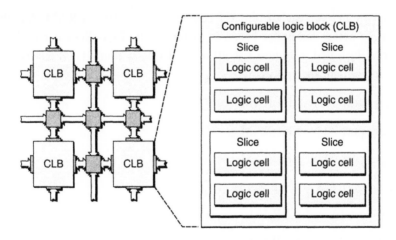

Figure 1.9: A CLB containing four slices (the number of slices depends on the FPGA family)

There is also some fast programmable interconnect within the CLB. This interconnect (not shown in Figure 1.9 for reasons of clarity) is used to connect neighboring slices.

> The point where a set of data or control signals enters or exits a logic function is commonly referred to as a "port."
>
> In the case of a single-port RAM, data is written in and read out of the function using a common data bus.

The reason for having this type of logic-block hierarchy—LC→Slice (with two LCs)→CLB (with four slices)—is that it is complemented by an equivalent hierarchy in the interconnect. Thus, there is fast interconnect between the LCs in a slice, then slightly slower interconnect between slices in a CLB, followed by the interconnect between CLBs. The idea is to achieve the optimum trade-off between making it easy to connect things together without incurring excessive interconnect-related delays.

In the case of a dual-port RAM, data is written into the function using one data bus (port) and read out using a second data bus (port). In fact, the read and write operations each have an associated address bus (used to point to a word of interest inside the RAM). This means that the read and write operations can be performed simultaneously.

1.6.5   Distributed RAMs and Shift Registers

We previously noted that each 4-bit LUT can be used as a 16 × 1 RAM. And things just keep on getting better and better because, assuming the four-slices-per-CLB configuration illustrated in Figure 1.9, all of the LUTs within a CLB can be configured together to implement the following:

- Single-port 16 × 8 bit RAM

- Single-port 32 × 4 bit RAM

- Single-port 64 × 2 bit RAM

- Single-port 128 × 1 bit RAM

- Dual-port 16 × 4 bit RAM

- Dual-port 32 × 2 bit RAM

- Dual-port 64 × 1 bit RAM

Alternatively, each 4-bit LUT can be used as a 16-bit shift register. In this case, there are special dedicated connections between the logic cells within a slice and between the slices themselves that allow the last bit of one shift register to be connected to the first bit of another without using the ordinary LUT output (which can be used to view the contents of a selected bit within that 16-bit register). This allows the LUTs within a single CLB to be configured together to implement a shift register containing up to 128 bits as required.

1.7   Fast Carry Chains

A key feature of modern FPGAs is that they include the special logic and interconnect required to implement fast carry chains. In the context of the CLBs introduced in the previous section, each LC contains special carry logic. This is complemented by dedicated interconnect between the two LCs in each slice, between the slices in each CLB, and between the CLBs themselves.

This special carry logic and dedicated routing boosts the performance of logical functions such as counters and arithmetic functions such as adders. The availability of these fast carry chains—in conjunction with features like the shift register incarnations of LUTs (discussed above) and embedded multipliers and the like (introduced below)— provided the wherewithal for FPGAs to be used for applications like DSP.

DSP is pronounced by spelling it out as "D-S-P."

1.8 Embedded RAMs

A lot of applications require the use of memory, so FPGAs now include relatively large chunks of embedded RAM called *e-RAM* or *block RAM*. Depending on the architecture of the component, these blocks might be positioned around the periphery of the device, scattered across the face of the chip in relative isolation, or organized in columns, as shown in Figure 1.10.

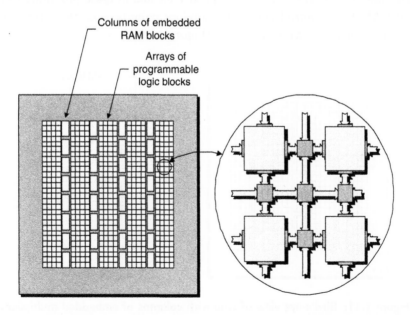

Figure 1.10: Bird's-eye view of chip with columns of embedded RAM blocks

Depending on the device, such a RAM might be able to hold anywhere from a few thousand to tens of thousands of bits. Furthermore, a device might contain anywhere from tens to hundreds of these RAM blocks, thereby providing a total storage capacity of a few hundred thousand bits all the way up to several million bits.

Each block of RAM can be used independently, or multiple blocks can be combined together to implement larger blocks. These blocks can be used for a variety of purposes, such as implementing standard single- or dual-port RAMs, *first-in first-out (FIFO)* functions, state machines, and so forth.

> FIFO is pronounced "fi" to rhyme with "hi," followed by "fo" to rhyme with "no" (like the "Hi-Ho" song in "Snow White and the Seven Dwarfs").

1.9   Embedded Multipliers, Adders, MACs, Etc.

Some functions, like multipliers, are inherently slow if they are implemented by connecting a large number of programmable logic blocks together. Since these functions are required by a lot of applications, many FPGAs incorporate special hardwired multiplier blocks. These are typically located in close proximity to the embedded RAM blocks introduced in the previous point because these functions are often used in conjunction with each other (Figure 1.11).

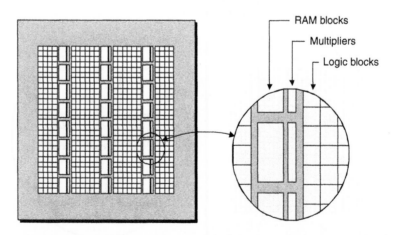

Figure 1.11: Bird's-eye view of chip with columns of embedded multipliers and RAM blocks

Similarly, some FPGAs offer dedicated adder blocks. One operation that is very common in DSP-type applications is called a *multiply-and-accumulate (MAC)* (Figure 1.12). As its name would suggest, this function multiplies two numbers together and adds the result to a running total stored in an accumulator.

1829: England. Sir Charles Wheatstone invents the concertina.

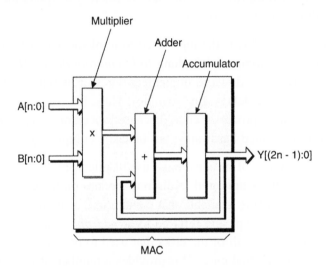

Figure 1.12: The functions forming a MAC

If the FPGA you are working with supplies only embedded multipliers, you will have to implement this function by combining the multiplier with an adder formed from a number of programmable logic blocks, while the result is stored in some associated flip-flops, in a block RAM, or in a number of distributed RAMs. Life becomes a little easier if the FPGA also provides embedded adders, and some FPGAs provide entire MACs as embedded functions.

1.10   Embedded Processor Cores (Hard and Soft)

Almost any portion of an electronic design can be realized in hardware (using logic gates and registers, etc.) or software (as instructions to be executed on a microprocessor). One of the main partitioning criteria is how fast you wish the various functions to perform their tasks:

1831: England. Michael Faraday creates the first electric dynamo.

- *Picosecond and nanosecond logic*: This has to run insanely fast, which mandates that it be implemented in hardware (in the FPGA fabric).

- *Microsecond logic*: This is reasonably fast and can be implemented either in hardware or software (this type of logic is where you spend the bulk of your time deciding which way to go).

- *Millisecond logic*: This is the logic used to implement interfaces such as reading switch positions and flashing *light-emitting diodes (LEDs)*. It's a pain slowing the hardware down to implement this sort of function (using huge counters to generate delays, for example). Thus, it's often better to implement these tasks as microprocessor code (because processors give you lousy speed—compared to dedicated hardware—but fantastic complexity).

The fact is that the majority of designs make use of microprocessors in one form or another. Until recently, these appeared as discrete devices on the circuit board. Of late, high-end FPGAs have become available that contain one or more embedded microprocessors, which are typically referred to as *microprocessor cores*. In this case, it often makes sense to move all of the tasks that used to be performed by the external microprocessor into the internal core. This provides a number of advantages, not the least being that it saves the cost of having two devices; it eliminates large numbers of tracks, pads, and pins on the circuit board; and it makes the board smaller and lighter.

1.10.1   Hard Microprocessor Cores

A hard microprocessor core is implemented as a dedicated, predefined block. There are two main approaches for integrating such a core into the FPGA. The first is to locate it in a strip (actually called "The Stripe") to the side of the main FPGA fabric (Figure 1.13).

In this scenario, all of the components are typically formed on the same silicon chip, although they could also be formed on two chips and packaged as a *multichip module (MCM)*. The main FPGA fabric would also include the embedded RAM blocks, multipliers, and the like introduced earlier, but these have been omitted from this illustration to keep things simple.

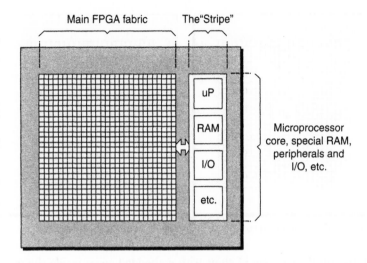

Figure 1.13: Birds-eye view of chip with embedded core outside of the main fabric

MCM is pronounced by spelling it out as "M-C-M."

One advantage of this implementation is that the main FPGA fabric is identical for devices with and without the embedded microprocessor core, which can help make things easier for the design tools used by the engineers. The other advantage is that the FPGA vendor can bundle a whole load of additional functions in the strip to complement the microprocessor core, such as memory, special peripherals, and so forth.

An alternative is to embed one or more microprocessor cores directly into the main FPGA fabric. One, two, and even four core implementations are currently available as I pen these words (Figure 1.14).

Once again, the main FPGA fabric would also include the embedded RAM blocks, multipliers, and the like introduced earlier, but these have been omitted from this illustration to keep things simple.

1831: England. Michael Faraday creates the first electrical transformer.

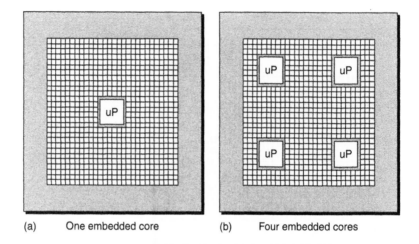

(a)   One embedded core   (b)   Four embedded cores

Figure 1.14: Bird's-eye view of chips with embedded cores inside the main fabric

In this case, the design tools have to be able to take account of the presence of these blocks in the fabric; any memory used by the core is formed from embedded RAM blocks, and any peripheral functions are formed from groups of general-purpose programmable logic blocks. Proponents of this scheme will argue that there are inherent speed advantages to be gained from having the microprocessor core in intimate proximity to the main FPGA fabric.

1.10.2   Soft Microprocessor Cores

As opposed to embedding a microprocessor physically into the fabric of the chip, it is possible to configure a group of programmable logic blocks to act as a microprocessor. These are typically called *soft cores*, but they may be more precisely categorized as either "soft" or "firm" depending on the way in which the microprocessor's functionality is mapped onto the logic blocks (see also the discussions associated with the hard IP, soft IP, and firm IP topics later in this chapter).

Soft cores are simpler (more primitive) and slower than their hard-core counterparts.[2] However, they have the advantage that you only need to implement a core if you need it and also that you can instantiate as many cores as you require until you run out of resources in the form of programmable logic blocks.

---

[2] A soft core typically runs at 30 to 50% of the speed of a hard core.

1831: England. Michael Faraday discovers magnetic lines of force.

1.11 Clock Trees and Clock Managers

All of the synchronous elements inside an FPGA—for example, the registers configured to act as flip-flops inside the programmable logic blocks—need to be driven by a clock signal. Such a clock signal typically originates in the outside world, comes into the FPGA via a special clock input pin, and is then routed through the device and connected to the appropriate registers.

1.11.1 Clock Trees

Consider a simplified representation that omits the programmable logic blocks and shows only the clock tree and the registers to which it is connected (Figure 1.15).

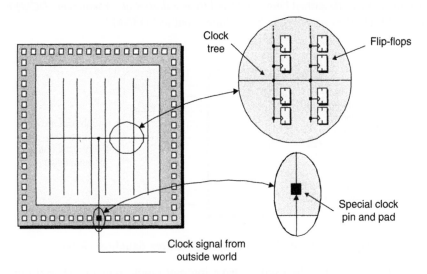

Figure 1.15: A simple clock tree

This is called a "clock tree" because the main clock signal branches again and again (the flip-flops can be considered to be the "leaves" on the end of the branches). This structure is used to ensure that all of the flip-flops see their versions of the clock signal as close together as possible. If the clock were distributed as a single long track driving all of the flip-flops one after another, then the flip-flop closest to the clock pin would see the clock signal much sooner than the one at the end of the chain. This is referred to as *skew*, and it can cause all sorts of problems (even when using a clock tree, there will be a certain amount of skew between the registers on a branch and also between branches).

The clock tree is implemented using special tracks and is separate from the general-purpose programmable interconnect. The scenario shown above is actually very simplistic. In reality, multiple clock pins are available (unused clock pins can be employed as general-purpose I/O pins), and there are multiple *clock domains* (clock trees) inside the device.

1.11.2   Clock Managers

Instead of configuring a clock pin to connect directly into an internal clock tree, that pin can be used to drive a special hard-wired function (block) called a *clock manager* that generates a number of *daughter clocks* (Figure 1.16).

A clock manager as described here is referred to as a *digital clock manager (DCM)* in the Xilinx world. DCM is pronounced by spelling it out as "D-C-M."

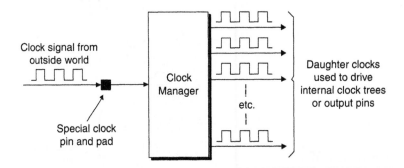

Clock signal from
outside world

Special clock
pin and pad

Clock
Manager

etc.

Daughter clocks
used to drive
internal clock trees
or output pins

Figure 1.16: A clock manager generates daughter clocks

These daughter clocks may be used to drive internal clock trees or external output pins that can be used to provide clocking services to other devices on the host circuit board. Each family of FPGAs has its own type of clock manager (there may be multiple clock manager blocks in a device), where different clock managers may support only a subset of the following features:

Jitter removal: For the purposes of a simple example, assume that the clock signal has a frequency of 1 MHz (in reality, of course, this could be much, much higher). In an ideal environment each clock edge from the outside world would arrive exactly one millionth of a second after its predecessor. In the real world, however, clock edges may arrive a little early or a little late.

> The term hertz was taken from the name of Heinrich Rudolf Hertz, a professor of physics at Karlsruhe Polytechnic in Germany, who first transmitted and received radio waves in a laboratory environment in 1888.

As one way to visualize this effect—known as *jitter*—imagine if we were to superimpose multiple edges on top of each other; the result would be a "fuzzy" clock (Figure 1.17).

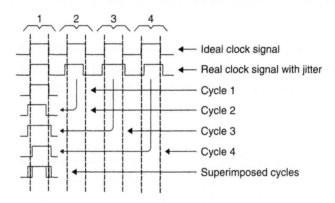

Figure 1.17: Jitter results in a fuzzy clock

> One hertz (Hz) equates to "one cycle per second," so MHz stands for megahertz or "million Hertz."

The FPGA's clock manager can be used to detect and correct for this jitter and to provide "clean" daughter clock signals for use inside the device (Figure 1.18).

Frequency synthesis: It may be that the frequency of the clock signal being presented to the FPGA from the outside world is not exactly what the design engineers wish for. In this case, the clock manager can be used to generate daughter clocks with frequencies that are derived by multiplying or dividing the original signal.

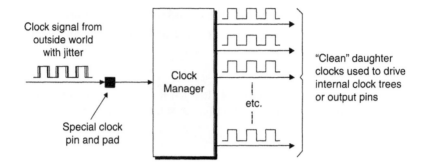

Figure 1.18: The clock manager can remove jitter

1831: England. Michael Faraday discovers the principle of electromagnetic induction.

As a really simple example, consider three daughter clock signals: the first with a frequency equal to that of the original clock, the second multiplied to be twice that of the original clock, and the third divided to be half that of the original clock (Figure 1.19).

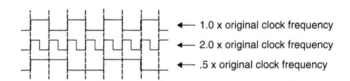

Figure 1.19: Using the clock manager to perform frequency synthesis

Once again, Figure 1.19 reflects very simple examples. In the real world, one can synthesize all sorts of internal clocks, such as an output that is four-fifths the frequency of the original clock.

Phase shifting: Certain designs require the use of clocks that are phase shifted (delayed) with respect to each other. Some clock managers allow you to select from fixed phase shifts of common values such as 120° and 240° (for a three-phase clocking scheme) or 90°, 180°, and 270° (if a four-phase clocking scheme is required). Others allow you to configure the exact amount of phase shift you require for each daughter clock.

For example, let's assume that we are deriving four internal clocks from a master clock, where the first is in phase with the original clock, the second is phase shifted by 90°, the third by 180°, and so forth (Figure 1.20).

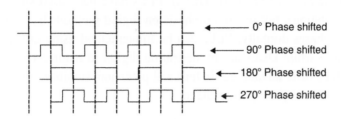

0° Phase shifted

90° Phase shifted

180° Phase shifted

270° Phase shifted

Figure 1.20: Using the clock manager to phase-shift the daughter clocks

Auto-skew correction: For the sake of simplicity, let's assume that we're talking about a daughter clock that has been configured to have the same frequency and phase as the main clock signal coming into the FPGA. By default, however, the clock manager will add some element of delay to the signal as it performs its machinations. Also, more significant delays will be added by the driving gates and interconnect employed in the clock's distribution. The result is that—if nothing is done to correct it—the daughter clock will lag behind the input clock by some amount. Once again, the difference between the two signals is known as *skew*.

Depending on how the main clock and the daughter clock are used in the FPGA (and on the rest of the circuit board), this can cause a variety of problems. Thus, the clock manager may allow a special input to feed the daughter clock. In this case, the clock manager will compare the two signals and specifically add additional delay to the daughter clock sufficient to realign it with the main clock (Figure 1.21).

PLL is pronounced by spelling it out as "P-L-L."

DLL is pronounced by spelling it out as "D-L-L."

To be a tad more specific, only the *prime* (zero phase-shifted) daughter clock will be treated in this way, and all of the other daughter clocks will be phase aligned to this prime daughter clock.

At this time, I do not know why digital delay-locked loop is not abbreviated to "DDLL."

Some FPGA clock managers are based on *phase-locked loops (PLLs)*, while others are based on *digital delay-locked loops (DLLs)*. PLLs have been used since the 1940s in analog implementations, but recent emphasis on digital methods has made it desirable to match signal phases digitally. PLLs can be implemented using either analog or digital techniques, while DLLs are by definition digital in nature. The proponents of DLLs say that they offer advantages in terms of precision, stability, power management, noise insensitivity, and jitter performance.

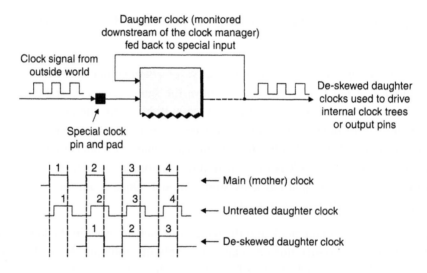

Figure 1.21: De-skewing with reference to the mother clock

1.12   General-Purpose I/O

Today's FPGA packages can have 1,000 or more pins, which are arranged as an array across the base of the package. Similarly, when it comes to the silicon chip inside the package, flip-chip packaging strategies allow the power, ground, clock, and I/O pins to be presented across the surface of the chip. Purely for the purposes of these discussions (and illustrations), however, it makes things simpler if we assume that all of the connections to the chip are presented in a ring around the circumference of the device, as indeed they were for many years.

1.12.1 Configurable I/O Standards

Let's consider for a moment an electronic product from the perspective of the architects and engineers designing the circuit board. Depending on what they are trying to do, the devices they are using, the environment the board will operate in, and so on, these guys and gals will select a particular standard to be used to transfer data signals. (In this context, "standard" refers to electrical aspects of the signals, such as their logic 0 and logic 1 voltage levels.)

1831: England. Michael Faraday discovers that a moving magnet induces an electric current.

The problem is that there is a wide variety of such standards, and it would be painful to have to create special FPGAs to accommodate each variation. For this reason, an FPGA's general-purpose I/O can be configured to accept and generate signals conforming to whichever standard is required. These general-purpose I/O signals will be split into a number of banks—we'll assume eight such banks numbered from 0 to 7 (Figure 1.22).

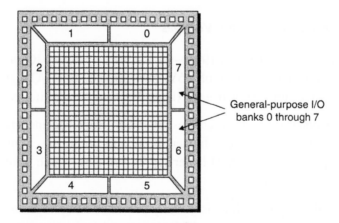

General-purpose I/O
banks 0 through 7

Figure 1.22: Bird's-eye view of chip showing general-purpose I/O banks

The interesting point is that each bank can be configured individually to support a particular I/O standard. Thus, in addition to allowing the FPGA to work with devices using multiple I/O standards, this allows the FPGA to actually be used to interface between different I/O standards (and also to translate between different protocols that may be based on particular electrical standards).

1832: England. Charles Babbage conceives the first mechanical computer, the Analytical Engine.

1.12.2 Configurable I/O Impedances

The signals used to connect devices on today's circuit board often have fast *edge rates* (this refers to the time it takes the signal to switch between one logic value and another). In order to prevent signals reflecting back (bouncing around), it is necessary to apply appropriate terminating resistors to the FPGA's input or output pins.

In the past, these resistors were applied as discrete components that were attached to the circuit board outside the FPGA. However, this technique became increasingly problematic as the number of pins started to increase and their *pitch* (the distance between them) shrank. For this reason, today's FPGAs allow the use of internal terminating resistors whose values can be configured by the user to accommodate different circuit board environments and I/O standards.

1.12.3 Core versus I/O Supply Voltages

In the days of yore—circa 1965 to 1995—the majority of digital ICs used a ground voltage of 0V and a supply voltage of +5V. Furthermore, their I/O signals also switched between 0V (logic 0) and +5V (logic 1), which made life really simple.

Over time, the geometries of the structures on silicon chips became smaller because smaller transistors have lower costs, higher speed, and lower power consumption. However, these processes demanded lower supply voltages, which have continued to fall over the years (Table 1.2).

The point is that this supply (which is actually provided using large numbers of power and ground pins) is used to power the FPGA's internal logic. For this reason, this is known as the *core voltage*. However, different I/O standards may use signals with voltage levels significantly different from the core voltage, so each bank of general-purpose I/Os can have its own additional supply pins.

1832: England Joseph Henry discovers self-induction or inductance.

Table 1.2: Supply voltages versus technology nodes

Year	Supply (Core Voltage (V))	Technology Node (nm)
1998	3.3	350
1999	2.5	250
2000	1.8	180
2001	1.5	150
2003	1.2	130

It's interesting to note that—from the 350 nm node onwards—the core voltage has scaled fairly linearly with the process technology. However, there are physical reasons not to go much below 1V (these reasons are based on technology aspects such as transistor input switching thresholds and voltage drops), so this "voltage staircase" might start to tail off in the not-so-distant future.

1.13 Gigabit Transceivers

The traditional way to move large amounts of data between devices is to use a *bus*, a collection of signals that carry similar data and perform a common function (Figure 1.23).

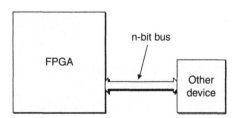

Figure 1.23: Using a bus to communicate between devices

Early microprocessor-based systems circa 1975 used 8-bit buses to pass data around. As the need to push more data around and to move it faster grew, buses grew to 16 bits in width, then 32 bits, then 64 bits, and so forth. The problem is that this requires a lot of pins on the device and a lot of tracks connecting the devices together. Routing these tracks so that they all have the same length and impedance becomes increasingly painful as boards grow in complexity. Furthermore, it becomes increasingly difficult to manage signal integrity issues (such as susceptibility to noise) when you are dealing with large numbers of bus-based tracks.

1833: England. Michael Faraday defines the laws of electrolysis.

For this reason, today's high-end FPGAs include special hard-wired gigabit transceiver blocks. These blocks use one pair of differential signals (which means a pair of signals that always carry opposite logical values) to *transmit (*TX*)* data and another pair to *receive (*RX*)* data (Figure 1.24).

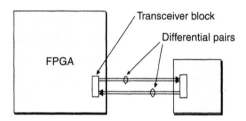

Figure 1.24: Using high-speed transceivers to communicate between devices

These transceivers operate at incredibly high speeds, allowing them to transmit and receive billions of bits of data per second. Furthermore, each block actually supports a number (say four) of such transceivers, and an FPGA may contain a number of these transceiver blocks.

1.14 Hard IP, Soft IP, and Firm IP

Each FPGA vendor offers its own selection of *hard, firm,* and *soft IP.* Hard IP comes in the form of preimplemented blocks such as microprocessor cores, gigabit interfaces, multipliers, adders, MAC functions, and the like. These blocks are designed to be as efficient as possible in terms of power consumption, silicon real estate, and performance. Each FPGA family will feature different combinations of such blocks together with various quantities of programmable logic blocks.

IP is pronounced by spelling it out as "I-P."

HDL is pronounced by spelling it out as "H-D-L."

VHDL is pronounced by spelling it out as "V-H-D-L."

RTL is pronounced by spelling it out as "R-T-L."

At the other end of the spectrum, soft IP refers to a source-level library of high-level functions that can be included to the users' designs. These functions are typically represented using a hardware description language, or HDL, such as Verilog or VHDL at the *register transfer level* (RTL) of abstraction. Any soft IP functions the design engineers decide to use are incorporated into the main body of the design—which is also specified in RTL—and subsequently synthesized down into a group of programmable logic blocks (possibly combined with some hard IP blocks like multipliers, etc.).

Holding somewhat of a middle ground is firm IP, which also comes in the form of a library of high-level functions. Unlike their soft IP equivalents, however, these functions have already been optimally mapped, placed, and routed into a group of programmable logic blocks (possibly combined with some hard IP blocks like multipliers, etc.). One or more copies of each predefined firm IP block can be instantiated (called up) into the design as required.

The problem is that it can be hard to draw the line between those functions that are best implemented as hard IP and those that should be implemented as soft or firm IP (using a number of general-purpose programmable logic blocks). In the case of functions like the multipliers, adders, and MACs discussed earlier in this chapter, these are generally useful for a wide range of applications. On the other hand, some FPGAs contain dedicated blocks to handle specific interface protocols like the PCI standard. It can, of course, make your life a lot easier if this happens to be the interface with which you wish to connect your device to the rest of the board. On the other hand, if you decide you need to use some other interface, a dedicated PCI block will serve only to waste space, block traffic, and burn power in your chip.

PCI is pronounced by spelling it out as "P-C-I."

Generally speaking, once FPGA vendors add a function like this into their device, they've essentially placed the component into a niche. Sometimes you have to do this to achieve the desired performance, but this is a classic problem because the next generation of the device is often fast enough to perform this function in its main (programmable) fabric.

1837: America. Samuel Finley Breese Morse exhibits an electric telegraph.

1.15 System Gates versus Real Gates

One common metric used to measure the size of a device in the ASIC world is that of *equivalent gates*. The idea is that different vendors provide different functions in their cell libraries, where each implementation of each function requires a different number of transistors. This makes it difficult to compare the relative capacity and complexity of two devices.

The answer is to assign each function an equivalent gate value along the lines of "Function A equates to five equivalent gates; function B equates to three equivalent gates ..." The next step is to count all of the instances of each function, convert them into their equivalent gate values, sum all of these values together, and proudly proclaim, "My ASIC contains 10 million equivalent gates, which makes it *much bigger* than your ASIC!"

Unfortunately, nothing is simple because the definition of what actually constitutes an equivalent gate can vary depending on whom one is talking to. One common convention is for a 2-input NAND function to represent one equivalent gate. Alternatively, some vendors define an equivalent gate as equaling an arbitrary number of transistors. And a more esoteric convention defines an ECL equivalent gate as being "one-eleventh the minimum logic required to implement a single-bit full adder" (who on earth came up with this one?). As usual, the best policy here is to make sure that everyone is talking about the same thing before releasing your grip on your hard-earned money.

And so we come to FPGAs. One of the problems FPGA vendors run into occurs when they are trying to establish a basis for comparison between their devices and ASICs. For example, if someone has an existing ASIC design that contains 500,000 equivalent gates and he wishes to migrate this design into an FPGA implementation, how can he tell if his design will fit into a particular FPGA? The fact that each 4-input LUT can be used to represent anywhere between one and more than twenty 2-input primitive logic gates makes such a comparison rather tricky.

1837: England Sir Charles Wheatstone and Sir William Fothergill Cooke patent the five-needle electric telegraph.

In order to address this issue, FPGA vendors started talking about *system gates* in the early 1990s. Some folks say that this was a noble attempt to use terminology that ASIC designers could relate to, while others say that it was purely a marketing ploy that didn't do anyone any favors.

Sad to relate, there appears to be no clear definition as to exactly what a system gate is. The situation was difficult enough when FPGAs essentially contained only generic programmable logic in the form of LUTs and registers. Even then, it was hard to state whether or not a particular ASIC design containing x equivalent gates could fit into an FPGA containing y system gates. This is because some ASIC designs may be predominantly combinatorial, while others may make excessively heavy use of registers. Both cases may result in a suboptimal mapping onto the FPGA.

The problem became worse when FPGAs started containing embedded blocks of RAM, because some functions can be implemented much more efficiently in RAM than in general purpose logic. And the fact that LUTs can act as distributed RAM only serves to muddy the waters; for example, one vendor's system gate count values now include the qualifier, "Assumes 20% to 30% of LUTs are used as RAM." And, of course, the problems are exacerbated when we come to consider FPGAs containing embedded processor cores and similar functions, to the extent that some vendors now say, "System gate values are not meaningful for these devices."

Is there a rule of thumb that allows you to convert system gates to equivalent gates and vice versa? Sure, there are lots of them! Some folks say that if you are feeling optimistic, then you should divide the system gate value by three (in which case three million FPGA system gates would equate to one million ASIC equivalent gates, for example). Or if you're feeling a tad more on the pessimistic side, you could divide the system gates by five (in which case three million system gates would equate to 600,000 equivalent gates).

1842: England. Joseph Henry discovers that an electrical spark between two conductors is able to induce magnetism in needles—this effect is detected at a distance of 30 meters.

However, other folks would say that the above is only true if you assume that the system gates value encompasses all of the functions that you can implement using both the general purpose programmable logic and the block RAMs. These folks would go on to say that if you remove the block RAMs from the equation, then you should divide the system gates value by ten (in which case, three million system gates would equate to only 300,000 equivalent gates), but in this case you still have the block RAMs to play with . . . arrggghhhh!

Ultimately, this topic spirals down into such a quagmire that even the FPGA vendors are trying desperately not to talk about system gates any more. When FPGAs were new on the scene, people were comfortable with the thought of equivalent gates and not so at ease considering designs in terms of LUTs, slices, and the like; however, the vast number of FPGA designs that have been undertaken over the years means that engineers are now much happier thinking in FPGA terms. For this reason, speaking as someone living in the trenches, I would prefer to see FPGAs specified and compared using only simple counts of:

- Number of logic cells or logic elements or whatever (which equates to the number of 4-input LUTs and associated flip-flops/latches)

- Number (and size) of embedded RAM blocks

- Number (and size) of embedded multipliers

- Number (and size) of embedded adders

- Number (and size) of embedded MACs

- etc.

Why is this so hard? And it would be really useful to take a diverse suite of real-world ASIC design examples, giving their equivalent gate values, along with details as to their flops/latches, primitive gates, and other more complex functions, then to relate each of these examples to the number of LUTs and flip-flops/latches required in equivalent FPGA implementations, along with the amount of embedded RAM and the number of other embedded functions.

1842: Scotland. Alexander Bail demonstrates first electromechanical means to capture, transmit, and reproduce an image.

Even this would be less than ideal, of course, because one tends to design things differently for FPGA and ASIC targets, but it would be a start.

1.16   FPGA Years

We've all heard it said that each year for a dog is equivalent to seven human years, the idea being that a 10-year-old pooch would be 70 years old in human terms. Thinking like this doesn't actually do anyone much good. On the other hand, it does provide a useful frame of reference so that when your hound can no longer keep up with you on a long walk, you can say, "Well, it's only to be expected because the poor old fellow is almost 100 years old" (or whatever).

Similarly, in the case of FPGAs, it may help to think that one of their years equates to approximately 15 human years. Thus, if you're working with an FPGA that was only introduced to the market within the last year, you should view it as a teenager. On the one hand, if you have high hopes for the future, he or she may end up with a Nobel Peace Prize or as the President of the United States. On the other hand, the object of your affections will typically have a few quirks that you have to get used to and learn to work around.

By the time an FPGA has been on the market for two years (equating to 30 years in human terms), you can start to think of it as reasonably mature and a good all-rounder at the peak of its abilities. After three years (45 years old), an FPGA is becoming somewhat staid and middle-aged, and by four years (60 years old). You should treat it with respect and make sure that you don't try to work it like a carthorse!

CHAPTER 2

Design Techniques, Rules, and Guidelines

Bob Zeidman

This chapter comes from the book Designing with FPGAs and CPLDs *by Bob Zeidman, ISBN: 9781578201129. Bob is a very interesting fellow and our paths have crossed many times over the years. Since the 1980s, Bob has designed numerous integrated circuits and circuit boards and has written software for many different systems.*

Bob's company, Zeidman Consulting (www.zeidmanconsulting.com) is a research and development contract firm specializing in digital hardware and software design. An offshoot of Zeidman Consulting—Zeidman Technologies (www.zeidman.biz)—develops hardware/software codesign tools for embedded system development that promise to change the way these systems are designed, simulated, and implemented. For example, Bob recently introduced me to his patented SynthOS™ technology, which can be used to automatically generate a real-time operating system (RTOS) that's targeted toward your unique embedded application.

But we digress . . . In this chapter, Bob examines in detail the issues that arise when designing a circuit that is to be implemented in a CPLD or FPGA. As Bob says: "These concepts are essential to designing a chip that functions correctly in your system and will be reliable throughout the lifetime of your product."

—Clive "Max" Maxfield

This chapter presents design techniques, rules, and guidelines that are critical to FPGA and CPLD design. By understanding these concepts, you increase your chances of producing a working, reliable device that will work for different chip vendor processes and continue to work for the lifetime of your system.

In certain sections of this chapter, I show a number of examples of incorrect or inefficient designs, and the equivalent function designed correctly. I assume that you are using, or will be using, a hardware description language (HDL) like Verilog or VHDL to design your CPLD or FPGA. Most, if not all, CPLDs and FPGAs are designed using HDLs. This chapter presents these design examples using schematics. Although HDLs are much more efficient for creating large designs, schematics are still preferred for illustrating small designs because they give a nice visual representation of what is going on.

This chapter focuses on the potential problems that an engineer must recognize when designing an FPGA or CPLD and the design techniques that are used to avoid these problems. More specifically, reading this chapter will help you:

- Learn the fundamental concepts of hardware description languages.

- Appreciate the process of top-down design and how it is used to organize a design and speed up the development time.

- Comprehend how FPGA and CPLD architecture and internal structures affect your design.

- Understand the concept of synchronous design, know how to spot asynchronous circuits, and how to redesign an asynchronous circuit to be synchronous.

- Recognize what problems floating internal nodes can cause and learn how to avoid these problems.

- Understand the consequences of bus contention and techniques for avoiding it.

- Comprehend one-hot state encoding for optimally creating state machines in FPGAs.

- Design testability into a circuit from the beginning and understand various testability structures that are available.

2.1 Hardware Description Languages

Design teams can use a hardware description language to design at any level of abstraction, from high-level architectural models to low-level switch models. These levels, from least amount of detail to most amount of detail, are as follows:

- Behavioral models
  - ○ Algorithmic
  - ○ Architectural
- Structural models
  - ○ Register transfer level (RTL)
  - ○ Gate level
  - ○ Switch level

These levels refer to the types of models that are used to represent a circuit design. The top two levels use what are called *behavioral models*, whereas the lower three levels use what are called structural models. Behavioral models consist of code that represents the behavior of the hardware without respect to its actual implementation. Behavioral models don't include timing numbers. Buses don't need to be broken down into their individual signals. Adders can simply add two or more numbers without specifying registers or gates or transistors. The two types of behavioral models are called *algorithmic models* and *architectural models*.

Algorithmic models simply represent algorithms that act on data. No hardware implementation is implied in an algorithmic model. So an algorithmic model is similar to what a programmer might write in C or Java to describe a function. The algorithmic model is coded to be fast, efficient, and mathematically correct. An algorithmic model of a circuit can be simulated to test that the basic specification of the design is correct.

Architectural models specify the blocks that implement the algorithms. Architectural models may be divided into blocks representing PC boards, ASICs, FPGAs, or other major hardware components of the system, but they do not specify how the algorithms are implemented in each particular block. These models of a circuit can be compared to an algorithmic model of the same circuit to discover if a chip's architecture is correctly implementing the algorithm. The design team can simulate the algorithmic model to find bottlenecks and inefficiencies before any of the low level design has begun.

Some sample behavioral level HDL code is shown in Listing 2.1. This sample shows a multiplier for two unsigned numbers of any bit width. Notice the very high level of description—there are no references to gates or clock signals.

Listing 2.1: Sample behavioral level HDL code

```
// *************************************************

// ***** Multiplier for two unsigned numbers *****

// *************************************************

// Look for the multiply enable signal

always @(posedge multiply_en) begin

    product <= a*b;

end
```

Structural models consist of code that represents specific pieces of hardware. RTL specifies the logic on a register level. In other words, the simplest RTL code specifies register logic. Actual gates are avoided, although RTL code may use Boolean functions that can be implemented in gates. The example RTL code in Listing 2.2 shows a multiplier for two unsigned 4-bit numbers. This level is the level at which most digital design is done.

Listing 2.2: Sample RTL HDL code

```
// *************************************************

// ***** Multiplier for two unsigned 4-bit numbers *****

// *************************************************

                                        // Look at the rising edge of the clock

always @(posedge clk) begin

        if (multiply_en == 1) begin         // Set up the multiplication

            count <= ~0;                    // Set count to its max value
```

```
            product <= 0;              // Zero the product

    end

    if (count) begin

        if (b[count]) begin

                            // If this bit of the multiplier is 1, shift

                            // the product left and add the multiplicand

            product <= (product << 1) + a;

        end

        else begin

                            // If this bit of the multiplier is 0,

                            // just shift the product left

            product <= product << 1;

        end
        count <= count - 1;        // Decrement the count

    end

end
```

Gate level modeling consists of code that specifies gates such as NAND and NOR gates (Listing 2.3) Gate level code is often the output of a synthesis program that reads the RTL level code that an engineer has used to design a chip and writes the gate level equivalent. This gate level code can then be optimized for placement and routing within the CPLD or FPGA. The code in Listing 2.3 shows the synthesized 4-bit unsigned multiplier where the logic has been mapped to individual CLBs of an FPGA. Notice that at this level all logic must be described in primitive functions that map directly to the CLB logic, making the code much longer.

(In fact, much of the code was removed for clarity. Buffers used to route signals and the Boolean logic for the lookup tables (LUTs) are not included in this code, even though they would be needed in a production chip.)

Listing 2.3: Sample gate level HDL code

```
// *****************************************************
// ***** Multiplier for two unsigned 4-bit numbers *****
// *****************************************************
module UnsignedMultiply (
        clk,
        a,
        b,
        multiply_en,
        product);


input clk;
input [3:0] a;
input [3:0] b;
input multiply_en;
output [7:0] product;


wire clk;
wire [3:0] a;
wire [3:0] b;
wire multiply_en;
wire [7:0] product;
wire [3:0] count;
wire [7:0] product_c;
wire [3:0] a_c;
```

```
wire [7:0] product_10;

wire [3:0] b_c;

wire clk_c;

wire count16;

wire un1_count_5_axb_1;

wire un1_count_5_axb_2;

wire un7_product_axb_1;

wire un7_product_axb_2;

wire un7_product_axb_3;

wire un7_product_axb_4;

wire un7_product_axb_5;

wire un1_un1_count16_i;

wire multiply_en_c;

wire un1_multiply_en_1_0;

wire product25_3_0_am;

wire product25_3_0_bm;

wire product25;

wire un7_product_axb_0;

wire un7_product_s_1;

wire un7_product_s_2;

wire un7_product_s_3;

wire un7_product_s_4;

wire un7_product_s_5;

wire un7_product_s_6;
```

```
wire un1_count_5_axb_0;

wire un1_count_5_axb_3;

wire un7_product_axb_6;

wire un1_count_5_s_1;

wire un1_count_5_s_2;

wire un1_count_5_s_3;

wire un7_product_cry_5;

wire un7_product_cry_4;

wire un7_product_cry_3;

wire un7_product_cry_2;

wire un7_product_cry_1;

wire un7_product_cry_0;

wire un1_count_5_cry_2;

wire un1_count_5_cry_1;

wire un1_count_5_cry_0;

LUT2_6 un1_count_5_axb_1_Z (

    .I0(count[1]),

    .I1(count16),

    .O(un1_count_5_axb_1));

LUT2_6 un1_count_5_axb_2_Z (

    .I0(count[2]),

    .I1(count16),

    .O(un1_count_5_axb_2));
```

```
LUT2_6 un7_product_axb_1_Z (
    .I0(product_c[1]),
    .I1(a_c[2]),
    .O(un7_product_axb_1));

LUT2_6 un7_product_axb_2_Z (
    .I0(product_c[2]),
    .I1(a_c[3]),
    .O(un7_product_axb_2));

LUT1_2 un7_product_axb_3_Z (
    .I0(product_c[3]),
    .O(un7_product_axb_3));

LUT1_2 un7_product_axb_4_Z (
    .I0(product_c[4]),
    .O(un7_product_axb_4));

LUT1_2 un7_product_axb_5_Z (
    .I0(product_c[5]),
    .O(un7_product_axb_5));

FDE \product_Z[7] (
    .Q(product_c[7]),
    .D(product_10[7]),
    .C(clk_c),
    .CE(un1_un1_count16_i));
```

```
FDE \product_Z[0] (
    .Q(product_c[0]),
    .D(product_10[0]),
    .C(clk_c),
    .CE(un1_un1_count16_i));

FDE \product_Z[1] (
    .Q(product_c[1]),
    .D(product_10[1]),
    .C(clk_c),
    .CE(un1_un1_count16_i));

FDE \product_Z[2] (
    .Q(product_c[2]),
    .D(product_10[2]),
    .C(clk_c),
    .CE(un1_un1_count16_i));

FDE \product_Z[3] (
    .Q(product_c[3]),
    .D(product_10[3]),
    .C(clk_c),
    .CE(un1_un1_count16_i));

FDE \product_Z[4] (
    .Q(product_c[4]),
```

```
    .D(product_10[4]),

    .C(clk_c),

    .CE(un1_un1_count16_i));

FDE \product_Z[5] (

    .Q(product_c[5]),

    .D(product_10[5]),

    .C(clk_c),

    CE(un1_un1_count16_i));

FDE \product_Z[6] (

    Q(product_c[6]),

    D(product_10[6]),

    C(clk_c),

    CE(un1_un1_count16_i));

LUT2_4 un1_multiply_en_1 (

    I0(count16),

    I1(multiply_en_c),

    O(un1_multiply_en_1_0));

MUXF5 product25_3_0 (

    I0(product25_3_0_am),

    I1(product25_3_0_bm),

    S(count[1]),

    O(product25));
```

```
LUT3_D8 product25_3_0_bm_Z (
    I0(count[0]),
    I1(b_c[3]),
    I2(b_c[2]),
    O(product25_3_0_bm));

LUT3_D8 product25_3_0_am_Z (
    I0(count[0]),
    I1(b_c[1]),
    I2(b_c[0]),
    O(product25_3_0_am));

LUT4_A280 \product_10_Z[1] (
    .I0(count16),
    .I1(product25),
    .I2(un7_product_axb_0),
    .I3(product_c[0]),
    .O(product_10[1]));

LUT4_A280 \product_10_Z[2] (
    .I0(count16),
    .I1(product25),
    .I2(un7_product_s_1),
    .I3(product_c[1]),
    .O(product_10[2]));
```

```
LUT4_A280 \product_10_Z[3] (
    .I0(count16),
    .I1(product25),
    .I2(un7_product_s_2),
    .I3(product_c[2]),
    .O(product_10[3]));

LUT4_A280 \product_10_Z[4] (
    .I0(count16),
    .I1(product25),
    .I2(un7_product_s_3),
    .I3(product_c[3]),
    .O(product_10[4]));

LUT4_A280 \product_10_Z[5] (
    .I0(count16),
    .I1(product25),
    .I2(un7_product_s_4),
    .I3(product_c[4]),
    .O(product_10[5]));

LUT4_A280 \product_10_Z[6] (
    .I0(count16),
    .I1(product25),
    .I2(un7_product_s_5),
    .I3(product_c[5]),
```

```
    .O(product_10[6]));

LUT4_A280 \product_10_Z[7] (
    .I0(count16),

    .I1(product25),

    .I2(un7_product_s_6),

    .I3(product_c[6]),

    .O(product_10[7]));

LUT2_6 un1_count_5_axb_0_Z (
    .I0(count[0]),

    .I1(count16),

    .O(un1_count_5_axb_0));

LUT2_6 un1_count_5_axb_3_Z (
    .I0(count[3]),

    .I1(count16),

    .O(un1_count_5_axb_3));

LUT2_6 un7_product_axb_0_Z (
    .I0(product_c[0]),

    .I1(a_c[1]),

    .O(un7_product_axb_0));

LUT1_2 un7_product_axb_6_Z (
    .I0(product_c[6]),

    .O(un7_product_axb_6));
```

```
LUT4_FFFE count16_Z (
    .I0(count[2]),
    .I1(count[3]),
    .I2(count[0]),
    .I3(count[1]),
    .O(count16));

LUT3_80 \product_10_Z[0] (
    .I0(count16),
    .I1(product25),
    .I2(a_c[0]),
    .O(product_10[0]));

LUT2_E un1_un1_count16_i_Z (
    .I0(multiply_en_c),
    .I1(count16),
    .O(un1_un1_count16_i));

FDS \count_Z[0] (
    .Q(count[0]),
    .D(un1_count_5_axb_0),
    .C(clk_c),
    .S(un1_multiply_en_1_0));

FDS \count_Z[1] (
    .Q(count[1]),
```

```
    .D(un1_count_5_s_1),

    .C(clk_c),

    .S(un1_multiply_en_1_0));


FDS \count_Z[2] (

    .Q(count[2]),

    .D(un1_count_5_s_2),

    .C(clk_c),

    .S(un1_multiply_en_1_0));


FDS \count_Z[3] (

    .Q(count[3]),

    .D(un1_count_5_s_3),

    .C(clk_c),

    .S(un1_multiply_en_1_0));


XORCY un7_product_s_6_Z (

    .LI(un7_product_axb_6),

    .CI(un7_product_cry_5),

    .O(un7_product_s_6));


XORCY un7_product_s_5_Z (

    .LI(un7_product_axb_5),

    .CI(un7_product_cry_4),

    .O(un7_product_s_5));
```

```
MUXCY_L un7_product_cry_5_Z (

    .DI(GND),

    .CI(un7_product_cry_4),

    .S(un7_product_axb_5),

    .LO(un7_product_cry_5));


XORCY un7_product_s_4_Z (

    .LI(un7_product_axb_4),

    .CI(un7_product_cry_3),

    .O(un7_product_s_4));


MUXCY_L un7_product_cry_4_Z (

    .DI(GND),

    .CI(un7_product_cry_3),

    .S(un7_product_axb_4),

    .LO(un7_product_cry_4));


XORCY un7_product_s_3_Z (

    .LI(un7_product_axb_3),

    .CI(un7_product_cry_2),

    .O(un7_product_s_3));


MUXCY_L un7_product_cry_3_Z (

    .DI(GND),

    .CI(un7_product_cry_2),

    .S(un7_product_axb_3),
```

```
      .LO(un7_product_cry_3));

  XORCY un7_product_s_2_Z (
      .LI(un7_product_axb_2),
      .CI(un7_product_cry_1),
      .O(un7_product_s_2));

  MUXCY_L un7_product_cry_2_Z (
      .DI(product_c[2]),
      .CI(un7_product_cry_1),
      .S(un7_product_axb_2),
      .LO(un7_product_cry_2));

  XORCY un7_product_s_1_Z (
      .LI(un7_product_axb_1),
      .CI(un7_product_cry_0),
      .O(un7_product_s_1));

  MUXCY_L un7_product_cry_1_Z (
      .DI(product_c[1]),
      .CI(un7_product_cry_0),
      .S(un7_product_axb_1),
      .LO(un7_product_cry_1));

  MUXCY_L un7_product_cry_0_Z (
      .DI(product_c[0]),
      .CI(GND),
```

```
        .S(un7_product_axb_0),

        .LO(un7_product_cry_0));


    XORCY un1_count_5_s_3_Z (

        .LI(un1_count_5_axb_3),

        .CI(un1_count_5_cry_2),

        .O(un1_count_5_s_3));


    XORCY un1_count_5_s_2_Z (

        .LI(un1_count_5_axb_2),

        .CI(un1_count_5_cry_1),

        .O(un1_count_5_s_2));


    MUXCY_L un1_count_5_cry_2_Z (

        .DI(count[2]),

        .CI(un1_count_5_cry_1),

        .S(un1_count_5_axb_2),

        .LO(un1_count_5_cry_2));


    XORCY un1_count_5_s_1_Z (

        .LI(un1_count_5_axb_1),

        .CI(un1_count_5_cry_0),

        .O(un1_count_5_s_1));


    MUXCY_L un1_count_5_cry_1_Z (

        .DI(count[1]),

        .CI(un1_count_5_cry_0),
```

```
    .S(un1_count_5_axb_1),

    .LO(un1_count_5_cry_1));


  MUXCY_L un1_count_5_cry_0_Z (

    .DI(count[0]),

    .CI(GND),

    .S(un1_count_5_axb_0),

    .LO(un1_count_5_cry_0));

  endmodule /* UnsignedMultiply */
```

Finally, the lowest level is that of a switch level model. A switch level model specifies the actual transistor switches that are combined to make gates. Digital design is never done at this level. Switch level code can be used for physical design of an ASIC and can also be used for the design of analog devices.

The advantage of HDLs is that they enable all of these different levels of modeling within the same language. This makes all the stages of design very convenient to implement. You don't need to learn different tools. You can easily simulate the design at a behavioral level, and then substitute various behavioral code modules with structural code modules. For system simulation, this allows you to analyze your entire project using the same set of tools. First, you can test and optimize the algorithms. Next, you can use the behavioral models to partition the hardware into boards, ASIC, and FPGAs. You can then write the RTL code and substitute it for behavioral blocks, one at a time, to easily test the functionality of each block. From that, you can synthesize the design, creating gate and switch level blocks that can be resimulated with timing numbers to get actual performance measurements. Finally, you can use this low-level code to generate a netlist for layout. All stages of the design have been performed using the same basic tool.

The main HDLs in existence today are Verilog and VHDL. Both are open standards, maintained by standards groups of the Institute of Electrical and Electronic Engineers (IEEE). VHDL is maintained as IEEE-STD-1076; Verilog is maintained as IEEE-STD-1364. Although some engineers prefer one language over the other, the differences are minor. As these standard languages progress with new versions, the differences become even fewer. Also, several languages, including C++, are being offered as a system level language, which would enable engineers to design and simulate an entire system consisting of multiple chips, boards, and software. These system level design languages are still evolving.

2.2 Top-Down Design

Top-down design is the design methodology whereby high-level functions are defined first, and the lower level implementation details are filled in later. A design can be viewed as a hierarchical tree, as shown in Figure 2.1. The top level block represents the entire chip. The next lower level blocks also represent the entire chip but divided into the major function blocks of the chip. Intermediate level blocks divide the functionality into more manageable pieces. The bottom level contains only gates and macrofunctions, which are vendor-supplied high level functions.

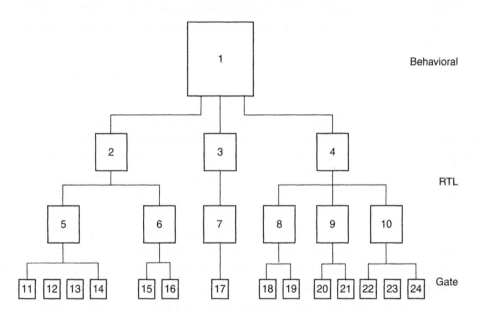

Figure 2.1: Top-down design

2.2.1 Use of Hardware Design Languages

Top-down design methodology lends itself particularly well to using HDLs, the generally accepted method of designing complex CPLDs and FPGAs. Each block in the design corresponds to the code for a self-contained module. The top-level blocks correspond to the behavioral models that comprise the chip. The intermediate levels correspond to the RTL models that will become input to the synthesis process. The lowest level of the hierarchy corresponds to gate level code which is output from the synthesis software and which directly represents logic structures within the chip.

2.2.2   Written Specifications

Top-down design methodology works hand in hand with a written specification that is an essential starting point for any design. The specification must include general aspects of the design, including the major functional blocks. The highest blocks of a top-down design are behavioral level models that correspond to the major functional blocks described in the specification. Thus, using a top-down design approach, the specification becomes a starting point for the actual HDL code. Specification changes can immediately be turned into HDL design changes, and design changes can be quickly and easily translated back to the specification, keeping the specification accurate and up to date.

2.2.3   Allocating Resources

These days, chips typically incorporate a large number of gates and a very high level of functionality. A top-down approach simplifies the design task and allows more than one engineer, when necessary, to design the chip. For example, the lead designer or the system architect may be responsible for the specification and the top-level block. Engineers in the design team may each be responsible for one or several intermediate blocks, depending on their strengths, experience, and abilities. An experienced ALU designer may be responsible for the ALU block and several other blocks. A junior engineer can work on a smaller block, such as a bus controller. Each engineer can work in parallel, writing code and simulating, until it is time to integrate the pieces into a single design. No one person can slow down the entire design. With a top-down design, your not-too-bright colleague in the next cubicle won't delay the entire project or make you look bad. That may be the single best reason for using this design methodology.

2.2.4   Design Partitioning

Even if you are the only engineer designing the chip, this methodology allows you to break the design into simpler functions that you (or others) can design and simulate independently from the rest of the design. A large, complex design becomes a series of independent smaller ones that are easier to design and simulate.

2.2.5 Flexibility and Optimization

Top-down design allows flexibility. Teams can remove sections of the design and replace them with higher-performance or optimized designs without affecting other sections of the design. Adding new or improved functionality involves simply redesigning one section of the design and substituting it for the current section.

2.2.6 Reusability

Reusability is an important topic in chip design these days. In the days when a CPLD consisted of a few small state machines, it was no big deal to design it from scratch. Nowadays, CPLDs and FPGAs contain so much logic that reusing any function from a previous design can save days, weeks, or months of design time. When one group has already designed a certain function, say a fast, efficient 64-bit multiplier, HDLs allow you to take the design and reuse it in your design. If you need a 64-bit multiplier, you can simply take the designed, verified code and plop it into your design. Or you can purchase the code from a third party. But it will only fit easily into your design if you have used a top-down approach to break the design into smaller pieces, one of which is a 64-bit multiplier.

2.2.7 Floorplanning

Floorplanning is another important topic in chip design these days. As chips become larger, it may be necessary to help the design tools place the various functions in the device. If you have used a top-down approach, you will be able to plan the placement of each block in the chip. The FBs or CLBs that implement the logic in each block can be placed in proximity to each other. The relationship between blocks will also be apparent, and so you can understand which blocks should be placed near each other.

2.2.8 Verification

Verification has become an extremely important aspect of the design process, but can be very resource-intensive and thus often needs to be optimized. Top-down design is one important means for improving verification. A top-down design approach allows each module to be simulated independently from the rest of the design. This is important for complex designs where an entire design can take weeks to simulate and days to debug. By using a top-down approach, design teams can efficiently perform behavioral, RTL, and gate level simulations and use the results to verify functionality at each level of design.

In summary, top-down design facilitates these good design practices:

- Use of hardware design languages

- Writing accurate and up-to-date specifications

- Allocation of resources for the design task

- Simplification and easy partitioning of the design task

- Flexibility in experimenting with different designs and optimizing the design

- Reusing previous designs

- Floorplanning

- Improved verification and less time spent on verification

2.2.9   Know the Architecture

Look at the particular architecture for the CPLD or FPGA that you are using to determine which logic devices fit best into it. You should choose a device with an architecture that fits well with your particular design. In addition, as you design, keep in mind the architecture of the device. For example, you may be using a CPLD that includes exclusive ORs. When you are deciding which kind of error detection to use, you could perform parity checking efficiently in this device. Similarly, if the device includes a fast carry chain, make sure that you are able to use it for any adders that you are designing.

Many FPGA and CPLD vendors now include specialized logic functions in their devices. For example, vendors may offer a device with a built-in digital signal processor (DSP). This device will not be useful, and is the wrong choice, if your design does not use a DSP. On the other hand, if you are implementing signal processing functions, you should make sure you use this DSP function as much as possible throughout the design.

The vendor will be able to offer advice about their device architecture and how to efficiently utilize it. Most synthesis tools can target their results to a specific FPGA or CPLD family from a specific vendor, taking advantage of the architecture to provide you with faster, more optimal designs.

2.3 Synchronous Design

One of the most important concepts in chip design, and one of the hardest to enforce on novice chip designers, is that of synchronous design. Once a chip designer uncovers a problem due to a design that is not synchronous (i.e., asynchronous) and attempts to fix it, he or she usually becomes an evangelical convert to synchronous design practices. This is because asynchronous design problems often appear intermittently due to subtle variations in the voltage, temperature, or semiconductor process. Or they may appear only when the vendor changes its semiconductor process. Asynchronous designs that work for years in one process may suddenly fail when the programmable part is manufactured using a newer process.

Unlike technologies like printed circuit boards, the semiconductor processes for creating FPGAs change very rapidly. Moore's Law, an observation about semiconductor technology improvements, currently says that the number of transistors per square inch doubles every 18 months. This doubling is due to rapid increases in semiconductor process technology and advances in the machinery used to create silicon structures. Due to these improvements, the FPGA or CPLD device that holds your design today will have different, faster timing parameters than the one that holds your design a year from now. The vendor will no doubt have improved its process by that time.

Even if you were certain that the semiconductor process for your programmable device would remain constant for each device in your system, each process has natural variations from chip to chip and even within a single chip. To add even more uncertainty, the exact timing for a programmable device depends on the specific routing and logic implementation. Essentially, you cannot determine exact delay numbers; you can only know timing ranges and relative delays. Synchronous design is a formal methodology for ensuring that your design will work correctly and within your speed requirements as long as the timing numbers remain within certain ranges and with delays that remain relatively controlled, if not absolutely controlled.

Synchronous design is not only more reliable than asynchronous design, but for the most part, EDA tools now assume that your design is synchronous. In the early days of

EDA software for digital circuits, the tools made no assumptions about the design. As chip designs grew, the software tools became more difficult to develop, the algorithms became more complex, and the tools became slower and less efficient. The EDA vendors finally realized that synchronous design was required anyway, for the reasons I gave previously. So the EDA vendors also began enforcing synchronous design rules, which made their algorithms simpler, the software complexity more manageable, and the tools faster and more efficient.

2.3.1   Five Rules of Synchronous Design

I use five rules to define synchronous design for a single clock domain. (A single clock domain means that all logic is clocked by a single clock signal.)

1.   All data is passed through combinatorial logic, and through delay elements, typically flip-flops, that are synchronized to a single clock.

2.   Delay is always controlled by delay elements, not combinatorial logic.

3.   No signal that is generated by combinatorial logic can be fed back to the same combinatorial logic without first going through a synchronizing delay element.

4.   Clocks cannot be gated; clocks must go directly to the clock inputs of the delay elements without going through any combinatorial logic.

5.   Data signals must go only to combinatorial logic or data inputs of delay elements.

Note that I use the term "delay elements." Typically, these elements will be flip-flops because those are the common delay element devices in use. Strictly speaking, the delay elements do not need to be flip-flops, they can be any element whose delay is predictable and synchronized to a clock signal.

A design may have multiple clocks and thus multiple clock domains. In other words, there will be logic clocked by one clock signal and logic clocked by another clock signal, but the design must treat all signals passed between the two domains as asynchronous signals. In Section 2.3.7, you will see how to deal with asynchronous signals.

The following sections cover common asynchronous design problems, what specific problems they can cause, and how to design the same functionality using synchronous logic. In my career, I have seen many of these problems in real designs and, unfortunately, I have had to debug many of them.

2.3.2 Race Conditions

Figure 2.2 shows an asynchronous race condition where a clock signal is connected to the asynchronous reset of a flip-flop. This violates rules 2 and either 4 or 5. It violates rule 2 because an asynchronous reset has a delay that is controlled by the internal design of the flip-flop, not by a delay element. It violates rule 4 if SIG2 is a clock signal, because it should not go to the CLR input. Otherwise, if SIG2 is a data signal, it should not go to the CLK input.

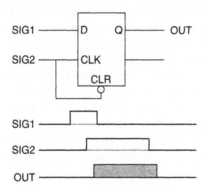

Figure 2.2: Asynchronous: Race condition. Note that OUT goes to an undefined state

Gate Count Controversy

What exactly is a gate count? The term comes from ASIC designs, specifically gate array ASICs, where designs are eventually reduced to the simplest elements consisting of logic gates—NANDs, NORs, buffers, and inverters. When FPGA vendors were courting ASIC designers, it made sense for them to compare the amount of logic that could be put into an FPGA with the amount that could be put into an ASIC. Because ASIC designers used gate counts, FPGA vendors started advertising gate counts for their devices.

The FPGA gate count had two problems. First, FPGAs don't have gates. They have larger grain logic such as flip-flops, and lookup tables that designers can use to implement Boolean equations that don't depend on gates. For example, the equation:

A = B & C & D & E & F

requires one 5-input AND gate in an ASIC or one 5-LUT in an FPGA. However, the equation:

A = ((B & C) | (D & E)) & ~F

requires five gates—three AND gates, one OR gate, and an inverter—in an ASIC, but still only one 5-LUT in an FPGA. So a gate count isn't an accurate measure of the logic a designer can fit into an FPGA.

The second problem is that utilization of the available logic in an FPGA is not nearly 100% and is very application dependent. Utilization percentages of 60 to 80 are much more common for any given design. So although an FPGA may be able to hold the equivalent of a 1 million-gate design, in theory, it is unlikely that a designer can actually fit and successfully route any particular 1 million-gate design in such a FPGA.

For this reason, the different FPGA vendors attacked their competitors' gate count numbers. Then, years ago, a nonprofit organization called PREP created what was called the PREP benchmarks. These benchmarks consisted of standard designs to be synthesized, placed, and routed into FPGAs from different vendors. The idea was that this would be a standard way of comparing the densities, routability, power consumption, and speed of these different FPGAs. This seemed like a better solution than the simple gate count. The different vendors, however, fought vehemently and many refused to participate in the benchmarks, claiming that some of the benchmark designs conformed to their competitors' architectures, producing deceptively better results. They also claimed that some synthesis and place and route tools used for benchmarking did a better job of optimizing their competitors' FPGAs, again making their competitors look better on these specific designs. Their arguments were not without merit and PREP eventually disbanded.

For some reason, though, gate count has come to be an accepted standard among FPGA vendors. They no longer complain, publicly at least, that their competitors are using misleading methods of counting available gates in their FPGAs. As a user of the FPGAs, however, you should understand that gate counts are a very rough estimate of capacity. Use them only for making rough determinations and rough comparisons.

How does this logic behave? When SIG2 is low, the flip-flop is reset to a low state. On the rising edge of SIG2, the designer wants the output, OUT, to change to reflect the current state of the input, SIG1. Unfortunately, because we do not know the exact internal timing of the flip-flop or the routing delay of the signal to the clock versus the routing delay of the reset input, we cannot know which signal will effectively arrive at the appropriate logic first—the clock or the reset. This is a race condition. If the clock rising edge arrives first, the output will remain low. If the reset signal arrives first, the output will go high. A slight change in temperature, voltage, or process may cause a chip that works correctly to suddenly work incorrectly because the order of arrival of the two signals changes.

My first step when creating a synchronous design, or converting an asynchronous design to a synchronous one, is to draw a state diagram. Although this may seem like overkill for such a small function, I find it useful to organize my thoughts and make sure that I've covered all of the possible conditions. The state diagram for this function is shown in Figure 2.3. From this diagram, it is easy to design the more reliable, synchronous solution shown in Figure 2.4. Here the flip-flop is reset synchronously on the rising edge of a fast clock. I've introduced a new signal, STATE, that together with the OUT signal, will uniquely identify the three states of the FSM. This circuit performs the correct function, and as long as SIG1 and SIG2 are produced synchronously—they change only after the rising edge of CLK—there is no race condition.

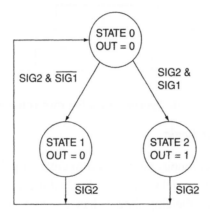

Figure 2.3: Synchronous state diagram

Now some people may argue that the synchronous design uses more logic, adding delay and using up expensive die space. They may also argue that the fast clock means that this design will consume more power. (This is especially true if it is implemented in CMOS, because CMOS devices consume power only while there is a logic transition. In this design, the flip-flops will consume power on every clock edge.) Finally, these people may argue that this design introduces extra signals that require more routing resources, add delay, and again, that consume precious die space. All of this is true. This design, however, will work reliably, and the previous design will not. End of argument.

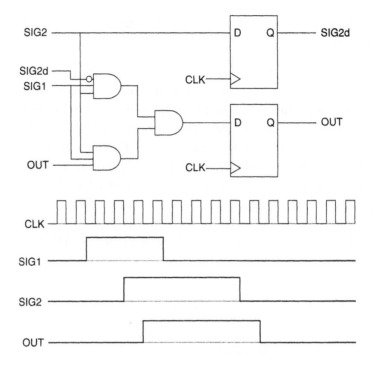

Figure 2.4: Synchronous: No race condition

2.3.3   *Delay Dependent Logic*

Figure 2.5 shows an asynchronous circuit used to create a pulse. The pulse width depends very explicitly on the delay of the individual logic gates. If the semiconductor process used to manufacture the chip should change, making the delay shorter, the pulse

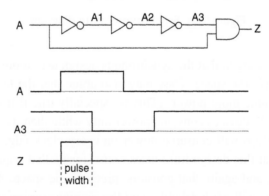

Figure 2.5: Asynchronous: Delay dependent logic

width will shorten also, to the point where the logic that it feeds may not recognize it at all. Because chip vendors are continually speeding up their processes, you can be certain that this type of design will eventually fail for some new batch of chips.

A synchronous version of a pulse generator is shown in Figure 2.6. This pulse depends only on the clock period. As our rule number 2 of synchronous design states, delay must always be controlled by delay elements. Changes to the semiconductor process will not cause any significant change in the pulse width for this design.

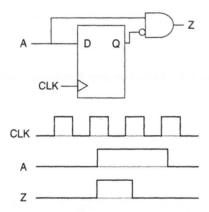

Figure 2.6: Synchronous: Delay independent logic

2.3.4   Hold Time Violations

Figure 2.7 shows an asynchronous circuit with a hold time violation. Hold time violations occur when data changes around the same time as the clock edge; it is uncertain which value will be registered by the clock—the value of the data input right before the clock edge or the value right after the clock edge. It all depends on the internal characteristics of the flip-flop. This can also result in a metastability problem, as discussed later.

The circuit in Figure 2.8 fixes this problem by putting both flip-flops on the same clock and using a flip-flop with an enable input. A pulse generator creates a pulse, signal Dp3, by ANDing signal D3 and a signal D3d, which is D3 delayed by a single clock cycle. The pulse D3p enables the flip-flop for one clock cycle.

The pulse generator also turns out to be very useful for synchronous design, when you want to clock data into a flip-flop after a particular event.

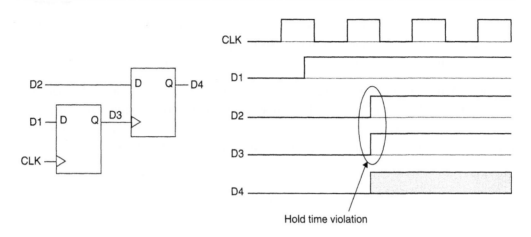

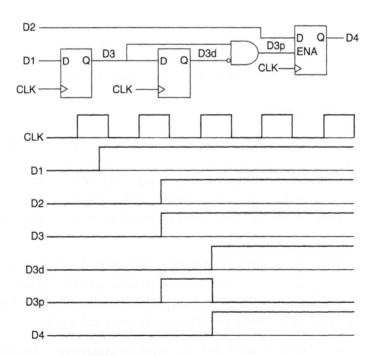

Figure 2.7: Asynchronous: Hold time violation

Figure 2.8: Synchronous: No hold time violation

2.3.5 Glitches

A glitch can occur due to small delays in a circuit, such as that shown in Figure 2.9. This particular example is one I like because the problem is not obvious at first. Here, a multiplexer switches between selecting two high inputs. It would appear, as it did to me when I was first shown this example, that the output would be high no matter what the value of the select input. One should be able to change the select input from low to high and back to low again and still get a high value out. In practice, though, the multiplexer produces a glitch when switching the select input. This is because of the internal design of the multiplexer, as shown in Figure 2.9. Due to the delay of the inverter on the select input, there is a short time when signals SEL and SELn are both low. Thus neither input is selected, causing the output to go low.

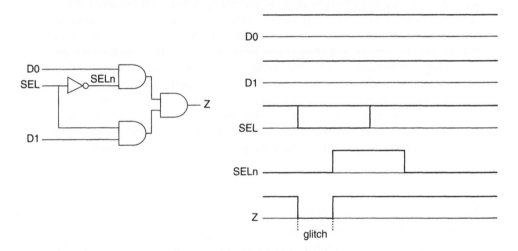

Figure 2.9: Asynchronous: Glitch

Changing Processes

Years ago I was working on a project designing some controller boards for a small client company. The vice president of manufacturing approached me and told me about a problem they were having and asked if I had any ideas. It seems that they had been shipping a particular board for about two years. Suddenly, every board would fail the preship tests they ran on it. They had assigned an engineer to look into it, but he couldn't find anything. What was particularly strange was that no part of the design had changed in two years.

I took time to look at the board and at the tests they were running on it. I narrowed the problem down to a particular FPGA and began examining the design. I found that there was one asynchronous circuit where a logic signal was being used to clock a flip-flop. I decided to call up the FPGA vendor and ask them whether they had recently changed their manufacturing process. They said that they had moved their devices over to a faster semiconductor process about two months ago. That corresponded exactly to the time when these boards started failing.

This illustrates a very important point about synchronous design. When you design synchronously, you are immune to process speedups because the chip vendor ensures that any speedups result with clock signals that are still much faster than data signals. However, if you have designed an asynchronous circuit, it works because the relationship between data signals has a specific timing relationship. In a new semiconductor process, these relationships between data signals may no longer hold.

Also, you will notice that FPGA vendors do not specify minimum delay times. This is because they want to have the ability to move older devices to newer, faster processes. When a semiconductor process is new, the bugs haven't been worked out, and the yields tend to be low. The vendor will charge more for denser, faster chips based on this process. Once the bugs are worked out and yields go up to a reasonable level, the vendor does not want to maintain two different processes because it is too expensive. Instead, the vendor will move the "slower" chips over to the faster process. So these so-called "slower" chips are now faster than before. As long as they have not specified the minimum times, the timing numbers for these "slower" chips are still within the specifications. And as long as you have designed synchronously, you will not have the problem that this client of mine did.

Synchronizing this output by sending it through a flip-flop, as shown in Figure 2.10, ensures that this glitch will not appear on the output and will not affect logic further downstream. As long as the timing calculations have been performed correctly, the entire design is synchronous, and the device is operated below the maximum clock frequency for the design, glitches such as this one will settle before the next clock edge.

2.3.6 Gated Clocking

Figure 2.11 shows an example of gated clocking. This violates the fourth and fifth rules of synchronous design because the circuit has data signals going to clock inputs and clock signals going to data inputs. This kind of clock gating will produce problems that will be particularly bad in FPGAs, because the GATE signal can easily be delayed so

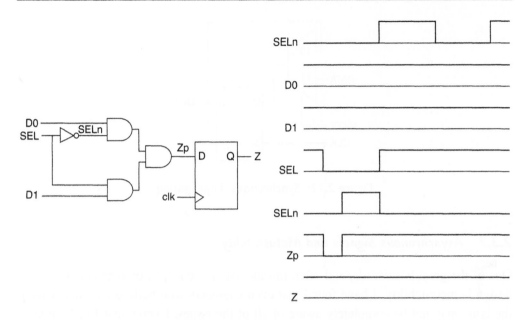

Figure 2.10: Synchronous: No glitch

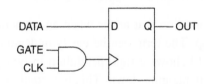

Figure 2.11: Asynchronous: Clock gating

that the clock signal rises before the GATE signal can prevent it. Data then gets clocked into the flip-flop on a cycle when it is not supposed to.

The correct way to enable and disable outputs is not by putting logic on the clock input, but by putting logic on the data input, as shown in Figure 2.12. Essentially, this circuit consists of an enable flip-flop that has a data signal, GATE, which enables and disables the flip-flop. In this synchronous design, the flip-flop is always being clocked directly by the CLK signal. The GATE input controls the mux on the input, to determine whether the new data gets clocked in or the old data gets clocked back in.

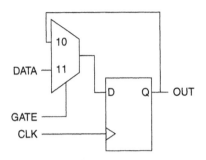

Figure 2.12: Synchronous: Logic gating

2.3.7   Asynchronous Signals and Metastability

One of the great buzzwords, and often-misunderstood concepts, of synchronous design is metastability. I have found that even engineers who believe they understand the issue may not be completely aware of all of the issues. I know that I had a good knowledge of metastability before I began writing about it. When I write about a topic, though, I make sure I understand it exactly. After doing some research on metastability, I realized that I still hadn't completely appreciated some issues.

Metastability refers to a condition that arises when an asynchronous signal is clocked into a synchronous flip-flop. The term can be easily understood from its equivalent in basic physics. Figure 2.14 shows a rigid pendulum in two possible stable states. In the first, the pendulum is hanging down. This is a stable state. The pendulum will tend to go into this state. If you tweak the pendulum, it will swing back and forth for a while and eventually end up back where it started.

The Enable Flip-Flop

The enable flip-flop is an important piece of logic for synchronous design. Many pieces of asynchronous logic can be turned into synchronous logic with the use of an enable flip-flop. The enable flip-flop allows data to be selectively clocked into a flip-flop rather than being clocked in on every clock edge.

The logic for an enable flip-flop is simple, as shown in Figure 2.13. It consists of a mux placed in the data path going to a normal flip-flop. When the enable signal is asserted, the data goes through the mux and is clocked into the D-input at the next clock edge. When the enable signal is not asserted, the output of the flip-flop is simply fed back, through the mux, into the data input of the flip-flop so that the data is continually clocked in on each clock edge.

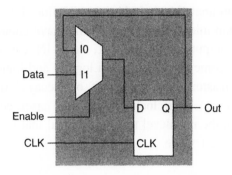

Figure 2.13: An enable flip-flop

The second part of the figure shows a rigid pendulum in a metastable state. If the pendulum is exactly in this state, upside-down, it will remain in this state indefinitely. However, if some small perturbation occurs—you tweak the pendulum in this state—the pendulum will swing down and eventually settle into the first, stable state. In other words, a metastable state is a precarious one in which an object will not remain for very long.

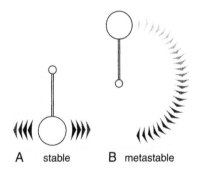

Figure 2.14: Stable states of a rigid pendulum

Although chip designers would prefer a completely synchronous world, the unfortunate fact is that signals coming into a chip will depend on a user pushing a button or an interrupt from a processor, or will be generated by a clock that is different from the one used by the chip. In these cases, designers must synchronize the asynchronous signal to the chip clock so that it can be used by the internal circuitry. The designer must do this carefully in order to avoid metastability problems. Figure 2.15 shows a circuit with potential metastability. If the ASYNC_IN signal goes high around the same time as the clock, it creates an unavoidable hold time violation. Flip-flop FF1 can go into a metastable state. If certain internal transistors do not have enough time to fully charge to the correct level, the output of flip-flop FF1, signal IN, can actually go to an undefined voltage level, somewhere between a logic 0 and logic 1. This "metalevel" will remain until the transistor voltage leaks off, or "decays," or until the next clock cycle when a good clean value gets clocked in. In some cases, depending on the internal structure of the flip-flop, the metalevel may be a signal that oscillates between a good 0 value and a good 1 value.

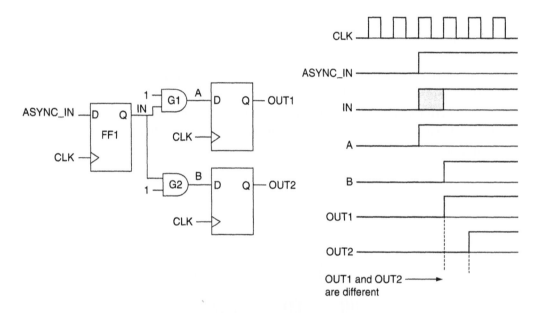

Figure 2.15: Metastability—the problem

So why is it that the large microprocessor manufacturers can use a technique that I am telling you not to use? Essentially they have a luxury that you do not have. First, the design teams at these companies have access to detailed circuit level simulations that can determine the timing more accurately than you can. These design teams work closely with the process engineers and the layout designers to calculate very exact timing numbers in order to avoid race conditions or other hazards due to asynchronous design.

Second, when the process engineers speed up a semiconductor process, changing the timing specifications, you can be sure that the design engineers know about it ahead of time. This gives them enough time to recalculate all of the critical timing and change the design and/or the layout to compensate for the new process.

On the other hand, when CPLD and FPGA vendors plan to change their process, it is unlikely that they will notify you beforehand. And if they did, would you really want to spend engineering time and resources redesigning your chip each time they change their processes? Definitely not. For this reason, the microprocessor manufacturers can gate their clocks in order to reduce power but you, unfortunately, should not.

This is where many engineers stop. This metalevel voltage on signal IN, though, is not really the problem. During the clock cycle, the gates driven by signal IN may interpret this metalevel differently. In the figure, the upper AND gate, G1, sees the level as a logic 1, whereas the lower AND gate, G2, sees it as a logic 0. This could occur because the two gates have different input thresholds because of process variations or power voltage variations throughout the die. It is possible that one gate is near the output of FF1, and the other is far away. Such differences in routing can cause the signal to change enough at the input to each gate to be interpreted differently.

In normal operation, OUT1 and OUT2 should always be the same value. Instead, we have created a condition that cannot occur according to the rules of logic. This condition is completely unpredictable. This is the problem with metastability—not that an output has a bad voltage, but that a single signal is interpreted differently by different pieces of logic. This problem will send the logic into an unexpected, unpredictable state from which it may never return. Metastability can permanently lock up your chip.

The "solution" to this metastability problem is shown in Figure 2.16. By placing a synchronizer flip-flop, S1, in front of the logic, the synchronized input, SYNC_IN, will be sampled by only one device, flip-flop FF1, and be interpreted only as a logic 0 or 1. The output of flip-flop FF1, IN, will be either a clean 1 or a clean 0 signal. The upper and lower gates, G1 and G2, will both sample the same logic level on signal IN.

There is still a very small but nonzero probability that a metastable signal SYNC_IN, will cause the output of flip-flop FF1, signal IN, to go metastable on the next clock edge, creating the same problem as before. Why would signal IN go metastable? A metalevel voltage on the input may cause a capacitor in flip-flop FF1 to change part way, creating a metalevel output. Of course, there is still an advantage with the synchronizer flip-flop. Because a device does not prefer to remain in a metastable state, there is a good chance that it has decayed into a stable state by the next clock edge. So this creates a time period—one clock period—during which the metastable device can stabilize before a problem occurs.

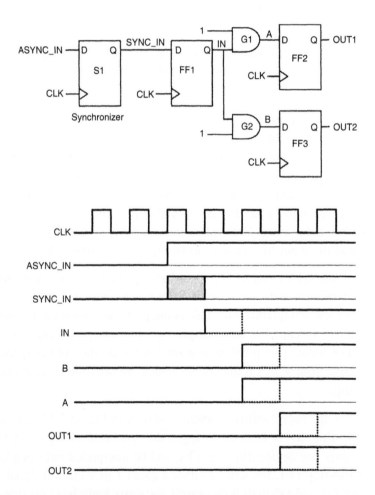

Figure 2.16: Metastability—the "solution"

It seems that, in either case, the metastability problem is solved. Or is it? I have put the word "solution" in quotation marks for a very good reason. Because this problem involves a probabilistic event, there is a possibility that the signal SYNC_IN will not decay to a valid logic level within one clock period. In this case, the next flip-flop will sample a metalevel value, and there is again a possibility that the output of flip-flop FF1, signal IN, will be indeterminate. At higher frequencies, this possibility increases.

Unfortunately, there is no certain solution to this problem. Some vendors provide special synchronizer flip-flops whose output transistors decay very quickly. Also, inserting more synchronizer flip-flops reduces the probability of metastability, but doing so will never reduce the probability to zero. The correct action involves discussing metastability problems with the vendor, and including enough synchronizing flip-flops to reduce the probability so that the problem is unlikely to occur within the lifetime of the product.

A good way to judge a CPLD or FPGA vendor is to ask them for metastability data and solutions. A good vendor will have run tests on their flip-flops and have characterized the probability of creating a metastable state. They may have special synchronizing flip-flops and they should have rules for the number of flip-flops needed at a given clock frequency, to reduce the odds of a problem due to metastability to a reasonable number.

Synchronizer Delay

Notice that each synchronizer flip-flop will delay the input signal by one clock cycle before it is recognized by the internal circuitry of the chip. This may at first seem to be a problem, but it is not. Given that the external signal is asynchronous, by definition the exact time that it is asserted will not be deterministic. You may need to respond to it within a set time period, but that time period should be orders of magnitude greater than several clock cycles. If this delay is a problem in your design, most likely this input should not be an asynchronous signal. Instead, you should generate this signal by logic that uses the same clock that is used in the rest of the chip, which will eliminate the metastability problem altogether.

2.3.8 Allowable Uses of Asynchronous Logic

Now that I've gone through a long argument against asynchronous design, I will tell you the few exceptions that I have found to this rule. I used to say that there were no exceptions. But experience always shows that every rule has an exception (except that rule). However, approach these exceptions with extreme caution and consideration.

2.3.8.1 Asynchronous Reset

Sometimes it is acceptable, or even preferable, to use an asynchronous reset throughout a design. If the vendor's library includes flip-flops with asynchronous reset inputs, designers can tie the reset input to a master reset in order to reduce the routing congestion and to reduce the logic required for a synchronous reset. FPGAs and CPLDs have master reset signals built into the architecture. Using these signals to reset state machines frees up interconnect for other uses. Because routing resources are often the limiting factor in the density of an FPGA design, you should take advantage of the asynchronous reset input to the flip-flops.

The rules to follow for an asynchronous reset are:

- Use it only to initialize the chip. Asynchronous resets should not occur during normal operation.

- Assert the reset signal for at least one clock cycle.

- After reset, ensure that the chip is in a stable state such that no flip-flops will change until an input changes. In other words, after reset every state machine should be in an idle state waiting for an input signal to change.

- The inputs to the chip should be stable and not change for at least one clock cycle after the reset is removed.

2.3.8.2 Asynchronous Latches on Inputs

Some buses, such as the VME bus, are designed to be asynchronous. In order to interface with these buses, designers need to use asynchronous latches to capture addresses or data. Once the data is captured, it must be synchronized to the internal clock. One suitable technique is to synchronize each incoming signal with its own synchronizing flip-flop. (This is what I suggested in the previous discussion on metastability.)

However, it is usually much more efficient to use asynchronous latches to capture the bus signals initially. Many buses have an address latch enable (ALE) signal to latch addresses and a data strobe (DSTROBE) to latch data. Unless your chip uses a clock that has a frequency much higher than that of the bus, attempting to synchronize all of these signals will cause a large amount of overhead and may actually create timing problems rather than eliminate them. Instead, you can latch the inputs asynchronously and then synchronize the signal that gets asserted last in the protocol, usually the data strobe, to the internal clock. By the time this late signal has been synchronized, you can be certain that all of the inputs are stable.

Synchronous VME Bus Is Not Synchronous

The VME bus has been in existence for a relatively long time now. It's a standard, a workhorse. In use since the early days of microprocessors, it's the granddaddy of all buses. And it's asynchronous, like most of the buses that came after it. Asynchronous buses are easier to implement over different media using various components. Synchronous buses require a lot of coordination between different devices, and they need carefully controlled timing.

When synchronous design became more recognized and more important for chip design for all of the reasons I discussed, synchronous buses such as PCI and SBUS started popping up. Synchronous buses have tighter timing requirements, but they can be much more efficient at transferring data. Some engineer or standards committee or perhaps a marketing person decided to make a synchronous VME bus by adding a clock to it. Unfortunately, no relationship exists between the signals of the bus and the clock. The clock was simply added to the existing asynchronous bus, resulting in an asynchronous bus with a clock.

I discovered this the hard way when I was asked to interface an FPGA with the "synchronous" VME bus. The overhead of logic and clock cycles required to synchronize this "synchronous" bus made data transfers very inefficient. My advice is this: When interfacing to a bus, look at the timing relationships to determine whether it is really synchronous. If these relationships are synchronous, a synchronous interface will be very efficient. Otherwise, despite the label that the bus may have, treat it as asynchronous and design the interface accordingly.

2.4 Floating Nodes

Floating nodes are internal nodes of a circuit that are not driven to a logic 0 or logic 1. They should always be avoided. An example of a potential floating node is shown in Figure 2.17. If signals SEL_A and SEL_B are both not asserted, signal OUT will float to an unknown level. Downstream logic may interpret OUT as a logic 1 or a logic 0, or the floating signal may create a metastable state. In particular, any CMOS circuitry that uses signal OUT as an input will use up power because CMOS dissipates power when the input is in the threshold region. The signal OUT will typically float somewhere in the threshold region. Also, even if downstream logic is not using this signal, the signal can bounce up and down, causing noise in the system and inducing noise in surrounding signals.

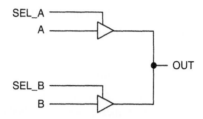

Figure 2.17: Floating nodes—the problem
NOTE: SEL_A and SEL_B are mutually exclusive.

Two solutions to the floating node problem are shown in Figure 2.18. At the top, signal OUT is pulled up using an internal pull-up resistor. This simple fix ensures that when both select signals are not asserted, OUT will be pulled to a good logic level. Note that the pull up represented in the picture may be a passive resistor, or it may be an active pull up circuit that can be faster and more power conservative.

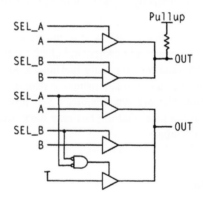

Figure 2.18: Floating nodes—solutions
NOTE: SEL_A and SEL_B are mutually exclusive.

The other solution, shown at the bottom of the figure, is to make sure that something is driving the output at all times. A third select signal is created that drives the OUT signal to a good level when neither of the other normal select signals are asserted.

2.5   Bus Contention

Bus contention occurs when two outputs drive the same signal at the same time, as shown in Figure 2.20. This reduces the reliability of the chip because it has multiple drivers fighting each other to drive a common output. If bus contention occurs regularly, even for short times, the possibility of damage to the drivers increases.

One place where this can occur, and that is often ignored, is during the turnaround of a bus. In a synchronous bus, when one device is driving the bus during one clock cycle and a different device is driving during the next clock cycle, there is a short time when both devices may be driving the bus, as shown in Figure 2.19.

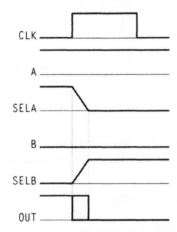

Figure 2.19: Contention during synchronous bus turnaround

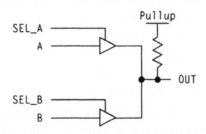

Figure 2.20: Bus contention—the problem
NOTE: SEL_A and SEL_B are not mutually exclusive.

To avoid contention problems, the designer must ensure that both drivers cannot be asserted simultaneously. This can be accomplished by inserting additional logic, as shown in Figure 2.21. The logic for each buffer enable has been modified so that a buffer is not turned on until its select line is asserted and all other select lines have been de-asserted. Due to routing delays, some contention may still occur, but this circuit has reduced it significantly. Of course, the best solution may be to find better implementations. For example, designers can use muxes instead of tri-state drivers, though muxes are often difficult to implement in FPGAs. Other solutions involve designing the system so that there is always a clock cycle where nothing is driving the bus. Of course, during those cycles, you want to be certain that the bus does not float by pulling it up.

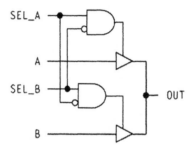

Figure 2.21: Bus contention—the solution
NOTE: SEL_A and SEL_B are not mutually exclusive.

2.6   One-Hot State Encoding

For a typical "large-grained" FPGA architecture, the normal method of designing state machines is not optimal. This is because the normal approach to FSMs tends to couple a few flip-flops that encode state to a large network of combinatorial logic that decodes the state. In an FPGA, though, large combinatorial networks must be built by aggregating several CLBs. Each single CLB may contain small lookup tables that can easily implement any eight or nine input combinatorial function. If you need a 10 input function, you need to spread the logic over two CLBs. Decoding the state representation used in classic FSM designs, can involve many CLBs. This means that routing becomes involved, significantly adding to the circuit delay, and slowing down the maximum clock speed for the design. An alternate design method, called one-hot state encoding, is better suited to FPGAs because it reduces the the number of inputs to the combinatorial network, thus reducing routing overhead and allowing better timing and thus faster clocks.

Figure 2.22 shows a small but typical state diagram for some simple state machine. Using the classic design methodology the four states would be represented as two state bits.

Figure 2.23 shows a typical design for this state machine, where the (S1, S0) states are defined as follows: IDLE (00), STATE1 (01), STATE2 (10), and STATE3 (11).

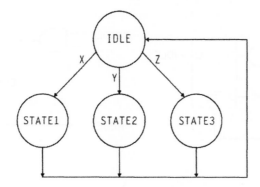

Figure 2.22: State diagram

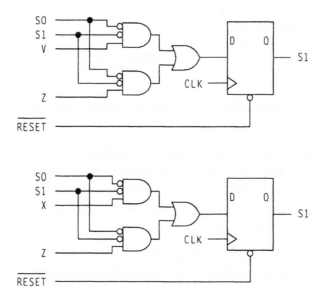

Figure 2.23: State machine: Usual method

Notice that although the number of flip-flops are minimized, the combinatorial logic is fairly large. As the number of states grows, the number of inputs needed for the combinatorial logic grows because the state representation must be decoded to determine the current state.

The better method of designing state machines for FPGAs is known as one-hot encoding, shown in Figure 2.24. Using this method, each state is represented by a single

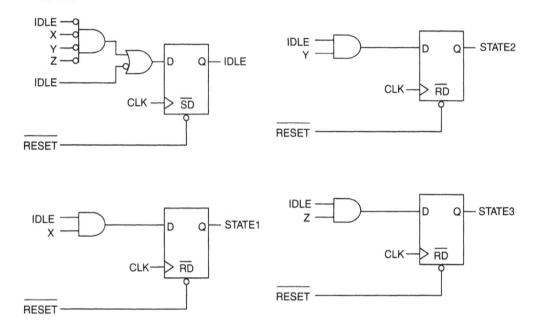

Figure 2.24: State machine: One-hot encoding

state bit, thus a single flip-flop, rather than encoded from several state bits. This greatly reduces the combinatorial logic, because designers need to check only one bit to determine the current state. Many synthesis tools now recognize the need for one-hot encoding for FPGAs, and can re-encode your state bits to produce optimal encoding for the particular CPLD or FPGA that you are using.

Note that each state bit flip-flop needs to be reset when initialized, except for the IDLE state flip-flop, which needs to be set so that the state machine begins in the IDLE state.

2.7   Design For Test (DFT)

The "design for test" philosophy stresses that testability should be a core design goal. Designed-in test logic plays two roles. The first role is to help debug a chip that has design flaws. These flaws are problems where the chip may perform the function for which it is designed, but that design will not operate properly in your system. The second role of test logic is to catch physical problems. Physical problems usually show up in production, but sometimes marginal problems appear only after the chip has been in the field for some time. Sometimes the same test logic can fill for both roles. Sometimes, the two roles require different kinds of test structures.

Both roles are particularly important for ASIC design because of the black box nature of ASICs, where internal nodes are simply not accessible when a problem occurs. These techniques are also applicable to CPLDs and FPGAs, many of which already have built-in test features. One difference between ASIC and FPGA/CPLD design is that for an ASIC design, you are expected to provide test structures and test vectors to use during production, to find any physical defects. With an FPGA, you can safely assume that the vendor has performed the appropriate production tests. However, some physical defects may show up only after prolonged use, so you may still want to design in test logic that allows you to check for physical defects while your chip is working in a system in the field. For each of the following tests, I note whether the test is applicable to physical defects, functional problems, or both, and whether you can use the test for the debug process, in the field, or both.

2.7.1 The 10/10 Rule of Testing

While test logic is intended to increase the testability and reliability of your FPGA, if the test logic becomes too large, it can actually decrease reliability. This is because the test logic can itself have problems that cause the FPGA to malfunction. A rule of thumb that I call the 10/10 rule is described in Figure 2.25.

- *Test circuitry should not make up more than 10 percent of the logic of the entire FPGA.*
- *You should not spend more than 10 percent of your time designing and simulating your test logic.*

Figure 2.25: The 10/10 rule of testing

The following sections describe DFT techniques that allow for better testing of a chip. While not all of these techniques need to be included in every design, those techniques that are needed should be included *during the design process* rather than afterwards. Otherwise, circuits can be designed that are later found to be difficult, if not impossible, to test.

2.8   Testing Redundant Logic

2.8.1   What Is Redundant Logic?

Tests for physical defects
It is used in the field

Redundant logic is used most often in systems that need to operate continuously without failure. Military systems and banking systems are two examples of systems that should not stop while in use. In these types of systems, logic will be duplicated. There will be a device following the redundant hardware that compares the outputs of the redundant hardware. Often, these systems will have three redundant blocks so that if one block fails, two blocks are still working, and the one bad block can be ignored. The comparison hardware is called "voting" logic because it compares signals from the three redundant blocks and decides that the majority of signals that agree have the correct value.

Most hardware is not designed to continue to operate with a physical failure, so redundant logic is not common. However, if you do use redundant logic in your design, you want to make sure that all of the redundant logic is working correctly. The idea is that redundant logic finds manufacturing faults that occur after the chip is in the field. The next section discusses how to functionally test redundant logic.

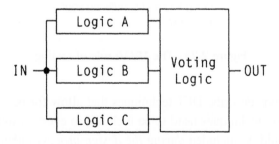

Figure 2.26: Redundant logic

Redundant Logic and the Battle Switch

Years ago, I worked for ROLM Corporation, the premiere Silicon Valley company of its day. ROLM was flying high, employing lots of people, and recording great revenue growth in the telecommunications industry. At the time, everyone wanted to work at ROLM. ROLM offered great salaries, stock option plans, and benefits. They epitomized the Silicon Valley workplace with large offices for engineers in spacious buildings on "campuses" that were nested among trees and artificial lakes. The cafeterias were subsidized by the company and served a great variety of foods in a quiet, elegant atmosphere. Most notably, the ROLM campus included a full health club with basketball courts, racquetball courts, weight room, swimming pool, Jacuzzi, and offered aerobics classes every day.

ROLM really defined the Silicon Valley company. It's sad that it grew too fast and didn't have the money to support the growth. It ended up being bought by IBM, split up, sold off in pieces, and eventually absorbed into several other companies. ROLM is a distant memory and most Silicon Valley engineers haven't even heard of it.

The reason I mention ROLM is that I worked in the Mil-Spec Computer Division, which manufactured computers for the military. We had to design fault tolerance and redundancy into the computers. On the side of our computers was a switch labeled the "battle switch," which no one really talked about. One day I asked a project leader about the function of this switch. He explained it this way. Throwing that switch turned off all error checking and fault tolerance. I asked him why. He replied that in a war, particularly a nuclear war, it's better to have a faulty computer fire off as many missiles as possible, even if some hit friendly targets, as long as one of them reaches the enemy. Pretty scary thought. I guess that's why no one discussed it, and I didn't either after that.

2.8.2   How to Test Redundant Logic

Tests for functional problems

It is used during debugging

Testing redundant logic is a separate issue. Figure 2.27 shows a circuit that has redundant logic. However, because the circuit is not testable, the effect is not as useful as it should be. If a design flaw exists, or a physical defect occurs before the chip is shipped, the redundant logic will hide the problem. If a defect occurs in the field, the chip will simply produce incorrect results—the incorrect redundant logic will not prevent failure. Thus, if it is to contribute meaningfully to reliability, redundant logic must properly tested before it is shipped.

The circuit in Figure 2.27 shows how to modify Figure 2.26 for testing. The extra test lines allow you to disable some of the redundant logic and test each piece independently during debug of the design. This allows you to be sure that each piece of redundant logic is working correctly and identically with its redundant counterparts.

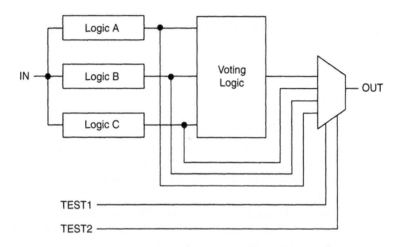

Figure 2.27: Testing redundant logic

2.9   Initializing State Machines

Tests for functional problems
It is used during debugging

It is important that all state machines, and in fact all registers in your design, can be initialized. This ensures that if a problem arises, testers can put the chip into a known state from which to begin debugging.

Also, for simulation purposes, simulation software needs clocked devices to start in a known state. I like to use the example of a divide-by-two counter—i.e., a flip-flop with its Qn output tied to its D input. The output of this flip-flop is a square wave that is half the frequency of the input clock signal. It may not matter in the design whether the flip-flop starts out high or low. In other words, the phase relationship between the input clock and the half-frequency output clock may not matter. But when you simulate the design, the output clock will start out undefined. On the next clock cycle, the output will change from undefined to the inverse of undefined, which is. . .undefined. The output clock will remain undefined for the entire simulation.

2.10 Observable Nodes

Tests for functional problems

It is used during debugging

It is a good idea to make internal nodes in your chip design observable. In other words, testers should be able to determine the values of these nodes by using the I/O pins of the chip. Figure 2.28a shows an unobservable state machine. In Figure 2.28b, the state machine has been made observable by routing each state machine output through a mux to an external pin. Test signals can be used to select which output is being observed.

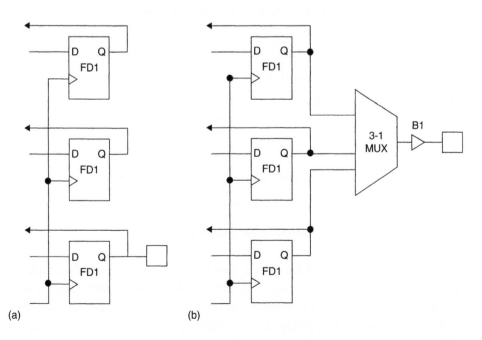

(a)　　　(b)

Figure 2.28: Observable nodes

If no pins are available, the state bits can be multiplexed onto an existing pin that, during testing, is used to observe the state machine. This configuration allows for much easier debugging of internal state machines. If your system includes a microprocessor, you can connect the microprocessor to the chip to allow it to read the internal nodes and assist with debugging.

2.11   Scan Techniques

Tests for physical defects or functional problems

It is used in the field or during debugging

Scan techniques sample the internal nodes of the chip serially so that they can be observed externally. Each flip-flop in the design gets replaced with a scan flip-flop, which is simply a flip-flop with a two-input mux in front of the data input as shown in Figure 2.29. The scan enable input (SE) is normally low, so that the normal data input gets clocked into the flip-flop. In scan mode, though, the scan enable input is high, causing the scan data (SD) to get clocked into the flip-flop.

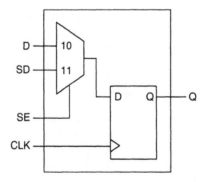

Figure 2.29: Scan flip-flop

The output of each flip-flop in the chip is then connected to the scan data input of another flip-flop, as shown in Figure 2.30. This creates a huge chain, called a scan chain. Automatic Test Pattern Generation (ATPG) software packages take an HDL description and convert all flip-flops to scan flip-flops and insert the scan chain into your design.

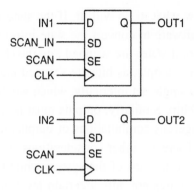

Figure 2.30: Scan chain

There are two main scan techniques—*full scan* and *boundary scan*. Full scan involves creating scan chains from every flip-flop in the design. Boundary scan involves using only flip-flops that are connected to I/O pins in the scan chains.

2.11.1 Full Scan

When performing full scan, the entire chip is put into scan mode, and the scan enable inputs to the scan flip-flops are asserted. Thus, testers can examine the state of each flip-flop in the design. Also, testers can put the chip into a completely predictable state by scanning a certain pattern into it. This technique of scanning patterns into and out of the chip is used for finding physical defects in ASICs after production. In fact, it is the only practical way to catch defects in today's very large ASICs.

Because FPGAs are not typically tested after production, except by the manufacturer, the use of scan is restricted to functional testing of FPGAs, if it is used at all. During debugging, if the chip malfunctions, it can be halted, put into scan mode, and the state of each flip-flop can be read via the scan. These bits can then be loaded into a simulation of the design to help figure out what went wrong. The simulation data can also be scanned back into the chip to put the chip into a known starting state. All of this allows the simulator and the physical chip to be used together to debug design problems—at least in theory.

The major problem with using this kind of technique for functional testing is that the scanning requires a lot of software development. Each flip-flop bit must be stored, and the software must know what to do with it. If the state is to be loaded into a simulator, there must be software to convert the state information to the simulator's format and back again. Also, if states are scanned into the chip, one must be careful not to scan in illegal states. For example, as bits are scanned into the chip, it is possible to turn on multiple drivers to a single net internally, which would normally not happen, but which would burn out the chip. Similarly, outputs must be disabled while the chip is being scanned because dangerous combinations of outputs may be asserted that can harm the attached system. There are other considerations, also, such as what to do with the clock and what to do with the rest of the system while the chip is being scanned. Avoiding these problems requires not only a certain level of sophistication in the software, but may also require extra hardware. Only very large, expensive systems can justify the cost of full scan for most designs. Also, scan requires that the chip be halted, which may not be practical or even allowable in certain systems, such as communication systems or medical devices.

2.11.2 Boundary Scan

Boundary scan is somewhat easier to implement and does not add as much logic to the design. Boundary scan reads only nodes around the boundary of the chip, not internal nodes. Limiting the scan to boundary nodes avoids internal contention problems, but not contention problems with the rest of the system. Boundary scan is also useful for testing the rest of your system, because testers can toggle the chip outputs and observe the effect on the rest of the system. Boundary scan can be used to check for defective solder joints or other physical connections between the chip and the printed circuit board or between the chip and other chips in the system.

The Institute of Electrical and Electronic Engineers (IEEE) has created a standard for boundary scan called JTAG, or IEEE 1149.1. It covers pin definitions and signaling. Most CPLDs and FPGAs support this standard in their architecture, without the need for making any changes to your design.

2.12 Built-In Self-Test (BIST)

Tests for functional problems
It is used during debugging

Another method of testing a chip is to put all of the test circuitry on the chip in such a way that the chip tests itself. This is called built-in self-test, or BIST. In this approach, some circuitry inside the chip can be activated by asserting a special input or combination of inputs. This circuitry then runs a series of tests on the chip. If the result of the tests does not match the expected result, the chip signals that a problem exists. The details of what type of tests to run and how to signal a good or bad chip are dependent on several factors, including the type of circuit to be tested, the amount of area to be devoted to test logic, and the amount of time that can be spent on testing.

BIST can be used for production testing or in-field testing. Because CPLD and FPGA production testing is done by the manufacturer, its use for these devices is for in-field testing. BIST allows you to periodically test a device for physical defects while it is running in the system, as long as there is an idle period where it is not in use. During those times, the chip can be commanded to run a test on itself. Commercial software is now available to add BIST circuitry to a design.

Figure 2.31 represents an entire chip, and shows, in general terms, how BIST is implemented. Your design is in the block labeled "circuit under test." Everything else is BIST logic. To put the chip into test mode, the external test signal is asserted. Then, all inputs to the circuit come from the test generator rather than the real chip inputs. The test generator circuit produces a series of test vectors that are applied to the circuit

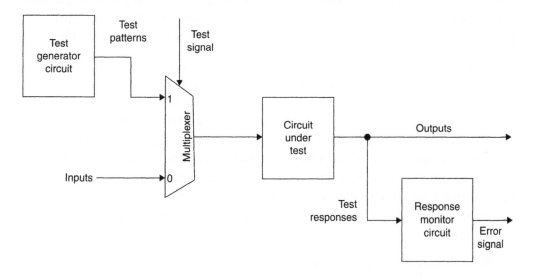

Figure 2.31: Built-in self-test

under test. The outputs of this circuit are then sent to a response monitor that compares the outputs to expected outputs. Any difference causes the error signal to be asserted.

One way to design the test generator is to store the simulation vectors in a memory and feed them out on each clock edge. The response monitor circuit would then also consist of a memory with all of the expected output vectors. The main problem, though, is that the memory to store the vectors would be too great, possibly even taking more silicon area than the original circuit. Another possibility would be to design logic to create the vectors and expected vectors, but that would require at least as much circuitry as the original chip design. The optimal solution is to use signature analysis, a form of BIST that is described in the next section.

2.13   Signature Analysis

Tests for physical defects

It is used in the field

A problem with the BIST technique, in general, is that the test generation and response monitoring circuitry can be extremely complex and take up a lot of silicon. Any test structure that is too large or complex ends up reducing the reliability of a design rather than increasing it, because now there is a significant chance that the test circuitry itself fails. Also, large test circuits increase the cost of a chip, another unwanted effect.

Signature analysis avoids these problems. BIST designs based on signature analysis use a test generation circuit that creates a pseudorandom sequence of test vectors. "Pseudorandom" means that the test vectors are distributed as though they were random, but they are actually predictable and repeatable. The response monitor takes the sequence of output vectors and compresses them into a single vector. This final compressed output vector is called the chip's signature. Although this type of testing is not exhaustive, if done right it can detect a very large number of failures. It is also possible that a nonworking chip will produce a correct signature, but if the number of bits in the signature is large, the probability of a bad device producing a good signature is very small.

Another significant advantage of signature analysis is that the pseudorandom number generation and the output vector compression can both be performed by a linear feedback shift register (LFSR), which is a very small device—simply a shift register and several XOR gates. Almost no silicon area is given up, and thus almost no additional cost is involved.

2.14   Summary

In this chapter, I have laid out specific design rules and guidelines that increase your chances of producing a working, reliable device—one that will work for different chip vendor processes and continue to work for the lifetime of your system. I have shown a number of examples of incorrect or inefficient designs, and the equivalent function designed correctly.

Specifically, after reading this chapter, you should be acquainted with these CPLD and FPGA design issues:

- Hardware description languages and different levels of design modeling.

- Top-down design, an approach to designing complex chips that allows you to better utilize or implement these important design capabilities:

  - Use of HDLs

  - Written specifications

  - Allocating resources

  - Design partitioning

  - Flexibility and optimization

  - Reusability

  - Floorplanning

  - Verification

- The significance of architectural differences and how they affect design trade-offs.

- Synchronous design—A requirement for creating devices that will function reliably over their entire lifetimes.

- Floating nodes—A condition that must be avoided in order to avoid excess noise and power consumption.

- Bus contention—Another condition that must be avoided to increase the long-term reliability of your chip.

- One-hot state encoding—A method of designing state machines that takes advantage of the architecture of FPGAs.

- Testing redundant logic—Why it's important and the typical design structures for implementing such tests.

- Initializing state machines—Designing a chip so that all internal state machines are initialized.

- Observable nodes—Bringing internal nodes to external pins so that the internal states of the chip can be observed.

- Scan techniques—Connecting flip-flops inside the chip (either boundary nodes, or both boundary and internal nodes) into one or more chains so that you can easily get 100% fault coverage to guard against manufacturing problems.

- Built-in self-test (BIST)—A method of including circuitry inside a chip that lets it test itself for physical problems.

- Signature analysis—A method using linear feedback shift registers (LFSRs) to simplify BIST circuitry.

The Linear Feedback Shift Register

The LFSR is an extremely useful function that, I believe, would be used even more often if there were more good references showing how to design them. Most textbooks that cover LFSRs (and there aren't many), discuss the mathematical theory of primitive polynomials. Unfortunately, this view doesn't help an electrical engineer design one from flip-flops and logic. The two books that do have practical design information are *Designus Maximus Unleashed!* by Clive "Max" Maxfield and my book, *Verilog Designer's Library.* The first is a very fun, off-the-wall book with lots of design tips. The second offers a library of tested Verilog functions, including just about any LFSR you would want to design.

The LFSR uses a shift register and an exclusive OR gate to create a sequence of numbers that are pseudorandom. This means they look as though they were chosen at random. They're not truly random because the sequence is predictable, which is a good thing because for any kind of testing, we need to repeat an exact sequence of patterns.

Figure 2.32 shows a 3-bit LFSR. Notice that the inputs to the XOR must be from specific outputs of the shift register to create a sequence that covers all seven possibilities. These outputs are known as taps. If the taps are not in the right places, the sequence will repeat early and some patterns will never appear in the sequence. Also note that the value of zero is never in the sequence. Starting with a zero in the shift register will always produce a sequence of all zeroes. When the LFSR is initialized, the starting patterns must be some value other than zero. If you need to put a zero into the sequence, you can insert it with extra logic, but then the sequence is not truly random.

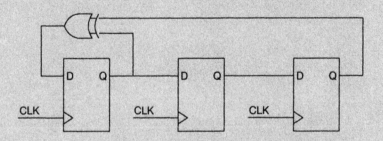

Figure 2.32: Linear feedback shift register

Using such an LFSR for the test generator and for the response monitor in a BIST circuit greatly simplifies the BIST circuit while still creating a chip signature that will flag most errors.

CHAPTER 3

A VHDL Primer: The Essentials

Peter R. Wilson

This chapter comes from the book Design Recipes for FPGAs by Peter Wilson, ISBN: 9780750668453. Peter's book is designed to be a desktop reference for engineers, students, and researchers who use FPGAs as their hardware platform of choice. The book is produced in the spirit of the "numerical recipe" series of books for various programming languages, where the intention is not to teach the language per se, but rather the philosophy and techniques required to make your application work. The rationale of Peter's book is similar in that the intention is to provide the methods and understanding to make the reader able to develop practical, operational VHDL that will run correctly on FPGAs.

Peter's book is divided into five main parts. In the introductory part of the book, primers are given into FPGAs, VHDL, and the standard design flow. In the second part of the book, a series of complex applications that encompass many of the key design problems facing designers today are worked through from start to finish in a practical way. In the third part of the book, important techniques are discussed, worked through, and explained from an example perspective, so you can see exactly how to implement a particular function. This part is really a toolbox of advanced specific functions that are commonly required in modern digital design. The fourth part on advanced techniques discusses the important aspect of design optimization, such as "How can I make my design faster?" or "How can I make my design more compact?" The fifth part investigates the details of fundamental issues that are implemented in VHDL. This final part is aimed at designers with a limited VHDL background, perhaps those looking for simpler examples to get started, or to solve a particular detailed issue.

Coming from the introductory section of Peter's book, this chapter provides a concise overview of important VHDL language constructs and usage. The chapter introduces the key concepts in VHDL and the important syntax required for most VHDL designs, particularly with reference to designs that are to be implemented in FPGAs.

—**Clive "Max" Maxfield**

3.1 Introduction

This chapter of the book is not intended as a comprehensive VHDL reference book—there are many excellent texts available that fit that purpose including Mark Zwolinski's *Digital System Design with VHDL*, Zainalabedin Navabi's *VHDL: Analysis and modeling of digital systems* or Peter Ashenden's *Designer's Guide to VHDL*. This section is designed to give concise and useful summary information on important language constructs and usage in VHDL—helpful and easy to use, but not necessarily complete.

This chapter will introduce the key concepts in VHDL and the important syntax required for most VHDL designs, particularly with reference to Field Programmable Gate Arrays (FPGAs). In most cases, the decision to use VHDL over other languages such as Verilog or SystemC will have less to do with designer choice, and more to do with software availability and company decisions. Over the last decade or so, a "war of words" has raged between the VHDL and Verilog communities about which is the best language, and in most cases it is completely pointless as the issue is more about design than syntax. There are numerous differences in the detail between VHDL and Verilog, but the fundamental philosophical difference historically has been the design context of the two languages. Verilog has come from a "bottom-up" tradition and has been heavily used by the IC industry for cell-based design, whereas the VHDL language has been developed much more from a "top-down" perspective. Of course, these are generalizations and largely out of date in a modern context, but the result is clearly seen in the basic syntax and methods of the two languages.

Without descending into a minute dissection of the differences between Verilog and VHDL, one important advantage of VHDL is the ability to use multiple levels of model with different architectures as shown in Figure 3.1.

This is not unique to VHDL, and in fact Verilog does have the concept of different behavior in a single "module"; however, it is explicitly defined in VHDL and is extremely useful in putting together practical multi-level designs in VHDL.

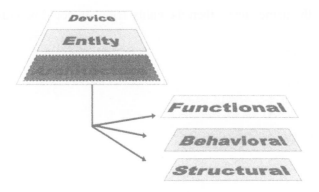

Figure 3.1: VHDL models with different architectures

The division of a model into its interface part (the "entity" in VHDL) and the behavior part (the "architecture" in VHDL) is an incredibly practical approach for modeling multiple behaviors for a single interface and makes model exchange and multiple implementations straightforward.

The remainder of this chapter will describe the key parts of VHDL, starting with the definition of a basic model structure using entities and architectures, discuss the important variable types, review the methods of encapsulating concurrent, sequential and hierarchical behavior and finally introduce the important fundamental data types required in VHDL.

3.2   Entity: Model Interface

3.2.1   Entity Definition

The entity defines how a design element described in VHDL connects to other VHDL models and also defines the name of the model. The entity also allows the definition of any parameters that are to be passed into the model using hierarchy. The basic template for an entity is as follows:

```
entity <name> is

....

entity <name>;
```

If the entity has the name "test", then the entity template could be either:

```
entity test is

end entity test;
```

or:

```
entity test is

end test;
```

3.2.2   Ports

The method of connecting entities together is using PORTS. These are defined in the entity using the following method:

```
port (

...list of port declarations...

);
```

The port declaration defines the type of connection and direction where appropriate. For example, the port declaration for an input bit called in1 would be as follows:

```
in1 : in bit;
```

And if the model had two inputs (in1 and in2) of type bit and a single output (out1) of type bit, then the declaration of the ports would be as follows:

```
port (

        in1, in2 : in bit;

        out1 : out bit

);
```

As the connection points between entities are effectively the same as those interprocess connections, they are effectively signals and can be used as such within the VHDL of the model.

3.2.3 Generics

If the model has a parameter, then this is defined using generics. The general declaration of generics is shown below:

```
generic (

...list of generic declarations...

);
```

In the case of generics, the declaration is similar to that of a constant with the form as shown below:

```
param1 : integer := 4;
```

Taking an example of a model that had two generics (gain (integer) and time_delay (time)), they could be defined in the entity as follows:

```
generic (

        gain : integer := 4;

        time_delay : time = 10 ns

);
```

3.2.4 Constants

It is also possible to include model specific constants in the entity using the standard declaration of constants method, for example:

```
constant : rpullup : real := 1000.0;
```

3.2.5 Entity Examples

To illustrate a complete entity, we can bring together the ports and generics examples previously and construct the complete entity for this example:

```
entity test is

        port (
```

```
            in1, in2 : in bit;

            out1 : out bit

      );

      generic (

            gain : integer := 4;

            time_delay : time := 10 ns

      );

      constant : rpullup : real := 1000.0;

   end entity test;
```

3.3   Architecture: Model Behavior

3.3.1   *Basic Definition of an Architecture*

While the entity describes the interface and parameter aspects of the model, the architecture defines the behavior. There are several types of VHDL architecture and VHDL allows different architectures to be defined for the same entity. This is ideal for developing behavioral, Register Transfer Level RTL and gate Level architectures that can be incorporated into designs and tested using the same test benches.

The basic approach for declaring an architecture could be as follows:

```
   architecture behaviour of test is

   ..architecture declarations

   begin

   ...architecture contents

   end architecture behaviour;
```

or

```
   architecture behaviour of test is

   ..architecture declarations
```

```
begin

...architecture contents

end behaviour;
```

3.3.2 Architecture Declaration Section

After the declaration of the architecture name and before the begin statement, any local signals or variables can be declared. For example, if there were two internal signals to the architecture called sig1 and sig2, they could be declared in the declaration section of the model as follows:

```
architecture behaviour of test is

        signal sig1, sig2 : bit;

begin
```

Then the signals can be used in the architecture statement section.

3.3.3 Architecture Statement Section

VHDL architectures can have a variety of structures to achieve different types of functionality. Simple combinatorial expressions use signal assignments to set new signal values as shown below:

```
out1 <= in1 and in2 after 10 ns;
```

Note that for practical design, the use of the "after 10 ns" is not synthesizable. In practice, the only way to ensure correct synthesizable design is to either make the design delay insensitive or synchronous. The design of combinatorial VHDL will result is additional delays due to the technology library gate delays, potentially resulting in glitches or hazards. An example of a multiple gate combinatorial architecture using internal signal declarations is given below:

```
architecture behavioural of test is

        signal int1, int2 : bit;

begin
```

```
        int1 <= in1 and in2;

        int2 <= in3 or in4;

        out1 <= int1 xor int2;

    end architecture behavioural;
```

3.4   Process: Basic Functional Unit in VHDL

The process in VHDL is the mechanism by which sequential statements can be
executed in the correct sequence, and with more than one process, concurrently. Each
process consists of a sensitivity list, declarations and statements. The basic process
syntax is given below:

```
process sensitivity_list is

    ... declaration part

begin

    ... statement part

end process;
```

The sensitivity list allows a process to be activated when a specific signal changes
value, for example a typical usage would be to have a global clock and reset signal to
control the activity of the process, for example:

```
process (clk, rst) is

begin

    ... process statements

end process;
```

In this example, the process would only be activated when either clk or rst changed
value. Another way of encapsulating the same behavior is to use a wait statement in the
process so that the process is automatically activated once, and then waits for activity
on either signal before running the process again. The same process could then be
written as follows:

```
process

begin

    ... process statements

    wait on clk, rst;

end process;
```

In fact, the location of the wait statement is not important, as the VHDL simulation cycle executes each process once during initialization, and so the wait statement could be at the start or the end of the process and the behavior would be the same in both cases.

In the declaration section of the process, signals and variables can be defined locally as described previously; for example, a typical process may look like the following:

```
process (a) is

    signal na : bit;

begin

    na <= not a;

end process;
```

With the local signal na and the process activated by changes on the signal a that is externally declared (with respect to the process).

3.5 Basic Variable Types and Operators

3.5.1 Constants

When a value needs to be static throughout a simulation, the type of element to use is a constant. This is often used to initialize parameters or to set fixed register values for comparison. A constant can be declared for any defined type in VHDL with examples as follows:

```
constant a : integer := 1;

constant b : real := 0.123;

constant c : std_logic := '0';
```

3.5.2 Signals

Signals are the link between processes and sequential elements within the processes. They are effectively "wires" in the design and connect all the design elements together. When simulating signals, the simulator will in turn look at updating the signal values and also checking the sensitivity lists in processes to see whether any changes have occurred that will mean that processes become active.

Signals can be assigned immediately or with a time delay, so that an event is scheduled for sometime in the future (after the specified delay). It is also important to recognize that signals are not the same as a set of sequential program code (such as in C), but are effectively concurrent signals that will not be able to be considered stable until the next time the process is activated.

Examples of signal declaration and assignment are shown below:

```
signal sig1 : integer := 0;

signal sig2 : integer := 1;

sig1 <= 14;

sig1 <= sig2;

sig1 <= sig2 after 10 ns;
```

3.5.3 Variables

While signals are the external connections between processes, variables are the internal values within a process. They are only used in a sequential manner, unlike the concurrent nature of signals within and between processes. Variables are used within processes and are declared and used as follows:

```
variable var1 : integer := 0;

variable var2 : integer := 1;

var1 := var2;
```

Notice that there is no concept of a delay in the variable assignment—if you need to schedule an event, it is necessary to use a signal.

3.5.4 Boolean Operators

VHDL has a set of standard Boolean operators built in, which are self-explanatory. The list of operators are and, or, nand, not, nor, xor. These operators can be applied to BIT, BOOLEAN or logic types with examples as follows:

```
out1 <= in1 and in2;

out2 <= in3 or in4;

out5 <= not in5;
```

3.5.5 Arithmetic Operators

There are a set of arithmetic operators built into VHDL which again are self-explanatory and these are described and examples provided (see Table 3.1).

Table 3.1: Arithmetic operators

Operator	Description	Example
+	Addition	out1 <= in1 + in2;
–	Subtraction	out1 <= in1 – in2;
*	Multiplication	out1 <= in1 * in2;
/	Division	out1 <= in1/in2;
abs	Absolute Value	absin1 <= abs(in1);
mod	Modulus	modin1 <= mod(in1);
rem	Remainder	remin1 <= rem(in1);
**	Exponent	out1 <= in1 ** 3;

3.5.6 Comparison Operators

VHDL has a set of standard comparison operators built in, which are self-explanatory. The list of operators are =, /=, <, <=, >, >=. These operators can be applied to a variety of types as follows:

```
in1 < 1

in1 /= in2

in2 >= 0.4
```

3.5.7   *Shifting Functions*

VHDL has a set of six built-in logical shift functions which are summarized in Table 3.2.

Table 3.2: Built-in logical shift functions

Operator	Description	Example
sll	Shift Left Logical	reg <= reg sll 2;
srl	Shift Right Logical	reg <= reg srl 2;
sla	Shift Left Arithmetic	reg <= reg sla 2;
sra	Shift Right Arithmetic	reg <= reg sra 2;
rol	Rotate Left	reg <= reg rol 2;
ror	Rotate Right	reg <= reg ror 2;

3.5.8   *Concatenation*

The concatenation function XE "VHDL:concatenation" in VHDL is denoted by the & symbol and is used as follows:

```
A <= '1111';

B <= '000';

out1 <= A & B & '1'; - out1 = '11110001';
```

3.6   Decisions and Loops

3.6.1   *If-Then-Else*

The basic syntax for a simple if statement is as follows:

```
if (condition) then

    ... statements

end if;
```

The condition is a Boolean expression, of the form a > b or a = b. Note that the comparison operator for equality is a single =, not to be confused with the double == used in some programming languages. For example, if two signals are equal, then set an output high would be written in VHDL as:

```
if ( a = b ) then

        out1 <= '1';

end if;
```

If the decision needs to have both the if and else options, then the statement is extended as follows:

```
if (condition) then

        ... statements

else

        ... statements

end if;
```

So in the previous example, we could add the else statements as follows:

```
if ( a = b ) then

        out1 <= '1';

else

        out1 <= '0';

end if;
```

And finally, multiple if conditions can be implemented using the general form:

```
if (condition1) then

        ... statements

elsif (condition2)

        ... statements

        ... more elsif conditions & statements
```

```
else

    ... statements

end if;
```

With an example:

```
if (a > 10) then

    out1 <= '1';

elsif (a > 5) then

    out1 <= '0';

else

    out1 <= '1';

end if;
```

3.6.2 Case

As we have seen with the IF statement, it is relatively simple to define multiple conditions, but it becomes a little cumbersome, and so the case statement offers a simple approach to branching, without having to use Boolean conditions in every case. This is especially useful for defining state diagrams or for specific transitions between states using enumerated types. An example of a case statement is:

```
case testvariable is

    when 1 =>

        out1 <= '1';

    when 2 =>

        out2 <= '1';

    when 3 =>

        out3 <= '1';

end case;
```

This can be extended to a range of values, not just a single value:

```
case test is

        when 0 to 4 => out1 <= '1';
```

It is also possible to use Boolean conditions and equations. In the case of the default option (i.e., when none of the conditions have been met), then the term when others can be used:

```
case test is

        when 0 => out1 <= '1';

        when others => out1 <= '0';

end case;
```

3.6.3   For

The most basic loop in VHDL is the FOR loop. This is a loop that executes a fixed number of times. The basic syntax for the FOR loop is shown below:

```
for loopvar in start to finish loop

        ... loop statements

end loop;
```

It is also possible to execute a loop that counts down rather than up, and the general form of this loop is:

```
for loopvar in start downto finish loop

        ... loop statements

end loop;
```

A typical example of a for loop would be to pack an array with values bit by bit, for example:

```
signal a : std_logic_vector(7 downto 0);
```

```
for i in 0 to 7 loop

    a(i) <= '1';

end loop;
```

3.6.4   *While and Loop*

Both the while and loop loops have an in-determinant number of loops, compared to the fixed number of loops in a FOR loop and as such are usually not able to be synthesized. For FPGA design, they are not feasible as they will usually cause an error when the VHDL model is compiled by the synthesis software.

3.6.5   *Exit*

The exit command allows a FOR loop to be exited completely. This can be useful when a condition is reached and the remainder of the loop is no longer required. The syntax for the exit command is shown below:

```
for i in 0 to 7 loop

    if ( i = 4 ) then

        exit;

    endif;

endloop;
```

3.6.6   *Next*

The next command allows a FOR loop iteration to be exited; this is slightly different from the exit command in that the current iteration is exited, but the overall loop continues onto the next iteration. This can be useful when a condition is reached and the remainder of the iteration is no longer required. An example for the next command is shown below:

```
for i in 0 to 7 loop

    if ( i = 4 ) then

        next;
```

```
    endif;

endloop;
```

3.7 Hierarchical Design

3.7.1 Functions

Functions are a simple way of encapsulating behavior in a model that can be reused in multiple architectures. Functions can be defined locally to an architecture or more commonly in a package, but in this section the basic approach of defining functions will be described. The simple form of a function is to define a header with the input and output variables as shown below:

```
function name (input declarations) return output_type is

    ... variable declarations

begin

    ... function body

end
```

For example, a simple function that takes two input numbers and multiplies them together could be defined as follows:

```
function mult (a,b : integer) return integer is

begin

    return a * b;

end;
```

3.7.2 Packages

Packages are a common single way of disseminating type and function information in the VHDL design community. The basic definition of a package is as follows:

```
package name is

... package header contents
```

```
end package;

package body name is

     ... package body contents

end package body;
```

As can be seen, the package consists of two parts, the header and the body. The header is the place where the types and functions are declared, and the package body is where the declarations themselves take place.

For example, a function could be described in the package body and the function is declared in the package header. Take a simple example of a function used to carry out a simple logic function:

```
and10 = and(a,b,c,d,e,f,g,h,i,j)
```

The VHDL function would be something like the following:

```
function and10 (a,b,c,d,e,f,g,h,i,j : bit) return bit is
begin
     return a and b and c and d and e and f and g and h
       and i and j;
end;
```

The resulting package declaration would then use the function in the body and the function header in the package header thus:

```
package new_functions is
function and10 (a,b,c,d,e,f,g,h,i,j : bit) return bit;
end;
package body new_functions is
     function and10 (a,b,c,d,e,f,g,h,i,j : bit) return
       bit is
```

```
     begin

          return a and b and c and d and e \

               and f and g and h and i and j;

     end;

end;
```

3.7.3 Components

While procedures, functions and packages are useful in including behavioral constructs generally, with VHDL being used in a hardware design context, often there is a need to encapsulate design blocks as a separate component that can be included in a design, usually higher in the system hierarchy. The method for doing this in VHDL is called a COMPONENT. Caution needs to be exercised with components as the method of including components changed radically between VHDL 1987 and VHDL 1993; as such, care needs to be taken to ensure that the correct language definitions are used consistently.

Components are a way of incorporating an existing VHDL entity and architecture into a new design without including the previously created model. The first step is to declare the component, in a similar way to declaring functions. For example, if an entity is called and4, and it has 4 inputs (a, b, c, d of type bit) and 1 output (q of type bit), then the component declaration would be of the form shown below:

```
component and4

     port ( a, b, c, d : in bit; q : out bit );

end component;
```

Then this component can be instantiated in a netlist form in the VHDL model architecture:

```
d1 : and4 port map ( a, b, c, d, q );
```

Note that, in this case, there is no explicit mapping between port names and the signals in the current level of VHDL; the pins are mapped in the same order as defined in the component declaration. If each pin is to be defined independent of the order of the pins, then the explicit port map definition needs to be used:

```
d1: and4 port map ( a => a, b => b, c => c, d => d, q =>

q);
```

The final thing to note is that this is called the default binding. The binding is the link between the compiled architecture in the current library and the component being used. It is possible, for example, to use different architectures for different instantiated components using the following statement for a single specific device:

```
for d1 : and4 use entity work.and4(behaviour) port map

  (a,b,c,d,q);
```

or the following to specify a specific device for all the instantiated components:

```
for all : and4 use entity work.and4(behaviour) port

  map (a,b,c,d,q);
```

3.7.4 Procedures

Procedures are similar to functions, except that they have more flexibility in the parameters, in that the direction can be in, out or inout. This is useful in comparison to functions where there is generally only a single output (although it may be an array) and avoids the need to create a record structure to manage the return value. Although procedures are useful, they should be used only for small specific functions. Components should be used to partition the design, not procedures, and this is especially true in FPGA design, as the injudicious use of procedures can lead to bloated and inefficient implementations, although the VHDL description can be very compact. A simple procedure to execute a full adder could be of the form:

```
procedure full_adder (a,b : in bit; sum, carry : out bit)

  is

begin

      sum := a xor b;

      carry := a and b;

end;
```

Notice that the syntax is the same as that for variables (NOT signals), and that multiple outputs are defined without the need for a return statement.

3.8   Debugging Models

3.8.1   Assertions

Assertions are used to check if certain conditions have been met in the model and are extremely useful in debugging models. Some examples:

```
assert value <= max_value

     report "Value too large";

assert clock_width >= 100 ns

     report "clock width too small"

     severity failure;
```

3.9   Basic Data Types

3.9.1   Basic Types

VHDL has the following standard types defined as built-in data types:

- BIT
- BOOLEAN
- BIT_VECTOR
- INTEGER
- REAL

3.9.2   Data Type: BIT

The BIT data type is the simple logic type built into VHDL. The type can have two legal values, "0" or "1". The elements defined as of type BIT can have the standard VHDL built-in logic functions applied to them. Examples of signal and variable declarations of type BIT follow:

```
signal ina : bit;

variable inb : bit := '0';

ina <= inb and inc;

ind <= '1' after 10 ns;
```

3.9.3   Data Type: Boolean

The Boolean data type is primarily used for decision-making, so the test value for "if" statements is a Boolean type. The elements defined as of type Boolean can have the standard VHDL built-in logic functions applied to them. Examples of signal and variable declarations of type Boolean follow:

```
signal test1 : Boolean;

variable test2 : Boolean := FALSE;
```

3.9.4   Data Type: Integer

The basic numeric type in VHDL is the integer and is defined as an integer in the range -2147483647 to $+2147483647$. There are obviously implications for synthesis in the definition of integers in any VHDL model, particularly the effective number of bits, and so it is quite common to use a specified range of integer to constrain the values of the signals or variables to within physical bounds. Examples of integer usage follow:

```
signal int1 : integer;

variable int2 : integer := 124;
```

There are two subtypes (new types based on the fundamental type) derived from the integer type which are integer in nature, but simply define a different range of values.

3.9.4.1   Integer Subtypes: Natural

The natural subtype is used to define all integers greater than and equal to zero. They are actually defined with respect to the high value of the integer range as follows:

```
natural values : 0 to integer'high
```

3.9.4.2 Integer Subtypes: Positive

The positive subtype is used to define all integers greater than and equal to one. They are actually defined with respect to the high value of the integer range as follows:

```
positive values : 1 to integer'high
```

3.9.5 Data Type: Character

In addition to the numeric types inherent in VHDL, there are also the complete set of ASCII characters available for designers. There is no automatic conversion between characters and a numeric value *per se*; however, there is an implied ordering of the characters defined in the VHDL standard (IEEE Std 1076-1993). The characters can be defined as individual characters or arrays of characters to create strings. The best way to consider characters is as an enumerated type.

3.9.6 Data Type: Real

Floating-point numbers are used in VHDL to define real numbers and the predefined floating-point type in VHDL is called real. This defines a floating-point number in the range $-1.0e38$ to $+10e38$. This is an important issue for many FPGA designs, as most commercial synthesis products do not support real numbers, precisely because they are floating point. In practice, it is necessary to use integer or fixed-point numbers that can be directly and simply synthesized into hardware. An example of defining real signals or variables is shown below:

```
signal realno : real;

variable realno : real := 123.456;
```

3.9.7 Data Type: Time

Time values are defined using the special time type. These not only include the time value, but also the unit—separated by a space. The basic range of the time type value is between -2147483647 and 2147483647, and the basic unit of time is defined as the femtosecond (fs). Each subsequent time unit is derived from this basic unit of the fs as shown below:

```
ps = 1000 fs;

ns = 1000 ps;

us = 1000 ns;

ms = 1000 us;

min = 60 sec;

hr = 60 min;
```

Examples of time definitions are shown below:

```
delay : time := 10 ns;

wait for 20 us;

y <= x after 10 ms;

z <= y after delay;
```

3.10 Summary

This chapter provides a very brief introduction to VHDL and is certainly not a comprehensive reference. It enables the reader, hopefully, to have enough knowledge to understand the syntax of the examples in this book. The author strongly recommends that anyone serious about design with VHDL should also obtain a detailed and comprehensive reference book on VHDL, such as Zwolinski (a useful introduction to digital design with VHDL—a common student textbook) or Ashenden (a more heavy duty VHDL reference that is perhaps more comprehensive, but less easy for a beginner to VHDL).

CHAPTER 4

Modeling Memories

Richard Munden

This chapter comes to us from the book ASIC and FPGA Verification *by Richard Munden, ISBN: 9780125105811. Richard has been using and managing CAE systems since 1987, and he has been concerned with simulation and modeling issues for just as long. Richard cofounded the Free Model Foundry (**http://eda.org/fmf/**) in 1995 and is its president and CEO. In addition to his real (daytime) job, Richard is a well-known contributor to several EDA users groups and industry conferences. His primary focus over the years has been verification of board-level designs.*

ASIC and FPGA Verification *covers the creation and use of models useful for verifying ASIC and FPGA designs and board-level designs that use off-the-shelf digital components. The intent of the book is show how ASICs and FPGAs can be verified in the larger context of a board or system.*

ASIC and FPGA Verification *is presented in four parts: Introduction, Resources and Standards, Modeling Basics, and Advanced Modeling. Each part covers a number of related modeling concepts and techniques, with individual chapters building upon previous material.*

"Memories are among the most frequently modeled components. How they are modeled can determine not just the performance, but the very practicality of board-level simulation.

"Many boards have memory on them, and FPGA designs frequently interface to one or more types of memory. These are not the old asynchronous static RAMs of more innocent times. These memories are pipelined zero bus turnaround (ZBT) synchronous static RAMs (SSRAM), multibanked synchronous dynamic RAMs (SDRAM), or double

> *data rate (DDR) DRAMs. The list goes on and complexities go up. Verify the interfaces or face the consequences.*
>
> *"Just as the memories have become complex, so have the models. There are several issues specific to memory models. How they are dealt with will determine the accuracy, performance, and resource requirements of the models."*
>
> **—Clive "Max" Maxfield**

Memories are among the most frequently modeled components. How they are modeled can determine not just the performance, but the very practicality of board-level simulation.

Many boards have memory on them, and FPGA designs frequently interface to one or more types of memory. These are not the old asynchronous static RAMs of more innocent times. These memories are pipelined zero bus turnaround (ZBT) synchronous static RAMs (SSRAM), multibanked synchronous dynamic RAMs (SDRAM), or double data rate (DDR) DRAMs. The list goes on and complexities go up. Verify the interfaces or face the consequences.

Just as the memories have become complex, so have the models. There are several issues specific to memory models. How they are dealt with will determine the accuracy, performance, and resource requirements of the models.

4.1   Memory Arrays

There are a number of ways memory arrays can be modeled. The most obvious and commonly used is an array of bits. This is the method that most closely resembles the way a memory component is constructed. Because the model's ports are of type `std_ulogic`, we can create an array of type `std_logic_vector` for our memory:

```
TYPE MemStore IS ARRAY (0 to 255) OF STD_LOGIC_VECTOR(7 DOWNTO 0);
```

This has the advantage of allowing reads and writes to the array without any type conversions.

However, using an `std_logic_vector` array is expensive in terms of simulation memory. `Std_logic` is a 9-value type, which is more values than we need or can use. A typical VHDL simulator will use 1B of simulation memory for each `std_logic` bit.

A 1 megabit memory array will consume about 1 MB of computer memory. At that rate, the amount of memory in a design can be too large to simulate.

In real hardware, memory can contain only 1s and 0s. That might suggest the use of type `bit_vector`, but the point of simulation is to debug and verify a design in an environment that makes it easier than debugging real hardware. It is useful if a read from a memory location that has never been written to gives a unique result. Although real hardware may contain random values, 'U's are more informative for simulation because they make it easy to see that an uninitialized location has been accessed. Likewise, if a timing violation occurs during a memory write, the simulation model usually emits a warning message. Ideally that location should also contain an invalid word that is recognizable as such. On reading that corrupt location, the user should see 'X's.

So it seems type `UX01` would provide all the values required. We could declare our memory array:

```
TYPE MemStore IS ARRAY (0 to 255) OF UX01_VECTOR(7 DOWNTO 0);
```

Unfortunately, because `UX01` is a subtype of `std_logic`, most simulators use the same amount of space to store type `UX01` as they do to store `std_logic`.

4.1.1   The Shelor Method

One option for modeling memories is the Shelor method. In 1996, Charles Shelor wrote an article in the *VHDL Times* as part of his "VHDL Designer" column [7], in which he discussed the problem of modeling large memories. He described several possible storage mechanisms. The method presented here is the one he favored.

Shelor noted that by converting the vector to a number of type natural we can store values in much less space. Of course, there are limitations. The largest integer guaranteed by the VHDL standard is $2^{31} - 1$, meaning a 30-bit word is the most you can safely model. It turns out this is not a problem. Few memory components use word sizes larger than 18 bits and most are either 8 or 9 bits wide.

So a range of 0 to 255 is sufficient for an 8-bit word, but that assumes every bit is either a 1 or a 0. It would be good to also allow words to be uninitialized or corrupted. To do so, just extend the range down to –2.

A generic memory array declaration is:

```
- Memory array declaration

TYPE MemStore IS ARRAY (0 to TotalLOC) OF INTEGER

                RANGE -2 TO MaxData;
```

where -2 is an uninitialized word, -1 is a corrupt word, and 0 to MaxData are valid data. This method does not lend itself to manipulation of individual bits, but that is rarely called for in a component model.

Simulators tested store integers in 4B, so each word of memory, up to 18 bits, will occupy 4B of simulator memory. This is a considerable improvement over using arrays of `std_logic_vector`.

4.1.2 The VITAL_Memory Package

Another option is to use the `VITAL_Memory` package released with VITAL2000. This package has an extensive array of features specific to memory modeling, including a method of declaring a memory array that results in a specific form of storage. In Figure 4.1, a memory array using the VITAL2000 memory package is declared.

```
- - VITAL Memory Declaration
VARIABLE Memdat : VitalMemoryDataType :=
  VitalDeclareMemory (
      NoOfWords              => TotalLOC,
      NoOfBitsPerWord        => DataWidth,
      NoOfBitsPerSubWord     => DataWidth,
      MemoryLoadFile         => MemLoadFileName,
      BinaryLoadFile         => FALSE
  );
```

Figure 4.1: VITAL2000 memory array declaration

In Figure 4.1, a procedure call is used to create the memory array.

The storage efficiency is very good—a 1B word occupies only 2B of memory—but this holds true only for 8-bit words. A 9-bit word occupies 4B of memory, which is the same as the Shelor method.

4.2 Modeling Memory Functionality

There are two distinct ways of modeling memory functionality. One is to use standard VHDL behavioral modeling methods. The other is to use the VITAL2000 memory package. Some features of the memory package can be used in behavioral models.

Let us look at how to model a generic SRAM using each method. The component modeled is a 4MB SRAM with an 8-bit word width.

4.2.1 Using the Behavioral (Shelor) Method

The model entity has the same general features as previous models. It begins with copyright, history, description, and library declarations:

```
-------------------------------------------------------------

--File Name: sram4m8.vhd

-------------------------------------------------------------

-- Copyright (C) 2001 Free Model Foundry;http://vhdl.org/fmf/

--

-- This program is free software; you can redistribute it and/or

-- modify it under the terms of the GNU General Public License

-- version 2 as published by the Free Software Foundation.

--

-- MODIFICATION HISTORY:

--

-- version:   |  author:   |  mod date:   |  changes made:

-- V1.0          R. Munden      01 MAY 27      Initial release

--
```

```
-------------------------------------------------------------------

-- PART DESCRIPTION:

--

-- Library:      MEM

-- Technology:   not ECL

-- Part:         SRAM4M8

--

-- Description: 4M X 8 SRAM

-------------------------------------------------------------------

LIBRARY IEEE;    USE IEEE.std_logic_1164.ALL;

                 USE IEEE.VITAL_timing.ALL;

                 USE IEEE.VITAL_primitives.ALL;

LIBRARY FMF;     USE FMF.gen_utils.ALL;

                 USE FMF.conversions.ALL;
```

The library declarations include the FMF conversions library. It is needed for converting between std_logic_vector and integer types.

The generic declarations contain the usual interconnect, path delay, setup and hold, and pulse width generics, as well as the usual control parameters:

```
-------------------------------------------------------------------

-- ENTITY DECLARATION

-------------------------------------------------------------------

ENTITY sram4m8 IS

  GENERIC (

    -- tipd delays : interconnect path delays

    tipd_OENeg     : VitalDelayType01 := VitalZeroDelay01;
```

```
tipd_WENeg      : VitalDelayType01 := VitalZeroDelay01;

tipd_CENeg      : VitalDelayType01 := VitalZeroDelay01;

tipd_CE         : VitalDelayType01 := VitalZeroDelay01;

tipd_D0         : VitalDelayType01 := VitalZeroDelay01;

tipd_D1         : VitalDelayType01 := VitalZeroDelay01;

tipd_D2         : VitalDelayType01 := VitalZeroDelay01;

tipd_D3         : VitalDelayType01 := VitalZeroDelay01;

tipd_D4         : VitalDelayType01 := VitalZeroDelay01;

tipd_D5         : VitalDelayType01 := VitalZeroDelay01;

tipd_D6         : VitalDelayType01 := VitalZeroDelay01;

tipd_D7         : VitalDelayType01 := VitalZeroDelay01;

tipd_A0         : VitalDelayType01 := VitalZeroDelay01;

tipd_A1         : VitalDelayType01 := VitalZeroDelay01;

tipd_A2         : VitalDelayType01 := VitalZeroDelay01;

tipd_A3         : VitalDelayType01 := VitalZeroDelay01;

tipd_A4         : VitalDelayType01 := VitalZeroDelay01;

tipd_A5         : VitalDelayType01 := VitalZeroDelay01;

tipd_A6         : VitalDelayType01 := VitalZeroDelay01;

tipd_A7         : VitalDelayType01 := VitalZeroDelay01;

tipd_A8         : VitalDelayType01 := VitalZeroDelay01;

tipd_A9         : VitalDelayType01 := VitalZeroDelay01;

tipd_A10        : VitalDelayType01 := VitalZeroDelay01;

tipd_A11        : VitalDelayType01 := VitalZeroDelay01;

tipd_A12        : VitalDelayType01 := VitalZeroDelay01;
```

```
tipd_A13            : VitalDelayType01 := VitalZeroDelay01;

tipd_A14            : VitalDelayType01 := VitalZeroDelay01;

tipd_A15            : VitalDelayType01 := VitalZeroDelay01;

tipd_A16            : VitalDelayType01 := VitalZeroDelay01;

tipd_A17            : VitalDelayType01 := VitalZeroDelay01;

tipd_A18            : VitalDelayType01 := VitalZeroDelay01;

tipd_A19            : VitalDelayType01 := VitalZeroDelay01;

tipd_A20            : VitalDelayType01 := VitalZeroDelay01;

tipd_A21            : VitalDelayType01 := VitalZeroDelay01;

-- tpd delays

tpd_OENeg_D0            : VitalDelayType01Z  := UnitDelay01Z;

tpd_CENeg_D0            : VitalDelayType01Z  := UnitDelay01Z;

tpd_A0_D0              : VitalDelayType01    := UnitDelay01;

-- tpw values: pulse widths

tpw_WENeg_negedge      : VitalDelayType      := UnitDelay;

tpw_WENeg_posedge      : VitalDelayType      := UnitDelay;

-- tsetup values: setup times

tsetup_D0_WENeg        : VitalDelayType      := UnitDelay;

tsetup_D0_CENeg        : VitalDelayType      := UnitDelay;

-- thold values: hold times

thold_D0_WENeg         : VitalDelayType      := UnitDelay;

thold_D0_CENeg         : VitalDelayType      := UnitDelay;

-- generic control parameters

InstancePath  : STRING := DefaultInstancePath;
```

```
TimingChecksOn   : BOOLEAN    := DefaultTimingChecks;

MsgOn            : BOOLEAN    := DefaultMsgOn;

Xon              : BOOLEAN    := DefaultXOn;

SeverityMode     : SEVERITY_LEVEL := WARNING;

-- For FMF SDF technology file usage

TimingModel      : STRING     := DefaultTimingModel
);
```

The port declarations are equally straightforward:

```
PORT (

    A0          : IN     std_ulogic := 'X';

    A1          : IN     std_ulogic := 'X';

    A2          : IN     std_ulogic := 'X';

    A3          : IN     std_ulogic := 'X';

    A4          : IN     std_ulogic := 'X';

    A5          : IN     std_ulogic := 'X';

    A6          : IN     std_ulogic := 'X';

    A7          : IN     std_ulogic := 'X';

    A8          : IN     std_ulogic := 'X';

    A9          : IN     std_ulogic := 'X';

    A10         : IN     std_ulogic := 'X';

    A11         : IN     std_ulogic := 'X';

    A12         : IN     std_ulogic := 'X';

    A13         : IN     std_ulogic := 'X';

    A14         : IN     std_ulogic := 'X';
```

```
        A15         : IN      std_ulogic := 'X';

        A16         : IN      std_ulogic := 'X';

        A17         : IN      std_ulogic := 'X';

        A18         : IN      std_ulogic := 'X';

        A19         : IN      std_ulogic := 'X';

        A20         : IN      std_ulogic := 'X';

        A21         : IN      std_ulogic := 'X';

        D0          : INOUT  std_ulogic := 'X';

        D1          : INOUT  std_ulogic := 'X';

        D2          : INOUT  std_ulogic := 'X';

        D3          : INOUT  std_ulogic := 'X';

        D4          : INOUT  std_ulogic := 'X';

        D5          : INOUT  std_ulogic := 'X';

        D6          : INOUT  std_ulogic := 'X';

        D7          : INOUT  std_ulogic := 'X';

        OENeg       : IN      std_ulogic := 'X';

        WENeg       : IN      std_ulogic := 'X';

        CENeg       : IN      std_ulogic := 'X';

        CE          : IN      std_ulogic := 'X'
    );

    ATTRIBUTE VITAL_LEVEL0 of sram4m8 : ENTITY IS TRUE;

END sram4m8;
```

Up to this point, the behavioral model and the VITAL2000 model are identical.
Next comes the VITAL_LEVEL0 architecture, beginning with the constant and signal
declarations:

```
-----------------------------------------------------------------
-- ARCHITECTURE DECLARATION
-----------------------------------------------------------------
ARCHITECTURE vhdl_behavioral of sram4m8 IS

    ATTRIBUTE VITAL_LEVEL0 of vhdl_behavioral : ARCHITECTURE IS TRUE;

    ---------------------------------------------------------------
    -- Note that this model uses the Shelor method of modeling large
    -- memory arrays. Data is stored as type INTEGER with the
    --  value -2 representing an uninitialized location and the
    -- value -1 representing a corrupted location.
    ---------------------------------------------------------------

    CONSTANT partID       : STRING := "SRAM 4M X 8";
    CONSTANT MaxData      : NATURAL := 255;
    CONSTANT TotalLOC     : NATURAL := 4194303;
    CONSTANT HiAbit       : NATURAL := 21;
    CONSTANT HiDbit       : NATURAL := 7;
    CONSTANT DataWidth    : NATURAL := 8;

    SIGNAL D0_ipd         : std_ulogic := 'U';
    SIGNAL D1_ipd         : std_ulogic := 'U';
    SIGNAL D2_ipd         : std_ulogic := 'U';
    SIGNAL D3_ipd         : std_ulogic := 'U';
    SIGNAL D4_ipd         : std_ulogic := 'U';
    SIGNAL D5_ipd         : std_ulogic := 'U';
    SIGNAL D6_ipd         : std_ulogic := 'U';
    SIGNAL D7_ipd         : std_ulogic := 'U';
```

```
      SIGNAL D8_ipd          : std_ulogic := 'U';

      SIGNAL A0_ipd          : std_ulogic := 'U';

      SIGNAL A1_ipd          : std_ulogic := 'U';

      SIGNAL A2_ipd          : std_ulogic := 'U';

      SIGNAL A3_ipd          : std_ulogic := 'U';

      SIGNAL A4_ipd          : std_ulogic := 'U';

      SIGNAL A5_ipd          : std_ulogic := 'U';

      SIGNAL A6_ipd          : std_ulogic := 'U';

      SIGNAL A7_ipd          : std_ulogic := 'U';

      SIGNAL A8_ipd          : std_ulogic := 'U';

      SIGNAL A9_ipd          : std_ulogic := 'U';

      SIGNAL A10_ipd         : std_ulogic := 'U';

      SIGNAL A11_ipd         : std_ulogic := 'U';

      SIGNAL A12_ipd         : std_ulogic := 'U';

      SIGNAL A13_ipd         : std_ulogic := 'U';

      SIGNAL A14_ipd         : std_ulogic := 'U';

      SIGNAL A15_ipd         : std_ulogic := 'U';

      SIGNAL A16_ipd         : std_ulogic := 'U';

      SIGNAL A17_ipd         : std_ulogic := 'U';

      SIGNAL A18_ipd         : std_ulogic := 'U';

      SIGNAL A19_ipd         : std_ulogic := 'U';

      SIGNAL A20_ipd         : std_ulogic := 'U';

      SIGNAL A21_ipd         : std_ulogic := 'U';

      SIGNAL OENeg_ipd       : std_ulogic := 'U';

      SIGNAL WENeg_ipd       : std_ulogic := 'U';
```

```
SIGNAL CENeg_ipd     : std_ulogic := 'U';

SIGNAL CE_ipd        : std_ulogic := 'U';
```

Manufacturers design memories in families, with the members differing only in depth and width. The constants declared here are used to enable the creation of models of memories within a family with a minimum amount of editing.

After the declarations, the architecture begins with the wire delay block:

```
BEGIN

   --------------------------------------------------------

   -- Wire Delays

   --------------------------------------------------------

   WireDelay : BLOCK

   BEGIN

      w_1: VitalWireDelay (OENeg_ipd, OENeg, tipd_OENeg);

      w_2: VitalWireDelay (WENeg_ipd, WENeg, tipd_WENeg);

      w_3: VitalWireDelay (CENeg_ipd, CENeg, tipd_CENeg);

      w_4: VitalWireDelay (CE_ipd, CE, tipd_CE);

      w_5: VitalWireDelay (D0_ipd, D0, tipd_D0);

      w_6: VitalWireDelay (D1_ipd, D1, tipd_D1);

      w_7: VitalWireDelay (D2_ipd, D2, tipd_D2);

      w_8: VitalWireDelay (D3_ipd, D3, tipd_D3);

      w_9: VitalWireDelay (D4_ipd, D4, tipd_D4);

      w_10: VitalWireDelay (D5_ipd, D5, tipd_D5);

      w_11: VitalWireDelay (D6_ipd, D6, tipd_D6);

      w_12: VitalWireDelay (D7_ipd, D7, tipd_D7);

      w_13: VitalWireDelay (A0_ipd, A0, tipd_A0);
```

```
w_14: VitalWireDelay (A1_ipd, A1, tipd_A1);

w_15: VitalWireDelay (A2_ipd, A2, tipd_A2);

w_16: VitalWireDelay (A3_ipd, A3, tipd_A3);

w_17: VitalWireDelay (A4_ipd, A4, tipd_A4);

w_18: VitalWireDelay (A5_ipd, A5, tipd_A5);

w_19: VitalWireDelay (A6_ipd, A6, tipd_A6);

w_20: VitalWireDelay (A7_ipd, A7, tipd_A7);

w_21: VitalWireDelay (A8_ipd, A8, tipd_A8);

w_22: VitalWireDelay (A9_ipd, A9, tipd_A9);

w_23: VitalWireDelay (A10_ipd, A10, tipd_A10);

w_24: VitalWireDelay (A11_ipd, A11, tipd_A11);

w_25: VitalWireDelay (A12_ipd, A12, tipd_A12);

w_26: VitalWireDelay (A13_ipd, A13, tipd_A13);

w_27: VitalWireDelay (A14_ipd, A14, tipd_A14);

w_28: VitalWireDelay (A15_ipd, A15, tipd_A15);

w_29: VitalWireDelay (A16_ipd, A16, tipd_A16);

w_30: VitalWireDelay (A17_ipd, A17, tipd_A17);

w_31: VitalWireDelay (A18_ipd, A18, tipd_A18);

w_32: VitalWireDelay (A19_ipd, A19, tipd_A19);

w_33: VitalWireDelay (A20_ipd, A20, tipd_A20);

w_34: VitalWireDelay (A21_ipd, A21, tipd_A21);

END BLOCK;
```

The model's ports have been declared as scalars to facilitate the model's use in a schematic capture environment and to provide a means of back-annotating individual interconnect delays. However, in modeling the function of the memory it is more convenient to use vectors. The mapping from scalar to vector and back is best done within a block.

The block opens with a set of port declarations:

```
------------------------------------------------------------

-- Main Behavior Block

------------------------------------------------------------

Behavior: BLOCK

    PORT (

            AddressIn   : IN    std_logic_vector(HiAbit downto 0);

            DataIn      : IN    std_logic_vector(HiDbit downto 0);

            DataOut     : OUT   std_logic_vector(HiDbit downto 0);

            OENegIn     : IN    std_ulogic := 'U';

            WENegIn     : IN    std_ulogic := 'U';

            CENegIn     : IN    std_ulogic := 'U';

            CEIn        : IN    std_ulogic := 'U'

    );
```

Note that the data ports that were of type INOUT in the entity are split into separate ports of type IN and type OUT in the block. Also note that the address and data buses are sized using the constants declared at the top of the architecture.

The next step is the port map:

```
PORT MAP (

  DataOut(0) => D0,

  DataOut(1) => D1,

  DataOut(2) => D2,

  DataOut(3) => D3,

  DataOut(4) => D4,
```

```
DataOut(5) => D5,

DataOut(6) => D6,

DataOut(7) => D7,

DataIn(0) => D0_ipd,

DataIn(1) => D1_ipd,

DataIn(2) => D2_ipd,

DataIn(3) => D3_ipd,

DataIn(4) => D4_ipd,

DataIn(5) => D5_ipd,

DataIn(6) => D6_ipd,

DataIn(7) => D7_ipd,

AddressIn(0) => A0_ipd,

AddressIn(1) => A1_ipd,

AddressIn(2) => A2_ipd,

AddressIn(3) => A3_ipd,

AddressIn(4) => A4_ipd,

AddressIn(5) => A5_ipd,

AddressIn(6) => A6_ipd,

AddressIn(7) => A7_ipd,

AddressIn(8) => A8_ipd,

AddressIn(9) => A9_ipd,

AddressIn(10) => A10_ipd,

AddressIn(11) => A11_ipd,

AddressIn(12) => A12_ipd,
```

```
    AddressIn(13) => A13_ipd,

    AddressIn(14) => A14_ipd,

    AddressIn(15) => A15_ipd,

    AddressIn(16) => A16_ipd,

    AddressIn(17) => A17_ipd,

    AddressIn(18) => A18_ipd,

    AddressIn(19) => A19_ipd,

    AddressIn(20) => A20_ipd,

    AddressIn(21) => A21_ipd,

    OENegIn => OENeg_ipd,

    WENegIn => WENeg_ipd,

    CENegIn => CENeg_ipd,

    CEIn    => CE_ipd
    );
```

In the port map, inputs to the block are associated with the delayed versions of the model's input ports. The block outputs are associated directly with the model's output ports. There is only one signal declaration in this block:

```
    SIGNAL D_zd   : std_logic_vector(HiDbit DOWNTO 0);
```

It is the zero delay data output.

The behavioral section of the model consists of a process. The process's sensitivity list includes all the input signals declared in the block:

```
    BEGIN

    ----------------------------------------------------------

    -- Behavior Process

    ----------------------------------------------------------
```

```
      Behavior : PROCESS (OENegIn, WENegIn, CENegIn, CEIn, AddressIn,
DataIn)

      -- Timing Check Variables

      VARIABLE   Tviol_D0_WENeg   : X01 := '0';

      VARIABLE   TD_D0_WENeg      : VitalTimingDataType;

      VARIABLE   Tviol_D0_CENeg   : X01 := '0';

      VARIABLE   TD_D0_CENeg      : VitalTimingDataType;

      VARIABLE   Pviol_WENeg      : X01 := '0';

      VARIABLE   PD_WENeg          : VitalPeriodDataType
                                     := VitalPeriodDataInit;
```

It is followed by the declarations for the timing check variables.

The memory array and functionality results variables are declared as follows:

```
   -- Memory array declaration
   TYPE MemStore IS ARRAY (0 to TotalLOC) OF INTEGER

                    RANGE -2 TO MaxData;

   -- Functionality Results Variables

   VARIABLE Violation   : X01 := '0';

   VARIABLE DataDrive   : std_logic_vector(HiDbit DOWNTO 0)

                          := (OTHERS => 'X');

   VARIABLE DataTemp    : INTEGER RANGE -2 TO MaxData := -2;

   VARIABLE Location    : NATURAL RANGE 0 TO TotalLOC := 0;

   VARIABLE MemData     : MemStore;
```

Again, these take advantage of the constants declared at the top of the architecture.

To avoid the need to test the control inputs for weak signals ('L' and 'H'), "no weak value" variables are declared for them:

```
-- No Weak Values Variables

VARIABLE OENeg_nwv  : UX01 := 'U';

VARIABLE WENeg_nwv  : UX01 := 'U';

VARIABLE CENeg_nwv  : UX01 := 'U';

VARIABLE CE_nwv     : UX01 := 'U';
```

and they are converted at the beginning of the process body:

```
BEGIN

    OENeg_nwv      := To_UX01 (s => OENegIn);

    WENeg_nwv      := To_UX01 (s => WENegIn);

    CENeg_nwv      := To_UX01 (s => CENegIn);

    CE_nwv         := To_UX01 (s => CEIn);
```

The timing check section comes near the top of the process body, as usual:

```
    -------------------------------------------------------------

    -- Timing Check Section

    -------------------------------------------------------------

IF (TimingChecksOn) THEN

    VitalSetupHoldCheck (

        TestSignal         => DataIn,

        TestSignalName     => "Data",

        RefSignal          => WENeg,

        RefSignalName      => "WENeg",

        SetupHigh          => tsetup_D0_WENeg,
```

```
          SetupLow            => tsetup_D0_WENeg,

          HoldHigh            => thold_D0_WENeg,

          HoldLow             => thold_D0_WENeg,

          CheckEnabled        => (CENeg ='0' and CE ='1'and OENeg ='1'),

          RefTransition       => '/',

          HeaderMsg           => InstancePath & PartID,

          TimingData          => TD_D0_WENeg,

          XOn                 => XOn,

          MsgOn               => MsgOn,

          Violation           => Tviol_D0_WENeg );

   VitalSetupHoldCheck (

          TestSignal          => DataIn,

          TestSignalName      => "Data",

          RefSignal           => CENeg,

          RefSignalName       => "CENeg",

          SetupHigh           => tsetup_D0_CENeg,

          SetupLow            => tsetup_D0_CENeg,

          HoldHigh            => thold_D0_CENeg,

          HoldLow             => thold_D0_CENeg,

          CheckEnabled        => (WENeg ='0' and OENeg ='1'),

          RefTransition       => '/',

          HeaderMsg           => InstancePath & PartID,

          TimingData          => TD_D0_CENeg,

          XOn                 => XOn,
```

```
    MsgOn                => MsgOn,

    Violation            => Tviol_D0_CENeg );

VitalPeriodPulseCheck (

    TestSignal           => WENegIn,

    TestSignalName       => "WENeg",

    PulseWidthLow        => tpw_WENeg_negedge,

    PeriodData           => PD_WENeg,

    XOn                  => XOn,

    MsgOn                => MsgOn,

    Violation            => Pviol_WENeg,

    HeaderMsg            => InstancePath & PartID,

    CheckEnabled         => TRUE );

    Violation := Pviol_WENeg OR Tviol_D0_WENeg OR
    Tviol_D0_CENeg;

    ASSERT Violation  = '0'

        REPORT InstancePath & partID & ": simulation may be" &

            "inaccurate due to timing violations"

        SEVERITY SeverityMode;

  END IF; -- Timing Check Section
```

This section is much shorter than it would have been if scalar signals had been used for the data bus instead of vectored signals.

An assertion statement is used to warn the user whenever a timing violation occurs. Corrupt data will also be written to memory when this happens, but the user could find it difficult to determine the source of the corruption without the assertion statement. The model's logic is in the functional section. The section begins by setting DataDrive, the output variable, to high impedance. Then, if the component is selected for either a read or a write operation the value of the address bus is translated to a natural and assigned to the variable `Location`:

```
-------------------------------------------------------------

-- Functional Section

-------------------------------------------------------------

DataDrive := (OTHERS => 'Z');

IF (CE_nwv = '1' AND CENeg_nwv = '0') THEN

    IF (OENeg_nwv = '0' OR WENeg_nwv = '0') THEN

      Location := To_Nat(AddressIn);

      IF (OENeg_nwv = '0' AND WENeg_nwv = '1') THEN

        DataTemp := MemData(Location);

        IF DataTemp >= 0 THEN

          DataDrive := To_slv(DataTemp, DataWidth);

        ELSIF DataTemp = -2 THEN

          DataDrive := (OTHERS => 'U');

        ELSE

          DataDrive := (OTHERS => 'X');

        END IF;

      ELSIF (WENeg_nwv = '0') THEN

        IF Violation = '0' THEN

          DataTemp := To_Nat(DataIn);

        ELSE

          DataTemp := -1;

        END IF;

        MemData(Location) := DataTemp;
```

```
        END IF;

    END IF;

  END IF;
```

If the operation is a read, the `Location` variable is used as an index to the memory array and the contents of that location assigned to `DataTemp`. `DataTemp` is tested to see if it contains a nonnegative number. If so, it is valid and assigned to `DataDrive`. If not, a –2 indicates the location is uninitialized and `'U's` are assigned to `DataDrive`. Anything else (–1) causes `'X's` to be assigned.

If the operation is a write and there is no timing violation, the value of the data bus is converted to a natural and assigned to `DataTemp`. If there is a timing violation, `DataTemp` is assigned a –1.

Finally, the element of the memory array indexed by Location is assigned the value of `DataTemp`.

At the end of the process, the zero delay signal gets the value of `DataDrive`:

```
  -----------------------------------------------------------

  -- Output Section

  -----------------------------------------------------------

  D_zd <=DataDrive;

END PROCESS;
```

The model concludes with the output path delay. Because the output is a bus (within the block), a generate statement is used to shorten the model:

```
  -----------------------------------------------------------

  -- Path Delay Processes generated as a function of data width

  -----------------------------------------------------------

  DataOut_Width : FOR i IN HiDbit DOWNTO 0 GENERATE

    DataOut_Delay : PROCESS (D_zd(i))

      VARIABLE D_GlitchData:VitalGlitchDataArrayType(HiDbit
      Downto 0);
```

```
    BEGIN

      VitalPathDelay01Z (

        OutSignal        =>DataOut(i),

        OutSignalName    => "Data",

        OutTemp          =>D_zd(i),

        Mode             =>OnEvent,

        GlitchData       =>D_GlitchData(i),

        Paths            =>(

          0 => (InputChangeTime    => OENeg_ipd'LAST_EVENT,

                PathDelay           => tpd_OENeg_D0,

                PathCondition       => TRUE),

          1 => (InputChangeTime    => CENeg_ipd'LAST_EVENT,

                PathDelay           => tpd_CENeg_D0,

                PathCondition       => TRUE),

          2 => (InputChangeTime    => AddressIn'LAST_EVENT,

                PathDelay => VitalExtendToFillDelay(tpd_A0_D0),

                PathCondition       => TRUE)

        )

      );

      END PROCESS;

    END GENERATE;

  END BLOCK;

END vhdl_behavioral;
```

Once again, one of the constants from the beginning of the architecture is used to control the size of the generate. Using a generate statement requires a separate process for the path delays, because although a process may reside within a generate, a generate statement may not be placed within a process.

4.2.2   Using the VITAL2000 Method

The VITAL2000 style memory model uses the same entity as the behavioral model, so the entity will not be repeated here. The first difference is the VITAL attribute in the architecture:

```
-------------------------------------------------------------

-- ARCHITECTURE DECLARATION

-------------------------------------------------------------

ARCHITECTURE vhdl_behavioral of sram4m8v2 IS

    ATTRIBUTE VITAL_LEVEL1_MEMORY of vhdl_behavioral :
    ARCHITECTURE IS TRUE;

    -------------------------------------------------------------

    -- Note that this model uses the VITAL2000 method of modeling

    -- large memory arrays.

    -------------------------------------------------------------
```

Here the attribute is `VITAL_LEVEL1_MEMORY`. It is required to get the full compiler benefits of a VITAL memory model.

From here the architecture is identical to the behavioral model for a while. The same constants and signals are declared. The wire delay block is the same. The behavior block is declared with the same ports and same port map. But the behavior process is completely different. It begins with the process declaration and sensitivity list:

```
BEGIN

    -------------------------------------------------------------

    -- Behavior Process

    -------------------------------------------------------------

MemoryBehavior : PROCESS (OENegIn, WENegIn, CENegIn, CEIn,
                         AddressIn, DataIn)
```

Then comes the declaration of a constant of type `VitalMemoryTableType`:

```
CONSTANT Table_generic_sram:    VitalMemoryTableType    :=    (

-----------------------------------------------------------

-- CE, CEN, OEN, WEN, Addr, DI, act, DO

-----------------------------------------------------------

-- Address initiated read

   ( '1', '0', '0', '1', 'G', '-', 's', 'm' ),

   ( '1', '0', '0', '1', 'U', '-', 's', 'l' ),

-- Output Enable initiated read

   ( '1', '0', 'N', '1', 'g', '-', 's', 'm' ),

   ( '1', '0', 'N', '1', 'u', '-', 's', 'l' ),

   ( '1', '0', '0', '1', 'g', '-', 's', 'm' ),

-- CE initiated read

   ( 'P', '0', '0', '1', 'g', '-', 's', 'm' ),

   ( 'P', '0', '0', '1', 'u', '-', 's', 'l' ),

-- CEN initiated read

   ( '1', 'N', '0', '1', 'g', '-', 's', 'm' ),

   ( '1', 'N', '0', '1', 'u', '-', 's', 'l' ),

-- Write Enable Implicit Read

   ( '1', '0', '0', 'P', '-', '-', 's', 'M' ),

-- Write Enable initiated Write

   ( '1', '0', '1', 'N', 'g', '-', 'w', 'S' ),

   ( '1', '0', '1', 'N', 'u', '-', 'c', 'S' ),
```

```
    -- CE initiated Write

        ( 'P', '0', '1', '0', 'g', '-', 'w', 'S' ),

        ( 'P', '0', '1', '0', 'u', '-', 'c', 'S' ),

    -- CEN initiated Write

        ( '1', 'N', '1', '0', 'g', '-', 'w', 'Z' ),

        ( '1', 'N', '1', '0', 'u', '-', 'c', 'Z' ),

    -- Address change during write

        ( '1', '0', '1', '0', '*', '-', 'c', 'Z' ),

        ( '1', '0', '1', 'X', '*', '-', 'c', 'Z' ),

    -- data initiated Write

        ( '1', '0', '1', '0', 'g', '*', 'w', 'Z' ),

        ( '1', '0', '1', '0', 'u', '-', 'c', 'Z' ),

        ( '1', '0', '-', 'X', 'g', '*', 'e', 'e' ),

        ( '1', '0', '-', 'X', 'u', '*', 'c', 'S' ),

    -- if WEN is X

        ( '1', '0', '1', 'r', 'g', '*', 'e', 'e' ),

        ( '1', '0', '1', 'r', 'u', '*', 'c', '1' ),

        ( '1', '0', '-', 'r', 'g', '*', 'e', 'S' ),

        ( '1', '0', '-', 'r', 'u', '*', 'c', 'S' ),

        ( '1', '0', '1', 'f', 'g', '*', 'e', 'e' ),

        ( '1', '0', '1', 'f', 'u', '*', 'c', '1' ),

        ( '1', '0', '-', 'f', 'g', '*', 'e', 'S' ),

        ( '1', '0', '-', 'f', 'u', '*', 'c', 'S' ),
```

```
        -- OEN is unasserted

        ( '-', '-', '1', '-', '-', '-', 's', 'Z' ),

        ( '1', '0', 'P', '-', '-', '-', 's', 'Z' ),

        ( '1', '0', 'r', '-', '-', '-', 's', '1' ),

        ( '1', '0', 'f', '-', '-', '-', 's', '1' ),

        ( '1', '0', '1', '-', '-', '-', 's', 'Z' )

        );
```

This table entirely defines the function of the memory model. Let us look at the table and see how it compares to the code in the behavioral model.

Columns CE, CEN, OEN, and WEN are the direct inputs. They are the control signals. Columns Addr and DI are the interpreted inputs. They represent the address and data buses, respectively. The act column specifies the memory action and the DO column specifies the output action. The table is searched from top to bottom until a match is found.

In the first section, an address initiated read is described:

```
        ---------------------------------------------------------------

        --CE, CEN, OEN, WEN, Addr, DI, act, DO

        ---------------------------------------------------------------

        -- Address initiated read

        ( '1', '0', '0', '1', 'G', '-', 's', 'm' ),

        ( '1', '0', '0', '1', 'U', '-', 's', '1' ),
```

For either line to be selected, CE and WEN, the chip enable and write enable, must be high. Write enable is an active low signal. In addition, CEN and OEN, also active low signals, must be low. If the address bus transitions to any good value (no 'X's), the memory location indexed by the address bus retains its previous value and the output bus gets the value of the memory location. Otherwise, if the address bus transitions to any unknown value (any bit is 'X'), the output gets a corrupt ('X') value.

If there is no match in the first section, the next section, describing an output enable initiated read, is searched:

```
------------------------------------------------------------

-- CE, CEN, OEN, WEN, Addr, DI, act, DO

------------------------------------------------------------

-- Output Enable initiated read

  ( '1', '0', 'N', '1', 'g', '-', 's', 'm' ),

  ( '1', '0', 'N', '1', 'u', '-', 's', 'l' ),

  ( '1', '0', '0', '1', 'g', '-', 's', 'm' ),
```

The two chip enables, CE and CEN, must be active and WEN inactive. If there is a falling edge or a low on OEN (1st and 3d lines) and the address bus has a good and steady (no transition) value, the value of the memory location is placed on the output. Otherwise, if OEN is falling (but steady) and the address is unknown (2nd line), the output is corrupted.

If there is no match in the previous sections, the next section, describing a chip enable initiated read is searched:

```
------------------------------------------------------------

-- CE, CEN, OEN, WEN, Addr, DI, act, DO

------------------------------------------------------------

-- CEN initiated read

  ( '1', 'N', '0', '1', 'g', '-', 's', 'm' ),

  ( '1', 'N', '0', '1', 'u', '-', 's', 'l' ),
```

CE and OEN must be active and WEN inactive. If there is a falling edge on CEN and the address bus has a good and steady (no transition) value, the value of the memory location is placed on the output. Otherwise, if the address is unknown (but steady), the output is corrupted.

The sections of the table just described correspond to the following lines of the behavioral model:

```
IF (CE_nwv = '1' AND CENeg_nwv = '0') THEN

   IF (OENeg_nwv = '0' OR WENeg_nwv = '0') THEN

      Location := To_Nat(AddressIn);

      IF (OENeg_nwv = '0' AND WENeg_nwv = '1') THEN

         DataTemp := MemData(Location);

         IF DataTemp >= 0 THEN

            DataDrive := To_slv(DataTemp, DataWidth);

         ELSIF DataTemp = -2 THEN

            DataDrive := (OTHERS => 'U');

         ELSE

            DataDrive := (OTHERS => 'X');

         END IF;
```

There are some differences. Although the VITAL model appears to be more complex, it does not require type conversions or special handling of uninitialized or corrupt locations. If a timing violation occurs during a read, the VITAL model will output `'X'`s. The behavioral model will send a warning message to the user but place valid data on the output bus. The VITAL modeling method provides more precise control of model behavior. How often that level of precision is required remains to be seen.

Continuing with the rest of the VITAL model process declarations, we have the following:

```
CONSTANT OENeg_D_Delay : VitalDelayArrayType01Z (HiDbit downto 0) :=

      (OTHERS => tpd_OENeg_D0);

CONSTANT CENeg_D_Delay : VitalDelayArrayType01Z (HiDbit downto 0) :=

      (OTHERS => tpd_CENeg_D0);
```

```
CONSTANT Addr_D_Delay : VitalDelayArrayType01 (175 downto 0) :=

        (OTHERS => tpd_A0_D0);
```

These constants are arrays of delays. A different delay could be assigned to every path from each input to each output. For the address to data out path, there are 22 inputs and 8 outputs. This method allows the assignment of 176 different delay values for address to data out. Of course, the same thing is possible in a behavior model; it would just take much more code. Although such detailed timing may be useful for memory embedded in an ASIC, the intended target of the VITAL_Memory package, it is rarely, if ever, required in component modeling.

The declaration for the timing check variables is the same as in the behavioral model:

```
-- Timing Check Variables

VARIABLE Tviol_D0_WENeg    : X01 := '0';

VARIABLE TD_D0_WENeg       : VitalTimingDataType;


VARIABLE Tviol_D0_CENeg    : X01 := '0';

VARIABLE TD_D0_CENeg       : VitalTimingDataType;


VARIABLE Pviol_WENeg       : X01 := '0';

VARIABLE PD_WENeg    : VitalPeriodDataType := VitalPeriodDataInit;
```

Although the VITAL_Memory package has its own set of timing check procedures, they are not used in this model. The more generic procedures are adequate in this case and easier to work with.

The memory declaration follows:

```
    -- VITAL Memory Declaration

    VARIABLE Memdat : VitalMemoryDataType :=

        VitalDeclareMemory (

            NoOfWords               => TotalLOC,

            NoOfBitsPerWord         => DataWidth,
```

```
         NoOfBitsPerSubWord         => DataWidth,

    -- MemoryLoadFile              => MemLoadFileName,

         BinaryLoadFile            => FALSE

    );
```

The `VITAL_Memory` package uses a procedure call for the memory array declaration. Included in the procedure is the ability to preload part or all of the memory array from a binary or ASCII file. Preloading memories is discussed later in this chapter.

The functionality results variables:

```
-- Functionality Results Variables

VARIABLE Violation : X01 := '0';

VARIABLE D_zd          : std_logic_vector(HiDbit DOWNTO 0);

VARIABLE Prevcntls     : std_logic_vector(0 to 3);

VARIABLE PrevData      : std_logic_vector(HiDbit downto 0);

VARIABLE Prevaddr      : std_logic_vector(HiAbit downto 0);

VARIABLE PFlag         : VitalPortFlagVectorType(0 downto 0);

VARIABLE Addrvalue     : VitalAddressValueType;

VARIABLE OENegChange   : TIME := 0 ns;

VARIABLE CENegChange   : TIME := 0 ns;

VARIABLE AddrChangeArray : VitalTimeArrayT(HiAbit downto 0);

VARIABLE D_GlitchData  : VitalGlitchDataArrayType
                         (HiDbit Downto 0);

VARIABLE DSchedData    : VitalMemoryScheduleDataVectorType
                         (HiDbit Downto 0);
```

include several that are specific to VITAL_Memory models.

The process begins with a timing check section similar to the one in the behavior model:

```
BEGIN

   ----------------------------------------------------------

   -- Timing Check Section

   ----------------------------------------------------------

   IF (TimingChecksOn) THEN

      VitalSetupHoldCheck (

         TestSignal      => DataIn,

         TestSignalName  => "Data",

         RefSignal       => WENeg,

         RefSignalName   => "WENeg",

         SetupHigh       => tsetup_D0_WENeg,

         SetupLow        => tsetup_D0_WENeg,

         HoldHigh        => thold_D0_WENeg,

         HoldLow         => thold_D0_WENeg,

         CheckEnabled    => (CENeg = '0' and CE = '1' and OENeg = '1'),

         RefTransition   => '/',

         HeaderMsg       => InstancePath & PartID,

         TimingData      => TD_D0_WENeg,

         XOn             => XOn,

         MsgOn           => MsgOn,

         Violation       => Tviol_D0_WENeg );

      VitalSetupHoldCheck (

         TestSignal      => DataIn,

         TestSignalName  => "Data",
```

```
        RefSignal        => CENeg,

        RefSignalName    => "CENeg",

        SetupHigh        => tsetup_D0_CENeg,

        SetupLow         => tsetup_D0_CENeg,

        HoldHigh         => thold_D0_CENeg,

        HoldLow          => thold_D0_CENeg,

        CheckEnabled     => (WENeg = '0' and OENeg = '1'),

        RefTransition    => '/',

        HeaderMsg        => InstancePath & PartID,

        TimingData       => TD_D0_CENeg,

        XOn              => XOn,

        MsgOn            => MsgOn,

        Violation        => Tviol_D0_CENeg );
    VitalPeriodPulseCheck (

        TestSignal       => WENegIn,

        TestSignalName   => "WENeg",

        PulseWidthLow    => tpw_WENeg_negedge,

        PeriodData       => PD_WENeg,

        XOn              => XOn,

        MsgOn            => MsgOn,

        Violation        => Pviol_WENeg,

        HeaderMsg        => InstancePath & PartID,

        CheckEnabled     => TRUE );
    Violation := Pviol_WENeg OR Tviol_D0_WENeg OR Tviol_D0_CENeg;

    ASSERT Violation = '0'
```

```
        REPORT InstancePath & partID & ": simulation may be" &

            "inaccurate due to timing violations"

        SEVERITY SeverityMode;

    END IF; -- Timing Check Section
```

The functionality section contains only a single call to the `VitalMemoryTable` procedure:

```
-------------------------------------------------------------

-- Functional Section

-------------------------------------------------------------

VitalMemoryTable (

        DataOutBus          => D_zd,

        MemoryData          => Memdat,

        PrevControls        => Prevcntls,

        PrevDataInBus       => Prevdata,

        PrevAddressBus      => Prevaddr,

        PortFlag            => PFlag,

        Controls            => (CEIn, CENegIn, OENegIn, WENegIn),

        DataInBus           => DataIn,

        AddressBus          => AddressIn,

        AddressValue        => Addrvalue,

        MemoryTable         => Table_generic_sram

    );
```

The model concludes with the output section. There are procedure calls to three different `VITAL_Memory` procedures in this section, which are the subject of the next section of this chapter:

```
------------------------------------------------------
-- Output Section

------------------------------------------------------

VitalMemoryInitPathDelay (

    ScheduleDataArray       =>      DSchedData,

    OutputDataArray         =>      D_zd

);

VitalMemoryAddPathDelay (       -- #11

    ScheduleDataArray       =>      DSchedData,

    InputSignal             =>      AddressIn,

    OutputSignalName        =>      "D",

    InputChangeTimeArray    =>      AddrChangeArray,

    PathDelayArray          =>      Addr_D_Delay,

    ArcType                 =>      CrossArc,

    PathCondition           =>      true

);

VitalMemoryAddPathDelay (       -- #14

    ScheduleDataArray       =>      DSchedData,

    InputSignal             =>      OENegIn,

    OutputSignalName        =>      "D",

    InputChangeTime         =>      OENegChange,

    PathDelayArray          =>      OENeg_D_Delay,

    ArcType                 =>      CrossArc,
```

```
            PathCondition           =>      true,

            OutputRetainFlag         =>      false

        );

        VitalMemoryAddPathDelay (        -- #14

            ScheduleDataArray        =>      DSchedData,

            InputSignal              =>      CENegIn,

            OutputSignalName         =>      "D",

            InputChangeTime          =>      CENegChange,

            PathDelayArray           =>      CENeg_D_Delay,

            ArcType                  =>      CrossArc,

            PathCondition            =>      true,

            OutputRetainFlag         =>      false

        );

        VitalMemorySchedulePathDelay (

            OutSignal                =>      DataOut,

            OutputSignalName         =>      "D",

            ScheduleDataArray        =>      DSchedData

        );

    END PROCESS;

    END BLOCK;

    END vhdl_behavioral;
```

The comment #14 is to remind the author and anyone who has to maintain the model which overloading of the `VitalMemoryAddPathDelay` procedure is being called.

4.3   VITAL_Memory Path Delays

The VITAL_Memory package has its own set of path delay procedures. There are three procedures that replace the VitalPathDelay01Z, and all three must be presented in the order shown.

The first is the VitalMemoryInitPathDelay. It is used to initialize the output delay data structure. It is called exactly once per output port. Output ports may be vectored. Outputs may also be internal signals rather than ports. The possible arguments to this procedure are shown in Table 4.1.

The second is the VitalMemoryAddPathDelay. It is used to add a delay path from an input to an output. There is one call to this procedure for each input to output path. Use of this procedure is analogous to the Paths parameter of the VitalPathDelay procedure. It is used for selecting candidate paths based on the PathCondition parameter. The procedure then updates the ScheduleDataArray structure.

Table 4.1: Arguments for VitalMemoryInitPathDelay

Name	Type	Description for Scalar Ports
ScheduleData	VitalMemorySchedule-DataType	Scalar form of the data structure used by VITAL_Memory path delay procedures to store persistent information for the path delay scheduling.
OutputData	STD_ULOGIC	Scalar form of the functional output value to be scheduled. **For Vectored Ports**
ScheduleDataArray	VitalMemorySchedule Data-VectorType	Vector form of the data structure used by VITAL_Memory path delay procedures to store persistent information for the path delay scheduling.
NumBitsPerSubWord	POSITIVE	Number of bits per memory subword. Optional.
OutputDataArray	STD_LOGIC_VECTOR	Vector form of the functional output value to be scheduled.

Table 4.2 shows the possible arguments to the `VitalMemoryAddPathDelay` procedure. The `VitalMemoryAddPathDelay` procedure is overloaded 24 ways. However, there are 64 possible parameter combinations, of which 40 will result in compiler errors that may or may not be informative. Therefore, it is recommended that you print out and read a copy of `memory_p_2000.vhd` if you intend to use this method. It will help in debugging your models.

Table 4.2: Arguments for VitalMemoryAddPathDelay

Name	Type	Description
ScheduleData	VitalMemorySchedule-DataType	Scalar form of the data structure used by VITAL_Memory path delay procedures to store persistent information for the path delay scheduling.
ScheduleDataArray	VitalMemoryScheduleData VectorType	Vector form of the data structure used by VITAL_Memory path delay procedures to store persistent information for the path delay scheduling.
InputSignal	STD_ULOGIC or STD_LOGIC_VECTOR	Scalar or vector input.
OutputSignalName	STRING	Name of output signal for use in messages.
InputChangeTime	TIME	Time since the last input change occurred.
PathDelay	VitalDelayType(01Z)	Path delay values used to delay the output values for scalar outputs.
PathDelayArray	VitalDelayArray- Type (01ZX)	Array of path delay values used to delay the output values for vector outputs.
ArcType	VitalMemoryArcType	Delay arc type between input and output.

(Continued)

Table 4.2: Arguments for VitalMemoryAddPathDelay (Cont'd)

Name	Type	Description
PathCondition	BOOLEAN	Condition under which the delay path is considered to be one of the candidate paths for propagation delay selection.
PathConditionArray	VitalBoolArrayT	Array of conditions under which the delay path is considered to be one of the candidate paths for propagation delay selection.
OutputRetainFlag	BOOLEAN	If TRUE, output retain (hold) behavior is enabled.
OutputRetain-Behavior	OutputRetainBehavior-Type	If value is BitCorrupt, output will be set to 'X' on a bit-by- bit basis. If WordCorrupt, entire word will be 'X' .

The third procedure is VitalMemorySchedulePathDelay. It is used to schedule the functional output value on the output signal using the selected propagation delay. This procedure is overloaded for scalar and vector outputs. It can also be used to perform result mapping of the output value using the Output-Map parameter. The possible arguments to the VitalMemorySchedulePathDelay are given in Table 4.3.

One of the most desirable features of the VITAL_Memory modeling path delay procedures is their support of output-retain behavior. Many memory components exhibit an output hold time that is greater than zero but less than the new output delay. This is shown in data sheets as a waveform similar to that in Figure 4.2. This behavior can be modeled using the behavioral style, but it is a nuisance to do so. The good news is that the VITAL_Memory path delay procedures can be used in a behavioral model to take advantage of this capability without requiring the use of other features of the package. They will work in a VITAL level 0 architecture.

Table 4.3: Arguments for VitalMemorySchedulePathDelay

Name	Type	Description
ScheduleData	VitalMemoryScheduleDataType	Scalar form of the data structure used by VITAL_Memory path delay procedures to store persistent information for the path delay scheduling.
ScheduleDataArray	VitalMemoryScheduleData-TypeVector	Vector form of the data structure used by VITAL_Memory path delay procedures to store persistent information for the path delay scheduling.
OutputSignal	STD_ULOGIC or STD_LOGIC_VECTOR	Scalar or vector output.
OutputSignalName	STRING	Name of output signal for use in messages.
OutputMap	VitalOutputMapType	Strength mapping of output values.

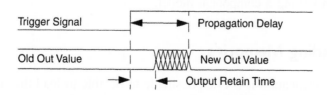

Figure 4.2: Output-retain waveform

4.4   VITAL_Memory Timing Constraints

The VITAL_Memory package has its own versions of two timing constraint checkers: VitalMemorySetupHoldCheck and VitalMemoryPeriodPulseCheck. The VitalMemorySetupHoldCheck procedure performs the same function as the VitalSetupHoldCheck procedure, with the following enhancements:

- Finer control over condition checking.

- Support for CrossArc, ParellelArc, and Subword timing relationships between the test signal and reference signal.

- Support for vector violation flags.

- Support for scalar and vector forms of condition in timing checks using CheckEnabled.

- Support of MsgFormat parameter to control the format of test/reference signals in the message.

- The VitalMemoryPeriodPulseCheck procedure is also similar to the VitalPeriodPulseCheck procedure, with the following differences:

- TestSignal is a vector rather than a scalar.

- The violation flag may be either a scalar or a vector.

- The MsgFormat parameter may be used to control the format of messages.

Although both of these procedures are required for accurate modeling of memories in an ASIC library environment, they are less useful for modeling off-the-shelf memory components. The VitalMemoryPeriodPulseCheck could be valuable should there be specification for minimum pulse width on an address bus. You will probably not find a need for the VitalMemorySetupHoldCheck procedure when writing a component model.

4.5   Preloading Memories

During system verification it is often desirable to be able to load the contents of a memory from an external file without going through the normal memory write process. Verilog has a system task, $readmemh, that can load a memory from a specified file at any time. It can be executed from the testbench if desired.

VHDL does not have an equivalent capability. However, that does not mean a memory cannot be preloaded in VHDL; it just takes a little more code. How it is done depends on the memory modeling style employed.

4.5.1 Behavioral Memory Preload

There must be a file to read. In a simple example, it may have the following format:

```
//format : @address

//          data -> address

//          data -> address+1

@1

1234

1235

@A

55AA
```

Lines beginning with / are comments and are ignored. A line beginning with @ indicates a new address. The following lines will contain the data starting at that address and incrementing the address with each new line. Address and data do not appear on the same line. For simple cases like this one, the format is compatible with that used by the Verilog $readmemh task.

Upon a triggering event, usually time zero, and assuming the feature is enabled, a file like that in the example is read and its contents loaded into the memory array as specified in the file. In a model, it all starts with the declaration of a generic:

```
-- memory file to be loaded

mem_file_name : STRING := "km416s4030.mem";
```

The value of the generic could be passed in from the schematic or the testbench. It will specify the name of the file to read for that particular instance of the model.

Further down in the model the line

```
FILE mem_file : text IS mem_file_name;
```

is required to declare that `mem_file` is an object of type `FILE`. Then somewhere in the same process that defines the memory array, the preload code is placed. The following is an example of some simple memory preload code:

```
------------------------------------------------------------

-- File Read Section

------------------------------------------------------------

IF PoweredUp'EVENT and PoweredUp and (mem_file_name /= "none") THEN
    ind := 0;
    WHILE (not ENDFILE (mem_file)) LOOP
        READLINE (mem_file, buf);
        IF buf(1) = '/' THEN
            NEXT;
        ELSIF buf(1) = '@' THEN
            ind := h(buf(2 to 5));
        ELSE
            MemData(ind) := h(buf(1 to 4));
            ind := ind + 1;
        END IF;
    END LOOP;
END IF;
```

This code waits for the triggering action, in this case an event on the signal `PoweredUp`. If the name of the memory load file is set to anything other than none, the memory preload code executes. It begins by initializing the index variable `ind`. Then it goes into a loop that will run until it reaches the end of the input file.

In this loop a line is read. If the line begins with the comment character (/) it is discarded and the next line is read. If the line begins with @ the following four characters, a hexadecimal number, are converted to a natural and assigned to ind, the array index. Otherwise, the first four characters on the line are converted to a natural and read into the memory array at the location indicated by ind, and then ind is incremented and the next line is read.

Memories with multiple words or multiple banks may be modeled. These may require slightly more complex memory preload files and file read sections. The model of a component that has four banks and is four words wide (32 bits) might have the following file read section shown in Figure 4.3.

```
-------------------------------------------------------------------
-- File Read Section
-------------------------------------------------------------------
IF PoweredUp'EVENT and PoweredUp and (mem_file_name /= "none") THEN
    ind := 0;
    WHILE (not ENDFILE (mem_file)) LOOP
        READLINE (mem_file, buf);
        IF buf(1) = '/' THEN
            NEXT;
        ELSIF buf(1) = '@' THEN
            file_bank := h(buf(2 to 2));
            ind := h(buf(4 to 8));
        ELSE
            MemData3(file_bank)(ind)  := h(buf(1 to 2));
            MemData2(file_bank)(ind)  := h(buf(3 to 4));
            MemData1(file_bank)(ind)  := h(buf(5 to 6));
            MemData0(file_bank)(ind)  := h(buf(7 to 8));
            ind := ind + 1;
        END IF;
    END LOOP;
END IF;
```

Figure 4.3: Memory preload for 4 bank memory

The corresponding preload file would have the following format:

```
// lines beginning with / are comments

// lines beginning with @set bank (0 to 3) and starting address

// other lines contain hex data values a 32-bit value

AAAAAAAA
```

```
55555555

00030003

@1 40000

FFFF2001

FFFE2001

FFFD2003

FFFC2003

@2 00000

2000FFFF

20012001

20022001

20032001
```

This format is not compatible with that of the Verilog $readmemh task because it includes a method for selecting memory banks. Verilog would require a separate file and a separate call to the system task for each bank.

4.5.2   VITAL_Memory Preload

The VitalDeclareMemory function, if given a file name for the MemoryLoadFile parameter, will cause the declared memory array to be initialized during elaboration.

It requires no additional code in the model. It is the only way to preload a VITAL_LEVEL1_MEMORY model. The preload file has the following format:

```
@8 aa

a5

bf

@a 00

01

02
```

The VITAL preload format does not allow comments. Address and data may be on the same line. A file may contain data in either binary or hexadecimal but not both. The address must always be in hex.

Although it is easy to implement, it has two drawbacks. Memory can be initialized only during elaboration. In some behavioral models, memory has been initialized by a reset signal improving verification efficiency.

If multiple memory arrays are declared, each must have its own preload file. In models of memories with multiple banks, it may be necessary to manage multiple preload files.

4.6 Modeling Other Memory Types

So far our discussion of memory models has centered around SRAMs. This is because SRAMs are the simplest type of memories to model. Now we will look at some more complex memory types. These models tend to be rather long, 2,000 to 4,000 lines each, so instead of presenting the entire models, only their defining features will be discussed. The complete models can be found on the Free Model Foundry Web site.

4.6.1 Synchronous Static RAM

The first model we will examine is for a pipelined zero bus turnaround SSRAM. Its IDT part number is IDT71V65803. Compatible parts are made by Micron and Cypress. This memory type is distinguished by its fast read-to-write turnaround time. This component has two 9-bit-wide bidirectional data buses with separate write enables and a common asynchronous output enable. Memory is modeled as two arrays, each holding 9 bits of data. They are 512K words deep.

The model includes three processes. The first is used for setup and runs only once, at time zero. The second describes the functionality of the component. The third contains a generate statement that generates the required number of `VitalPathDelay` calls to drive the output ports.

The first distinguishing feature we find in this model is a generic:

```
SeverityMode : SEVERITY_LEVEL := WARNING;
```

It is used to control the severity of some assertion statements.

This model contains a state machine with five states. They are chip deselect (`desel`), begin read (`begin_rd`), begin write (`begin_wr`), burst read (`burst_rd`), and burst write (`burst_wr`):

```
-- Type definition for state machine

TYPE mem_state IS (desel,

                   begin_rd,

                   begin_wr,

                   burst_rd,

                   burst_wr

                   );

SIGNAL state      : mem_state;
```

In burst mode, reads and writes may be either sequential or interleaved. The interleaved order is defined by a table:

```
TYPE sequence IS ARRAY (0 to 3) OF INTEGER RANGE -3 to 3;

TYPE seqtab IS ARRAY (0 to 3) OF sequence;

CONSTANT il0  : sequence   := (0, 1, 2, 3);

CONSTANT il1  : sequence   := (0, -1, 2, -1);

CONSTANT il2  : sequence   := (0, 1, -2, -1);

CONSTANT il3  : sequence   := (0, -1, -2, -3);

CONSTANT il   : seqtab     := (il0, il1, il2, il3);

CONSTANT ln0  : sequence   := (0, 1, 2, 3);

CONSTANT ln1  : sequence   := (0, 1, 2, -1);

CONSTANT ln2  : sequence   := (0, 1, -2, -1);

CONSTANT ln3  : sequence   := (0, -3, -2, -1);

CONSTANT ln   : seqtab     := (ln0, ln1, ln2, ln3);

SIGNAL Burst_Seq : seqtab;
```

The `il` constants are for the interleaved burst sequences. The `ln` constants are for the sequential (linear) bursts.

The burst mode for this component can be set only at power-up time. A special process is used to initialized the burst sequence:

```
Burst_Setup : PROCESS

BEGIN

   IF (LBONegIn = '1') THEN

      Burst_Seq <= il;

   ELSE

      Burst_Seq <= ln;

   END IF;

   WAIT; -- Mode can be set only during power up

END PROCESS Burst_Setup;
```

It is run once at time zero.

This component defines four commands. They are declared in the main behavior process:

```
-- Type definition for commands

TYPE command_type is (ds,

                      burst,

                      read,

                      write

                      );
```

On the rising edge of the clock, when the component is active

```
IF (rising_edge(CLKIn) AND CKENIn = '0' AND ZZIn = '0') THEN
```

each control input is checked for a valid value:

```
ASSERT (not(Is_X(RIn)))

    REPORT InstancePath & partID & ": Unusable value for R"

    SEVERITY SeverityMode;
```

If an invalid value is found the user is notified. Depending on how the user has modified the value of SeverityMode, the assertion might stop simulation. Assuming all the control inputs are valid, the command is decoded with an IF-ELSIF statement:

```
-- Command Decode
IF ((ADVIn = '0') AND (CE1NegIn = '1' OR CE2NegIn = '1' OR

    CE2In = '0')) THEN

  command := ds;
ELSIF (CE1NegIn = '0' AND CE2NegIn = '0' AND CE2In = '1' AND

    ADVIn = '0') THEN

  IF (RIn = '1') THEN

    command := read;

  ELSE

    command := write;

  END IF;
ELSIF (ADVIn = '1') AND (CE1NegIn = '0' AND CE2NegIn = '0' AND

    CE2In = '1') THEN

  command := burst;
ELSE

  ASSERT false

    REPORT InstancePath & partID & ": Could not decode "
```

```
                    & "command."
            SEVERITY SeverityMode;
      END IF;
```

Model behavior is controlled by the state machine. The state machine is built using a two-deep nesting of CASE statements:

```
      -- The State Machine

      CASE state IS

          WHEN desel =>

              CASE command IS

                  WHEN ds =>

                      OBuf1 := (others => 'Z');

                  WHEN read =>

...

                  WHEN write =>

...

                  WHEN burst =>

...

              END CASE;

          WHEN begin_rd =>

              Burst_Cnt := 0;

              CASE command IS

                  WHEN ds =>

...

                  WHEN read =>

...
```

The outer CASE statement uses the current state value. The inner CASE statement uses the current decoded command. The appropriate action is then taken.

Pipelining is affected using chained registers. For example, on the first clock of a read,

```
OBuf1(8 downto 0) := to_slv(MemDataA(MemAddr),9);
```

will execute. On the second clock,

```
OBuf2 := OBuf1;
```

and finally,

```
IF (OENegIn = '0') THEN

  D_zd <= (others => 'Z'), OBuf2 AFTER 1 ns;

END IF;
```

puts the output value on the zero delay output bus, ready to be used by the VitalPathDelay01Z procedure.

4.6.2 DRAM

DRAMs can be built with higher density and lower cost than SRAMs. That has made them very popular for use in computers and other memory-intensive devices. Their primary drawback is their inability to store data for more than a few tens of milliseconds without being refreshed. The refresh requirement makes them more complex than SRAM, both in their construction and their use.

The DRAM model we examine here is for Micron part number MT4LC4M16R6. It is also sourced by OKI and Samsung. This component is 64Mb memory organized as 4M words with a 16-bit-wide data bus. However, the data are also accessible as 8-bit bytes for both reads and writes. The memory is modeled as two arrays of 8-bit words.

The model includes three processes. The first describes the functionality of the component. The other two contain generate statements for generating the required VitalPathDelay calls.

The distinguishing features of this model begin with its generics. The component has a constraint for maximum time between refresh cycles. Because this time is not related to any ports, a `tdevice` generic is employed:

```
-- time between refresh

tdevice_REF          : VitalDelayType          := 15_625 ns;
```

It is important to initialize the generic to a reasonable default value so the model can be used without backannotation.

The next generic is for power-up initialization time. The component may not be written to until a period of time has passed after power is applied.

```
-- tpowerup: Power up initialization time. Data sheets say
100-200 us.

-- May be shortened during simulation debug.

tpowerup        : TIME          := 100 us;
```

By using a generic for this parameter, we allow the user to shorten the time during early phases of design debug.

Because a `tdevice` generic is used, there must be a `VITAL_Primitive` associated with it:

```
-- Artificial VITAL primitives to incorporate internal delays

REF : VitalBuf (refreshed_out, refreshed_in, (UnitDelay,
tdevice_REF));
```

Even though this primitive is not actually used in the model, it must be present to satisfy the `VITAL_Level1` requirements. In this model, `tdevice_REF` is used directly to time the rate of refresh cycles:

```
IF (NOW > Next_Ref AND PoweredUp = true AND Ref_Cnt > 0) THEN

    Ref_Cnt := Ref_Cnt - 1;

    Next_Ref := NOW + tdevice_REF;

END IF;
```

The component requires there be 4,096 refresh cycles in any 64 ms period. Every 15,625 nanosecond, the code decrements the value of the variable Ref_Cnt. Each time a refresh cycle occurs, the value of Ref_Cnt is incremented:

```
IF (falling_edge(RASNegIn)) THEN

    IF (CASHIn = '0' AND CASLIn = '0') THEN

      IF (WENegIn = '1') THEN

        CBR := TRUE;

        Ref_Cnt := Ref_Cnt + 1;
```

Should the value of Ref_Cnt ever reach zero,

```
-- Check Refresh Status

    IF (written = true) THEN

      ASSERT Ref_Cnt > 0

        REPORT InstancePath & partID &

          ": memory not refreshed (by ref_cnt)"

        SEVERITY SeverityMode;

      IF (Ref_Cnt < 1) THEN

        ready := FALSE;

    END IF;

  END IF;
```

a message is sent to the user and a flag, ready, is set to false. This flag will continue to warn the user if further attempts are made to store data in the component:

```
-- Early Write Cycle

ELSE

    ASSERT ready

      REPORT InstancePath & partID & ": memory is not ready for"
```

```
& "use - must be powered up and refreshed"

    SEVERITY SeverityMode;
```

Another interesting aspect of DRAMs is that they usually have multiplexed address buses. For this component there is a 12-bit column address and a 10-bit row address, but only a single 12-bit external address bus. Our memory is modeled as two linear arrays of naturals,

```
-- Memory array declaration

TYPE MemStore IS ARRAY (0 to 4194303) OF NATURAL

                RANGE 0 TO 255;

VARIABLE MemH        : MemStore;

VARIABLE MemL        : MemStore;
```

one for the high byte and one for the low byte.

The internal address bus is viewed as being 22 bits wide to accommodate both the row and column addresses. However, the index into the memory arrays is defined as a natural:

```
VARIABLE MemAddr       : std_logic_vector(21 DOWNTO 0)

                      := (OTHERS => 'X');

VARIABLE Location      : NATURAL RANGE 0 TO 4194303 := 0;
```

The row and column addresses must be read separately:

```
        MemAddr(21 downto 10) := AddressIn;

...

    MemAddr(9 downto 0) := AddressIn(9 downto 0);
```

then converted to a natural for use as an array index:

```
Location := to_nat(MemAddr);
```

4.6.3 SDRAM

SDRAMs are DRAMs with a synchronous interface. They often include pipelining as a means of improving their bandwidth. Refresh requirements are the same as for ordinary DRAMs.

The model we examine here covers the KM432S2030 SDRAM from Samsung. The equivalent Micron part is numbered MT48LC2M32B2. This component features all synchronous inputs, programmable burst lengths, and selectable CAS latencies. It is organized as 512K words, 32 bits wide. Internal memory is divided into four addressable banks. The 32-bit output is divided into four 8-bit words that can be individually masked during read and write cycles.

To accommodate the byte masking, each memory bank is modeled as an array of four 8-bit-wide memories. This makes byte access considerably simpler than using a single 32-bit memory array. It also solves the problem of having to deal with 32-bit integers on a 32-bit machine.

This SDRAM model is composed of two processes. One models the component functionality, the other generates the `VitalPathDelay` calls.

Distinguishing features of this model begin with its generics. This component has selectable CAS latency. This means the clock-to-output delays will depend on an internal register value. To annotate two sets of delays, two generics are needed:

```
-- tpd delays

tpd_CLK_DQ2            : VitalDelayType01Z    := UnitDelay01Z;

tpd_CLK_DQ3            : VitalDelayType01Z    := UnitDelay01Z;
```

The two CAS latencies also affect the maximum clock speed of the component.

Therefore, there must be two values annotated for period constraint checking:

```
-- CAS latency = 2

tperiod_CLK_posedge   : VitalDelayType       := UnitDelay;

-- CAS latency = 3

tperiod_CLK_negedge   : VitalDelayType       := UnitDelay;
```

In either of these cases the requirements could not have been met using conditional delay and constraint generics.

The SDRAM has a somewhat more complex state machine than the DRAM. In this case there are 16 states described in the data sheet. The state names are as follows:

```
-- Type definition for state machine

TYPE mem_state IS (pwron,

               precharge,

               idle,

               mode_set,

               self_refresh,

               self_refresh_rec,

               auto_refresh,

               pwrdwn,

               bank_act,

               bank_act_pwrdwn,

               write,

               write_suspend,

               read,

               read_suspend,

               write_auto_pre,

               read_auto_pre

               );
```

On careful examination of the data sheets from all the vendors, it becomes apparent that each bank has its own state machine. To reduce the amount of code in the model, it was decided to treat them as an array:

```
TYPE statebanktype IS array (hi_bank downto 0) of mem_state;

SIGNAL statebank : statebanktype;
```

It was decided to make statebank a signal rather than a variable. The reason was to delay any change of state till the next delta cycle. Had a variable been used, all state changes would be instantaneous.

In modeling a component as complex as this one, it is helpful to look at the data sheets from all the vendors producing compatible products. In the case of this component, one of four vendors studied, NEC, included a state diagram in their data sheet [8]. It is shown in Figure 4.4. Having an accurate state diagram greatly facilitates modeling a complex state machine.

A signal is needed for tracking the CAS latency:

```
SIGNAL CAS_Lat : NATURAL RANGE 0 to 3 := 0;
```

Like the DRAM, the SDRAM excepts a number of commands. An enumerated type is defined for them:

```
-- Type definition for commands

TYPE command_type is (desl,

                nop,

                bst,

                read,

                writ,

                act,

                pre,

                mrs,

                ref

                );
```

Memory arrays are declared in the behavior process:

```
-- Memory array declaration

TYPE MemStore IS ARRAY (0 to depth) OF INTEGER

            RANGE -2 TO 255;
```

3. Simplified State Diagram

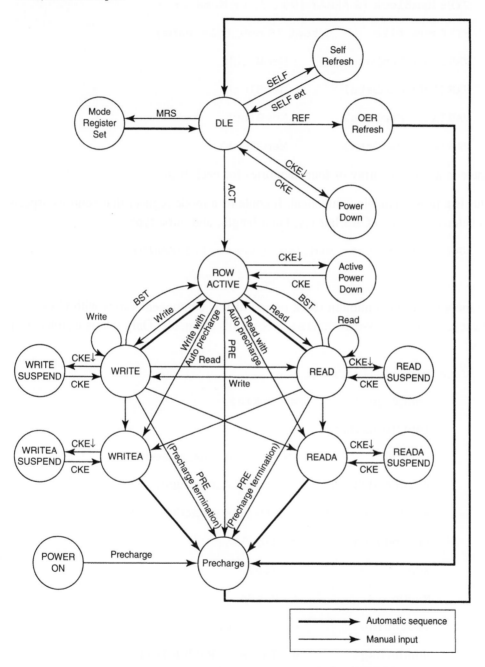

Figure 4.4: State diagram for SDRAM state machine

```
TYPE MemBlock IS ARRAY (0 to 3) OF MemStore;

FILE mem_file        : text IS mem_file_name;

VARIABLE MemData0        : MemBlock;

VARIABLE MemData1        : MemBlock;

VARIABLE MemData2        : MemBlock;

VARIABLE MemData3        : MemBlock;
```

There is a separate array of four memories for each bank.

This is a programmable component. It contains a mode register that controls aspects of its behavior, such as CAS latency, burst length, and burst type:

```
VARIABLE ModeReg : std_logic_vector(10 DOWNTO 0)

                    := (OTHERS => 'X');
```

As mentioned, the minimum clock period of this component varies with CAS latency. Two VitalPeriodPulseCheck calls are made. Only one is enabled at any time:

```
VitalPeriodPulseCheck (

    TestSignal          => CLKIn,

    TestSignalName      => "CLK",

    Period              => tperiod_CLK_posedge,

    PulseWidthLow       => tpw_CLK_negedge,

    PulseWidthHigh      => tpw_CLK_posedge,

    PeriodData          => PD_CLK,

    XOn                 => XOn,

    MsgOn               => MsgOn,

    Violation           => Pviol_CLK,

    HeaderMsg           => InstancePath & PartID,

    CheckEnabled        => CAS_Lat = 2 );
```

```
VitalPeriodPulseCheck (

    TestSignal          => CLKIn,

    TestSignalName      => "CLK",

    Period              => tperiod_CLK_negedge,

    PulseWidthLow       => tpw_CLK_negedge,

    PulseWidthHigh      => tpw_CLK_posedge,

    PeriodData          =>  PD_CLK,

    XOn                 => XOn,

    MsgOn               => MsgOn,

    Violation           => Pviol_CLK,

    HeaderMsg           => InstancePath & PartID,

    CheckEnabled        => CAS_Lat = 3 );
```

After the command is decoded (as in the DRAM), which bank it applies to must be determined. This is done with a CASE statement:

```
-- Bank Decode

CASE BAIn IS

    WHEN "00" => cur_bank := 0; BankString := " Bank-0 ";

    WHEN "01" => cur_bank := 1; BankString := " Bank-1 ";

    WHEN "10" => cur_bank := 2; BankString := " Bank-2 ";

    WHEN "11" => cur_bank := 3; BankString := " Bank-3 ";

    WHEN others =>

        ASSERT false

            REPORT InstancePath & partID & ": Could not decode bank"

                & " selection - results may be incorrect."
```

```
        SEVERITY SeverityMode;

    END CASE;
```

Next comes the state machine itself:

```
    -- The Big State Machine

    IF (rising_edge(CLKIn) AND CKEreg = '1') THEN
```

It contains a FOR loop that causes four passes through the state machine code for each active clock. That is one pass for each bank:

```
        banks : FOR bank IN 0 TO hi_bank LOOP

        CASE statebank(bank) IS

            WHEN pwron =>

    ...

                IF (command = bst) THEN

                    statebank(bank) <= bank_act;

                    Burst_Cnt(bank) := 0;
```

The state machine code is over 800 lines. The complete model may be found and downloaded from the FMF Web site.

Finally, the output delay is determined by CAS latency:

```
    VitalPathDelay01Z (

        OutSignal          => DataOut(i),

        OutSignalName      => "Data",

        OutTemp            => D_zd(i),

        Mode               => OnEvent,

        GlitchData         => D_GlitchData(i),

        Paths              => (

          1 => (InputChangeTime => CLKIn'LAST_EVENT,
```

```
        PathDelay => tpd_CLK_DQ2,

        PathCondition => CAS_Lat = 2),

    2 => (InputChangeTime => CLKIn'LAST_EVENT,

        PathDelay => tpd_CLK_DQ3,

        PathCondition => CAS_Lat = 3)

    )

  );
```

4.7   Summary

The efficiency with which memory is modeled can determine whether or not a board simulation is possible within the memory resources of your workstation. The most efficient ways of modeling large memories are the Shelor method and the `VITAL_memory` package's `VitalDeclareMemory` function.

Functionality may be modeled using the behavioral method or the VITAL2000 method. Which you choose will depend on the complexity of the model and how comfortable you are writing tables. Some functionality may be easier to describe with one method or the other.

Path delays in memory models may be written using the path delay procedures provided by the `VITAL_Timing` package or the those provided by the VITAL2000 memory package. If the part being modeled has output-retain behavior, the added complexity of the memory package procedures may be worth the effort. Otherwise, you will want to use the procedures from the timing package.

The VITAL2000 memory package has its own setuphold and periodpulsewidth checks. They are of limited value for component modeling.

Adding preload capability to a memory model is not difficult and well worth the additional effort required in a behavioral model. In a VITAL2000 model the capability, though somewhat restricted, comes for free.

CHAPTER 5

Introduction to Synchronous State Machine Design and Analysis

Richard Tinder

This chapter comes from the book Electrical Engineering Design *by Richard Tinder,*
ISBN: 9780126912951. This text emphasizes the successful engineering design of digital
devices and machines from first principles. A special effort has been made not to
"throw" logic circuits at the reader so that questions remain as to how the circuits
came about or whether or not they will function correctly.

From an engineering point of view, the design of a digital device or machine is of little or
no value unless it performs the intended operation(s) correctly and reliably. Both the
basics and background fundamentals are presented in this text. But it goes well beyond
the basics to provide significant intermediate-to-advanced coverage of digital design
material, some of which is covered by no other text.

The aim of the book is to provide the reader with the tools necessary for the successful
design of relatively complex digital systems from first principles. As part of this, the book
contains more than 600 figures and tables along with more than 1000 examples, exercises,
and problems. Also, VHDL behavioral and architectural descriptions of various
machines—combinational and sequential—are provided at various points in the text.

The text is divided into 16 relatively small chapters to provide maximum versatility in its
use. These chapters range from introductory remarks to advanced topics in asynchronous
systems.

In the case of this particular chapter, synchronous sequential machines and their design,
analysis, and operation are the subjects covered. Treatment begins with a discussion of
the models used for these machines. This is followed by a discussion of an important
type of graphic that is used to represent the sequential behavior of sequential machines

and by a detailed development of the devices used for their memory. The chapter ends with the design and analysis of relatively simple state machines (the intricacies of design are numerous and require detailed consideration; for this reason they are discussed later in the book).

—Clive "Max" Maxfield

5.1   Introduction

Up to this point only combinational logic machines have been considered, those whose outputs depend solely on the present state of the inputs. Adders, decoders, MUXs, PLAs, ALUs, and many other combinational logic machines are remarkable and very necessary machines in their own right to the field of logic design. However, they all suffer the same limitation. They cannot perform operations sequentially. A ROM, for example, cannot make use of its present input instructions to carry out a next-stage set of functions, and an adder cannot count sequentially without changing the inputs after each addition. In short, combinational logic devices lack *true memory*, and so lack the ability to perform sequential operations. Yet their presence in a sequential machine may be indispensable.

We deal with sequential devices all the time. In fact, our experience with such devices is so commonplace that we often take them for granted. For example, at one time or another we have all had the experience of being delayed by a modern four-way traffic control light system that is vehicle actuated with pedestrian overrides and the like. Once at the light we must wait for a certain sequence of events to take place before we are allowed to proceed. The controller for such a traffic light system is a fairly complex digital sequential machine.

Then there is the familiar elevator system for a multistory building. We may push the button to go down only to find that upward-bound stops have priority over our command. But once in the elevator and downward bound, we are likely to find the elevator stopping at floors preceding ours in sequence, again demonstrating a sequential priority. Added to these features are the usual safety and emergency overrides, and a motor control system that allows for the carrier to be accelerated or decelerated at some reasonable rate. Obviously, modern elevator systems are controlled by rather sophisticated sequential machines.

The list of sequential machines that touch our daily lives is vast and continuously growing. As examples, the cars we drive, the homes we live in, and our places of employment all use sequential machines of one type or another. Automobiles use digital sequential machines to control starting, braking, fuel injection, cruise control, and safety features. Most homes have automatic washing machines, microwave ovens, sophisticated audio and video devices of various types, and, of course, computers. Some homes have complex security, energy, and climate control systems. All of these remarkable and now commonplace gifts of modern technology are made possible through the use of digital sequential machines.

The machines just mentioned are called *sequential machines*, or simply *state machines*, because they possess true memory and can issue time-dependent sequences of logic signals controlled by present and past input information. These sequential machines may also be *synchronous* because the data path is controlled by a *system clock*. In synchronous sequential machines, input data are introduced into the machine and are processed sequentially according to some algorithm, and outputs are generated—all regulated by a system clock. Sequential machines whose operation is clock independent (i.e., self-timed) are called *asynchronous sequential machines*.

Synchronous sequential machines and their design, analysis, and operation are the subjects covered in this chapter. Treatment begins with a discussion of the models used for these machines. This is followed by a discussion of an important type of graphic that is used to represent the sequential behavior of sequential machines and by a detailed development of the devices used for their memory. The chapter ends with the design and analysis of relatively simple state machines.

5.1.1 A Sequence of Logic States

Consider that a synchronous sequential machine has been built by some means and that it is represented by the block symbol in Figure 5.1a. Then suppose the voltage waveforms from its three outputs are detected (say with a waveform analyzer) and displayed as in Figure 5.1b. From these physical waveforms the positive logic waveforms are constructed and a sequence of logic states is read in the order *ABC* as shown in Figure 5.1c. A group of logic waveforms such as these is commonly known as a *timing diagram*, and the sequence of logic states derived from these waveforms is seen to be a binary count sequence. Here, *A, B*, and *C* are called *state variables* because their values collectively define the *present state* of the machine at some point in time. Knowing the *present state* in a sequence of states also reveals the *next state*. Thus, in Figure 5.1c, if state 101 is the present state, then 110 is the next state. This short discussion evokes the following definition:

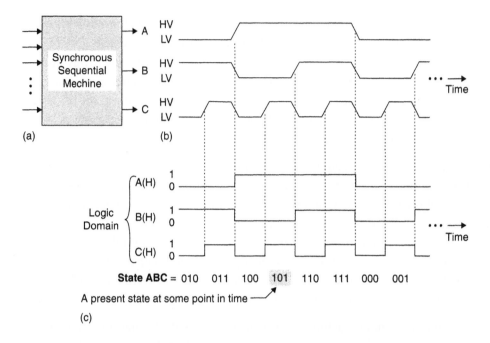

Figure 5.1: A sequence of logic events from a synchronous state machine (a) Block diagram symbol and (b) output voltage waveforms (c) Timing diagram representing the positive logic interpretation of the voltage waveforms and showing a sequence of logic states

A *logic state* is a unique set of binary values that characterize the logic status of a sequential machine at some point in time.

A sequential machine always has a finite number of states and is therefore called a *finite state machine* (*FSM*) or simply *state machine*. Thus, if there are N state variables, there can be no more than 2^N states in the FSM and no fewer than 2. That is, for any FSM,

$$2 \le (\text{number of states}) \le 2^N$$

For example, a two-state FSM requires one state variable, a three- or four-state FSM requires two state variables, five- to eight-state FSMs require three state variables, etc. More state variables can be used for an FSM than are needed to satisfy the 2^N requirement, but this is done only rarely to overcome certain design problems or limitations. The abbreviation FSM will be used frequently throughout the remainder of this text.

To help understand the meaning of the various models used in the description and design of FSMs, four binary sequences of states are given in Figure 5.2, each presenting a different feature of the sequence. The simple ascending binary sequence (a) is the same as that in Figure 5.1. This sequence and the remaining three will be described as they relate to the various models that are presented in the following section.

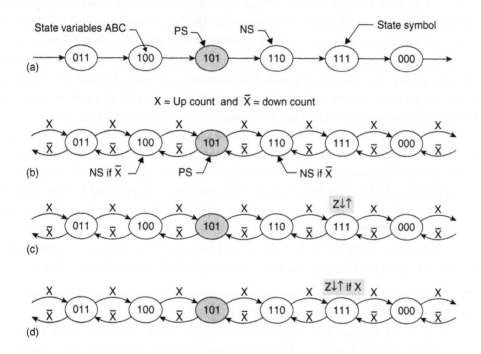

Figure 5.2: A sequence of states with present and next states based on Figure 5.1 (a) A simple ascending binary sequence showing present state (PS) and next state (NS) (b) A bidirectional (up/down) binary sequence showing PS and NS depending on logic level of input X (c) A bidirectional binary sequence with output Z in state 111 (d) A bidirectional sequence with output Z in state 111 conditional on input X (up-count)

5.2 Models for Sequential Machines

Models are important in the design of sequential machines because they permit the design process to be organized and standardized. The use of models also provides a means of communicating design information from one person to another. References can be made to specific parts of a design by using standard model nomenclature. In this section the general model for sequential machines will be developed beginning with the most elemental forms.

Notice that each state in the sequence of Figure 5.2a becomes the *present state* (*PS*) at some point in time and has associated with it a *next state* (*NS*) that is predictable given the PS. Now the question is: What logic elements of an FSM are required to do what is required in Figure 5.2a? To answer this question, consider the thinking process we use to carry out a sequence of events each day. Whether the sequence is the daily routine of getting up in the morning, eating breakfast, and going to work, or simply giving our telephone number to someone, we must be able to remember our present position in the sequence to know what the next step must be. It is no different for a sequential machine. There must be a memory section, as in Figure 5.3a, which generates the present state. And there must be a *next state logic* section, as in Figure 5.3b, which has been programmed to know what the next state must be, given the present state from the memory. Thus, an FSM conforming to the model of Figure 5.3b is capable of performing the simple ascending binary sequence represented by Figure 5.2a. However, to carry out the bidirectional binary sequence of Figure 5.2b, a machine conforming to the *basic model* of Figure 5.3c is required to have external input capability. As in the case of Figure 5.2b, an input X would force the FSM to count up in binary, while $\overline{X}$ would cause it to count down in binary.

If it is necessary that the FSM issue an output on arrival in any given state, output-forming logic must be added as indicated in Figure 5.4. This model has become known as *Moore's model* and any FSM that conforms to this model is often called a *Moore machine* in honor of E. F. Moore, a pioneer in sequential circuit design. For example, an FSM that can generate the bidirectional binary sequence in Figure 5.2c is called a Moore FSM, since an output Z is unconditionally activated on arrival in state 111 (up arrow, ↑) and is deactivated on exiting this state (down arrow, ↓); hence the double arrow (↓↑) for the output symbol Z↓↑. Such an output could be termed a *Moore output*, that is, an output that is issued as a function of the PS only. The functional relationships for a Moore FSM are:

$$\left\{ \begin{array}{l} PS = f(NS) \\ NS = f'(IP, PS) \\ OP = f''(PS) \end{array} \right\}, \tag{5.1}$$

where *IP* represents the external inputs and *OP* the outputs.

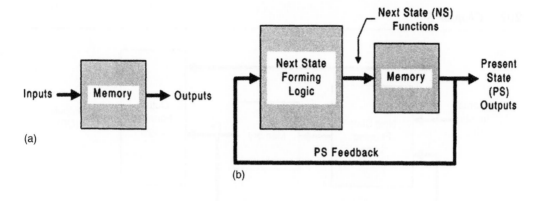

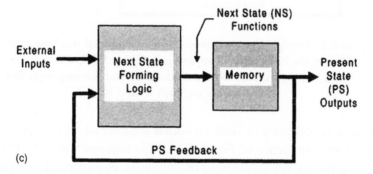

(a)

(b)

(c)

Figure 5.3: Development of the basic model for sequential machines (a) The memory section only (b) Model for an FSM capable of performing the sequence in Figure 5.2a, showing the memory section and NS-forming logic (c) The basic model for an FSM capable of performing the sequence in Figure 5.2b when the external input is X

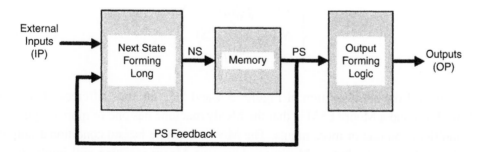

Figure 5.4: Moore's model for a sequential machine capable of performing the bidirectional binary sequence in Figure 5.2c, showing the basic model in Figure 5.3c with the added output-forming logic that depends only on the PS

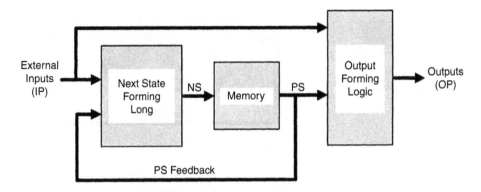

Figure 5.5: Mealy's (general) model for a sequential machine capable of performing the bidirectional sequence in Figure 5.2d, showing the basic model in Figure 5.3c with the added output-forming logic that depends on both IP and PS

Now suppose it is necessary to issue an output conditional on an input as in the bidirectional binary sequence of Figure 5.2d. This requires a model whose outputs not only depend on the PS, but also depend on the inputs, as illustrated in Figure 5.5. Such a model is the *most general model* for state machines and is known as *Mealy's model* after G. H. Mealy, another pioneer in the field of sequential machines. Thus, an FSM that conforms to this model can be called a *Mealy machine* and would be capable of generating the bidirectional binary sequence of Figure 5.2d, where the output is issued in state 111 but only if X is active (i.e., on an up count). Such an output could be termed a *Mealy output*, that is, an output that is issued conditional on an input. The functional relationships for a Mealy FSM are:

$$\left\{ \begin{array}{l} PS = f(NS) \\ NS = f'(IP, PS) \\ OP = f''(IP, PS) \end{array} \right\}. \tag{5.2}$$

As is evident from an inspection of Figures 5.4 and 5.5, the only difference between a Mealy FSM and a Moore FSM is that the Mealy machine has one or more outputs that are conditional on one or more inputs. The Moore machine has no conditional outputs. Hereafter, reference made to a Mealy machine or a Moore machine will imply this difference. Similarly, outputs that are referred to as Mealy outputs will be those that are issued conditionally on one or more inputs, and outputs referred to as Moore outputs will be those that are issued unconditionally.

5.3 The Fully Documented State Diagram

In Figure 5.2 a single input X is used to influence the sequential behavior of a binary sequence. A more complex example might involve several inputs that control the sequential behavior of the FSM. Such an FSM might be caused to enter one of several possible sequences (or routines), each with subroutines and outputs, all controlled by external inputs whose values change at various times during the operation of the FSM. Obviously, some means must be found by which both simple and complex FSM behavior can be represented in a precise and meaningful way. The *fully documented state diagram* discussed in this section is one means of representing the sequential behavior of an FSM.

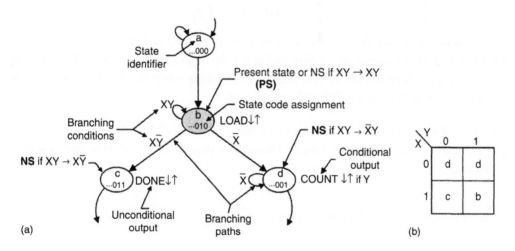

**Figure 5.6: (a) Features of the fully documented state diagram section
(b) The input/state map for state *b***

Presented in Figure 5.6a is a portion of a state diagram showing the important features used in its construction. Attention is drawn to states identified as a, b, c, and d. Here, state b is the present state (PS) at some point in time and is given the state code assignment $\cdots$ 010. Notice that state b branches to itself under the branching condition XY, the *holding condition* for that state, and that the next state (NS) depends on which input, X or Y, changes first. If X changes first, hence $XY \rightarrow \overline{X}Y$, the FSM will transit to the next state d, where it will hold on the input condition $\overline{X}$, where $\overline{X} = \overline{X}Y + X\overline{Y}$. Or if Y changes first, $XY \rightarrow X\overline{Y}$, the FSM will transit to state c, where there is no holding condition.

The output notation is straightforward. There are two types of outputs that can be represented in a fully documented state diagram. Referring to state b in Figure 5.6a, the output:

$$\text{LOAD } \downarrow\uparrow$$

is an *unconditional (Moore) output* issued any time the FSM is in state b. The down/up arrows ($\downarrow\uparrow$) signify that LOAD becomes active (up arrow, $\uparrow$) when the FSM enters state b and becomes inactive (down arrow, $\downarrow$) when the FSM leaves that state. The order in which the arrows are placed is immaterial as, for example, up/down. The output DONE in state c is also an unconditional or Moore output. The second type of output, shown in state d of Figure 5.6a and indicated by:

$$\text{COUNT } \downarrow\uparrow \text{ if } Y,$$

is a *conditional output* that is generated in state d but only if Y is active—hence, COUNT is a *Mealy output* according to the Mealy model in Figure 5.5. Thus, if input Y should toggle between active and inactive conditions while the FSM resides in state d, so also would the output COUNT toggle with Y.

The Sum Rule There are certain rules that must "normally" be followed for proper construction of state diagrams. One of these rules is called the *sum rule* and is stated as follows:

The Boolean sum of all branching conditions from a given state must be logic 1.

With reference to Figure 5.7, this rule is expressed mathematically as:

$$\sum_{i=0}^{n-1} f_{i \leftarrow j} = 1, \tag{5.3}$$

where $f_{i \leftarrow j}$ represents the branching condition from the jth to the ith state and is summed over n states as indicated in Figure 5.7b. For example, if the sum rule is applied to state b in Figure 5.6a, the result is:

$$XY + \overline{X} + X\overline{Y} = 1,$$

since according to the absorptive law, $\overline{X} + X\overline{Y} = \overline{X} + \overline{Y}$, which is the complement of XY. The graphical representation of the sum rule, as applied to state b, is shown in

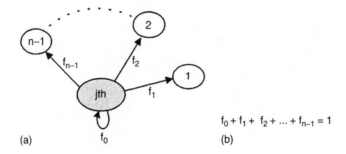

$$f_0 + f_1 + f_2 + \dots + f_{n-1} = 1$$

(a)   (b)

**Figure 5.7: Application of the sum rule given by equation (5.3)
(a) State diagram segment showing branching conditions relative to the
jth state (b) Application of the sum rule to the jth state in the
state diagram segment**

Figure 5.6b and is called the *input/state map*. Had the sum rule not been satisfied, one or more branching conditions would not be accounted for and one or more of the cells in the input/state map of Figure 5.6b would be vacant. If applied, unaccounted-for branching conditions can cause an FSM to malfunction.

The Mutually Exclusive Requirement While satisfying the sum ($\Sigma = 1$) rule is a necessary condition for state branching accountability in a fully documented state diagram, it is not sufficient to ensure that the branching conditions are nonoverlapping. The meaning of this can be stated as follows:

> Each possible branching condition from a given state must be associated with no more than one branching path.

With reference to Figure 5.8a, this condition is expressed mathematically as

$$f_{i \leftarrow j} = \overline{\sum_{\substack{k=0 \\ k \neq i}}^{n-1} f_{k \leftarrow j}}, \tag{5.4}$$

where each branching condition is seen to be the complement of the Boolean sum of those remaining as indicated in Figure 5.8b. When applied to state b in Figure 5.6, equation (5.4) gives the results $XY = \overline{\overline{X} + X\overline{Y}} = \overline{\overline{X} + \overline{Y}} = XY$ and $\overline{X} = \overline{XY + X\overline{Y}} = \overline{X}$, etc., clearly indicating that both the mutually exclusive requirement and the sum rule are satisfied.

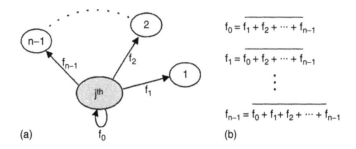

(a)

(b)

$$f_0 = \overline{f_1 + f_2 + \cdots + f_{n-1}}$$

$$f_1 = \overline{f_0 + f_2 + \cdots + f_{n-1}}$$

$$\vdots$$

$$f_{n-1} = \overline{f_0 + f_1 + f_2 + \cdots + f_{n-1}}$$

Figure 5.8: Application of the mutually exclusive requirement given by Eq. (5.4)
(a) State diagram segment showing branching conditions relative to the *j*th state
(b) Application of the mutually exclusive requirement to the
state diagram segment in (a)

Now consider the case shown in Figure 5.9a, where the sum rule is obeyed but not the mutually exclusive requirement. In this case, the branching condition XY is associated with both the $a \rightarrow b$ and the $a \rightarrow c$ branching paths. Thus, if $\overline{XY} \rightarrow XY$ while in state a, malfunction of the FSM is likely to occur. In Figure 5.9b is the input/state map showing violation of equation (5.4) under input condition XY shared by branching paths $a \rightarrow b$ and $a \rightarrow c$. Thus, if the mutually exclusive requirement is to hold for a given state, the input/state map must not have cells containing more than one state identifier.

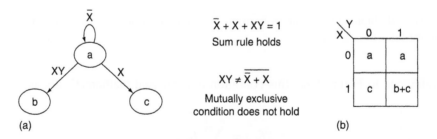

(a)

$$\overline{X} + X + XY = 1$$
Sum rule holds

$$XY \neq \overline{\overline{X} + X}$$
Mutually exclusive
condition does not hold

(b)

Figure 5.9: (a) A portion of a state diagram for which the mutually exclusive condition does not hold (b) Input/state map showing violation of the mutually exclusive requirement as applied to state *a* under branching condition *XY*

When Rules Can Be Broken There are conditions under which violation of the sum rule or of the mutual exclusion requirement is permissible. Simply stated, these conditions are as follows: If certain branching conditions can never occur or are never permitted to occur, they can be excluded from the sum rule and from the mutually exclusive requirement. This means that equations (5.3) and (5.4) need not be satisfied for the FSM to operate properly. As an example, suppose that in Figure 5.6 the branching condition is $\overline{Y}$ for branching path $b \rightarrow c$. Thus, the sum rule holds since $XY + \overline{X} + \overline{Y} = 1$. However, the branching condition $\overline{X}\overline{Y}$ is common to both the $b \rightarrow c$ and $b \rightarrow d$ branching paths with branching conditions $\overline{Y}$ and $\overline{X}$, respectively. Clearly, the mutually exclusive requirement of equation (5.4) is not satisfied, which is of no consequence if the input condition $\overline{X}\overline{Y}$ can never occur. But if the input condition $\overline{X}\overline{Y}$ is possible, then branching from state b under $\overline{X}\overline{Y}$ is ambiguous, leading to possible FSM malfunction.

5.4   The Basic Memory Cells

Developing the concept of memory begins with the basic building block for memory called the *basic memory cell* or simply *basic cell*. A basic cell plays a major role in designing a memory device (element) that will remember a logic 1 or a logic 0 indefinitely or until it is directed to change to the other value. In this section two flavors of the basic cell will be heuristically developed and used later in the design of important memory elements called *flip-flops*.

5.4.1   The Set-Dominant Basic Cell

Consider the wire loop in Figure 5.10a consisting of a fictitious lumped path delay (LPD) memory element Δt and two inverters whose function it is to maintain an imaginary signal. The LPD memory element is the path delay for the entire wire loop including inverters concentrated (lumped) in Δt, hence the meaning of the word "fictitious." But since there is no external access to this circuit, introduction of such a signal into this circuit is not possible. This can be partially remedied by replacing one of the inverters with an NAND gate performing the OR operation as in Figure 5.10b. Now, a Set $(0 \rightarrow 1)$ can be introduced into the circuit if $S(L) = 1(L)$. This can be further understood by an inspection of the Boolean expression $Q_{t+1} = S + Q_t$ for the circuit in Figure 5.10b, where the following definitions apply:

$$Q_{t+1} = \text{Next state}$$

$$Q_t = \text{Present state.}$$

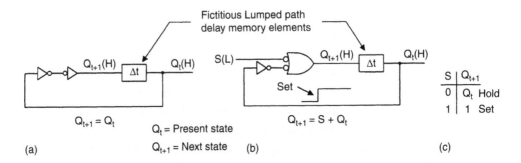

Figure 5.10: Development of the concept of Set (a) Wire loop with a fictitious lumped path delay memory element and two inverters used to restore an imaginary signal (b) Wire loop with one inverter replaced by a NAND gate used to introduce a Set condition (c) Truth table obtained from the logic expression for Q_{t+1} in (b) showing the Hold and Set conditions

The truth table in Figure 5.10c is constructed by introducing the values $\{0, 1\}$ for S into this equation and is another means of representing the behavior of the circuit in Figure 5.10b. The *hold condition* $Q_{t+1} = Q_t$ occurs any time the next state is equal to the present state, and the Set condition occurs any time the next state is a logic 1, i.e., $Q_{t+1} = 1$.

The circuit of Figure 5.10b has the ability to introduce a Set condition as shown, but no means of introducing a *Reset* $(1 \rightarrow 0)$ condition is provided. However, this can be done by replacing the remaining inverter with an NAND gate performing the AND operation as shown in Figure 5.11a. Then, if $R(L) = 1(L)$ when $S(L) = 0(L)$, a Reset condition is introduced into the circuit. Thus, both a Set and Reset condition can be introduced into the circuit by external means. This basic memory element is called the *set-dominant basic cell* for which the logic circuit in Figure 5.11a is but one of seven ways to represent its character, as discussed in the following paragraphs.

Reading the circuit in Figure 5.11a yields the following SOP expression for the next state function:

$$Q_{t+1} = S + \bar{R}Q_t. \tag{5.5}$$

When this expression is plotted in an EV K-map, Figure 5.11b results, where minimum cover is indicated by shaded loops. From this expression or from the EV K-map, it is clear that a set condition is introduced any time $S = 1$, and that a reset condition results

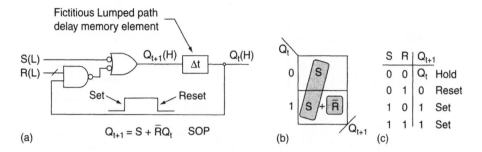

Figure 5.11: The set-dominant basic cell represented in different ways (a) Logic circuit showing the Set and Reset capability, and the Boolean equation for the next state function, Q_{t+1} (b) EV K-map with minimum cover indicated by shaded loops (c) Operation table for the set-dominant basic cell showing the Hold, Set, and Reset conditions inherent in the basic memory cell

only if $R = 1$ and $S = 0$. However, if both inputs are inactive, that is, if $S = R = 0$, it follows that $Q_{t+1} = Q_t$, which is the hold condition for the basic cell. The Hold, Set, and Reset conditions are easily observed by inspection of the *operation table* for the set-dominant basic cell given in Figure 5.11c. The basic cell is called *set-dominant* because there are two input conditions, $S\overline{R}$ and SR, that produce the Set condition as indicated by the operation table in Figure 5.11c. Notice that Figure 5.11 represents four ways of representing the set-dominant basic cell: logic circuit, NS function, NS K-map, and operation table.

By using the operation table in Figure 5.11c, the state diagram for the set-dominant basic cell can be constructed as given in Figure 5.12a. To clarify the nomenclature associated with any fully documented state diagram, the following definitions apply to the state variable changes and will be used throughout this text:

$$\left.\begin{cases} 0 \rightarrow 0 = Reset\ Hold \\ 0 \rightarrow 1 = Set \\ 1 \rightarrow 0 = Reset \\ 1 \rightarrow 1 = Set\ Hold \end{cases}\right\}. \tag{5.6}$$

Thus, for the state diagram in Figure 5.12a, $\overline{S}$ is the Reset Hold branching condition, S is the Set branching condition, $\overline{S}\,R$ is the Reset branching condition, and $S + \overline{R}$ is the Set Hold branching condition. The output Q is issued (active) only in the Set state (state 1), not in the Reset state (state 0). Notice that for each of the two states the

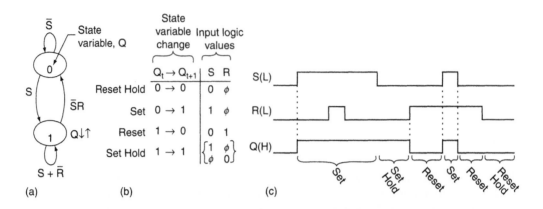

Figure 5.12: The set-dominant basic cell (cont'd.) (a) State diagram derived from the operation table in Figure 5.11c (b) Excitation table derived from the state diagram (c) Timing diagram illustrating the operation of the set-dominant basic cell

sum rule ($\Sigma = 1$) holds as it must. But all branching conditions are easily deduced from an inspection of the operation table. For example, the Set condition is the Boolean sum $\overline{S}\overline{R} + S\overline{R} = S$, or the Set Hold condition is the sum $\overline{S}\overline{R} + S\overline{R} = S\overline{R} = S + \overline{R}$, which is simply the complement of the Reset branching condition $\overline{S}\overline{R} = S + \overline{R}$ in agreement with the sum rule.

From the state diagram of Figure 5.12a another important table is derived, called the *excitation table*, and is presented in Figure 5.12b. Notice that a don't care ϕ is placed in either the S or R column of the excitation table for the basic cell to indicate an unspecified input branching condition. For example, the Set branching condition S requires that a 1 be placed in the S column while a ϕ is placed in the R, column indicating that R is not specified in the branching condition for Set. Similarly, for the Set Hold branching path $1 \rightarrow 1$, the branching condition $S + \overline{R}$ requires a 1 and ϕ to be placed in the S and R columns for the S portion of the branching condition, and that a ϕ and 0 to be placed in the S and R columns for the $\overline{R}$ portion, respectively. Thus, the excitation table specifies the input logic values for each of the four corresponding state variable changes in the state diagram as indicated.

As a seventh and final means of representing the behavior of the set-dominant basic cell, a timing diagram can be constructed directly from the operation table in Figure 5.11c. This timing diagram is given in Figure 5.12c, where the operating conditions Set, Set Hold, Reset, and Reset Hold are all represented—at this point no account is taken of the path delay through the gates. Notice that the set-dominant character is exhibited by the $S, R = 1, 0$ and $S, R = 1, 1$ input conditions in both the operation table and timing diagram.

5.4.2 The Reset-Dominant Basic Cell

By replacing the two inverters in Figure 5.10a with NOR gates, there results the logic circuit for the *reset-dominant basic cell* shown in Figure 5.13a. Now, the Set and Reset inputs are presented active high as $S(H)$ and $R(H)$. Reading the logic circuit yields the POS logic expression for the next state,

$$Q_{t+1} = \overline{R}(S + Q_t), \tag{5.7}$$

which is plotted in the EV K-map in Figure 5.13b with minimum cover indicated by shaded loops. The operation table for the reset-dominant basic cell is constructed directly from the Boolean expression for Q_{t+1} and is given in Figure 5.13c, where input conditions for Hold, Reset, and Set are depicted. Notice that the Set condition is introduced only when $S\overline{R}$ is active, whereas the Reset condition occurs any time R is active, the reset-dominant character of this basic memory element.

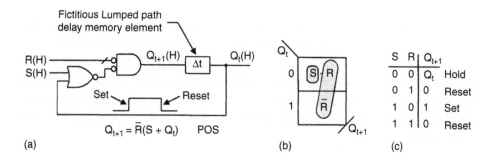

(a)   (b)   (c)

Figure 5.13: The reset-dominant basic cell represented in different ways (a) Logic circuit showing the Set and Reset capability, and the Boolean equation for the next state function, Q_{t+1} (b) EV K-map with minimum cover indicated by shaded loops (c) Operation table for the reset-dominant basic cell showing the Hold, Set, and Reset conditions inherent in the basic memory cell

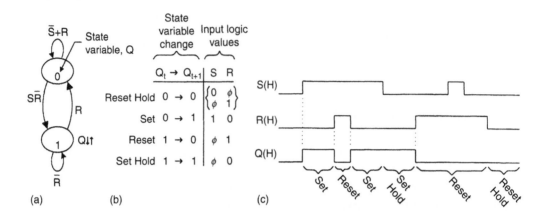

Figure 5.14: The reset-dominant basic cell (contd.) (a) State diagram derived from the operation table in Figure 5.13c (b) Excitation table derived from the state diagram (c) Timing diagram illustrating the operation of the reset-dominant basic cell

The state diagram for the reset-dominant basic cell is constructed from the operation table in Figure 5.13c with the result shown in Figure 5.14a. Here, the Set condition $S\overline{R}$ is placed on the $0 \rightarrow 1$ branching path. Thus, it follows that the Reset Hold condition is $\overline{S} + R$, which can be read from the operation table as $\overline{S}\overline{R} + \overline{S}R + SR = \overline{S} + R$, or is simply the complement of the Set input condition $\overline{S\overline{R}} = \overline{S} + R$, a consequence of the sum rule. The remaining two branching conditions follow by similar reasoning.

The excitation table for the reset-dominant basic cell is obtained directly from the state diagram in Figure 5.14(a) and is presented in Figure 5.14b. Again, a don't care ϕ is placed in either the S or R column of the excitation table for the basic cell to indicate an unspecified input branching condition, as was done in the excitation table for the set-dominant basic cell of Figure 5.12b. The nomenclature presented to the left of the excitation table follows the definitions for state variable change given by equations (5.6).

The seventh and final means of representing the reset-dominant basic cell is the timing diagram constructed in Figure 5.14c with help of the operation table in Figure 5.13c. Again, no account is taken at this time of the gate propagation delays. Notice that the reset-dominant character is exhibited by the $S, R = 0, 1$ and $S, R = 1, 1$ input conditions in both the operation table and the timing diagram.

At this point the reader should pause to make a comparison of the results obtained for the set-dominant and reset-dominant basic cells. Observe that there are some similarities, but there are also some basic differences that exist between the two basic memory elements. Perhaps these similarities and differences are best dramatized by observing the timing diagrams in Figs. 5.12c and 5.14c. First, notice that the S and R inputs arrive active low to the set-dominant basic cell but arrive active high to the reset-dominant cell. Next, observe the difference in the $Q(H)$ waveform for these two types of basic cells. Clearly, the set-dominant character is different from the reset-dominant character with regard to the $S, R = 1, 1$ input condition. This difference may be regarded as the single most important difference between these two memory cells and will play a role in the discussion that follows.

5.4.3 Combined Form of the Excitation Table

The excitation table for a memory element has special meaning and utility in state machine design. In subsequent sections it will be shown that the basic memory cell plays a major role in the design and analysis of flip-flops, the memory elements used in synchronous state machine design. Two such excitation tables have been identified so far: one associated with the set-dominant basic cell and the other associated with the reset-dominant cell. For purposes of flip-flop design these two excitation tables are inappropriate because of the different way they behave under the $S, R = 1, 1$ input condition. To overcome this difference, the two may be combined to give the single *generic (combined) excitation table* as shown in Figure 5.15. Here, common S, R input conditions for the two excitation tables in Figures. 5.15a and 5.15b are identified for each of the four branching paths given and are brought together to form the combined excitation table in Figure 5.15c. The important characteristic of the combined excitation is that *the S, R = 1, 1 condition is absent*. This leads to the following important statements:

- Because the S, R = 1, 1 condition is not present in the combined excitation table, it is applicable to either the set-dominant basic cell or the reset-dominant basic cell.

- Throughout the remainder of this text only the combined excitation table will be used in the design of other state machines, including other memory elements called flip-flops.

Thus, the individual excitation tables for the set-dominant and reset-dominant basic cells will be of no further use in the discussions of this text.

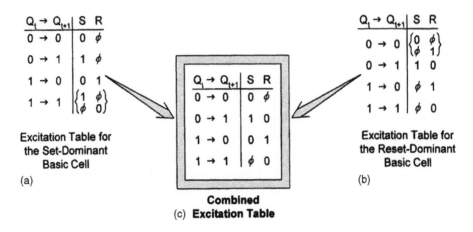

Figure 5.15: The excitation table for the basic cell (a) Excitation table for the set-dominant (NAND-based) basic cell (b) Excitation table for the reset-dominant (NOR-based) basic cell (c) Generic (combined) excitation table applicable to either of the basic cells since the S, R = 1, 1 condition is absent

5.4.4 Mixed-Rail Outputs of the Basic Cells

There are subtle properties of the basic cells, yet to be identified, that are essential to the design of other memory elements. These properties deal with the output character of the basic cells. Referring to the logic circuit in Figure 5.11a, only one output is identified for the set-dominant basic cell. However, by removing the fictitious lumped path delay (LPD) memory element Δt and arranging the conjugate NAND gate forms one above the other, there results the well-known "cross-coupled" NAND gate configuration shown in Figure 5.16a. There is but one feedback path for the basic cell (indicated by the heavy line), though it may appear to the reader as though there are two.

The mixed-logic output expression from each of the two conjugate NAND gate forms in the set-dominant basic cell is read and presented as shown in Figure 5.16a. Using these two output expressions, the truth table in Figure 5.16b is constructed. In this table it is observed that all input conditions except $S, R = 1, 1$ generate what are called *mixed-rail outputs* from the two conjugate NAND gate forms. This means that when a 0(H) is produced from the OR form, a 0(L) appears on the output of the AND form. Or when the former is 1(H), the latter is 1(L). The S, R = 1, 1 input condition, meaning $S(L) = R(L) = 1(L)$, produces outputs that are 1(H) and 0(L) = 1(H) from the OR and AND

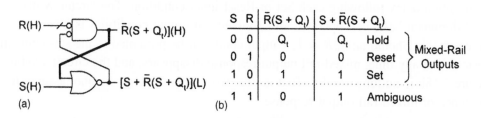

S	R	$S + \bar{R}Q_t$	$\bar{R}\cdot[S + \bar{R}Q_t]$	
0	0	Q_t	Q_t	Hold
0	1	0	0	Reset
1	0	1	1	Set
1	1	1	0	Ambiguous

Figure 5.16: Mixed-rail outputs of the set-dominant basic cell (a) Logic circuit showing the mixed-logic output expressions from the two gates (b) Truth table indicating the input conditions required for mixed-rail outputs

forms, respectively, and are not mixed-rail outputs—the NAND gate outputs are ambiguous, since they cannot be labeled as either Set or Reset.

A similar procedure is used in defining the mixed-rail outputs from the reset-dominant basic cell. Shown in Figure 5.17a are the "cross-coupled" NOR gates where the fictitious LPD memory element Δt has been removed, and outputs from the two conjugate NOR gate forms are given in mixed-logic notation. Notice again that only one feedback path exists as indicated by the heavy line.

The input conditions required to generate mixed-rail outputs from the reset-dominant basic cell are presented in the truth table of Figure 5.17b. This table is obtained from the logic circuit and mixed-logic expressions in Figure 5.17a. Notice that all input conditions except the $S, R = 1, 1$ condition generate *mixed-rail outputs* from the two conjugate NOR gate forms, similar to the case of the set-dominant basic cell in Figure 5.16b. Thus, again, the $S, R = 1, 1$ condition produces an ambiguous output, since the outputs from the conjugate NOR gates are neither a Set nor a Reset.

S	R	$\bar{R}(S + Q_t)$	$S + \bar{R}(S + Q_t)$	
0	0	Q_t	Q_t	Hold
0	1	0	0	Reset
1	0	1	1	Set
1	1	0	1	Ambiguous

Figure 5.17: Mixed-rail outputs of the reset-dominant basic cell (a) Logic circuit showing the mixed-logic output expressions from the two conjugate gate forms (b) Truth table indicating the input conditions required to produce mixed-rail output conditions

Clearly, the mixed-rail outputs of the two types of basic memory cells and the combined excitation table representing both basic cells all have something in common. From the results of Figs. 5.15(c), 5.16(b), and 5.17(b), the following important conclusion is drawn:

> The mixed-rail output character of the set- and reset-dominant basic cells is inherent in the combined excitation table of Figure 5.15c, since the S, R = 1, 1 input condition is absent.

Use of this fact will be made later in the design of the memory elements, called flip-flops, where the basic cells will serve as the memory. Thus, if the $S, R = 1, 1$ condition is never allowed to happen, mixed-rail output response is ensured. But how is this output response manifested? The answer to this question is given in the following subsection.

5.4.5   Mixed-Rail Output Response of the Basic Cells

From subsection 5.4.4, one could gather the impression that a mixed-rail output response from the conjugate gate forms of a basic cell occurs simultaneously. Actually, it does not. To dramatize this point, consider the set-dominant basic cell and its mixed-rail output response to nonoverlapping Set and Reset input conditions shown in Figure 5.18a. It is observed that the active portion of the waveform from the ANDing operation is symmetrically set inside of that from the ORing (NAND gate) operation by an amount τ on each edge. Here, it is assumed that $\tau_1 = \tau_2 = \tau$ is the propagation delay of a two-input NAND gate. Thus, it is evident that the mixed-rail output response of the conjugate gate forms does not occur simultaneously but is delayed by a gate propagation delay following each Set or Reset input condition. The circuit symbol for a set-dominant basic cell operated under mixed-rail output conditions is given in Figure 5.18b. Should an $S, R = 1, 1$ input condition be presented to the set-dominant basic cell at any time, mixed-rail output response disappears, and the circuit symbol in Figure 5.18b is no longer valid. That is, the two Q's in the logic symbol assume the existence of mixed-rail output response.

In a similar manner, the mixed-rail output response of the reset-dominant basic cell to nonoverlapping Set and Reset input conditions is illustrated in Figure 5.18c. Again, it is observed that the active portion of the waveform from the ANDing (NOR gate)

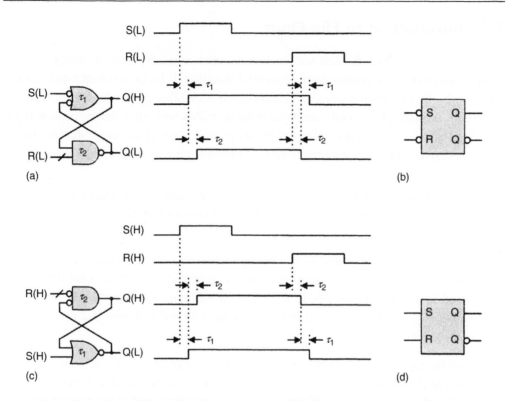

**Figure 5.18: Mixed-rail output response of the basic cells and circuit symbols
(a) Logic circuit and mixed-rail output response for the set-dominant basic cell
(b) Circuit symbol for the set-dominant basic cell (c) Logic circuit and mixed-rail
output response for the reset-dominant basic cell (d) Circuit symbol for the
reset-dominant basic cell**

operation is symmetrically set within that of the ORing operation by an amount equal to $\tau = \tau_1 = \tau_2$, the propagation delay of a NOR gate. The circuit symbol for the reset-dominant basic cell operated under mixed-rail output conditions is given in Figure 5.18d. The difference in circuit symbols for set- and reset-dominant basic cells is indicative of the fact that the former requires active low inputs while the latter requires active high inputs. As is the case for the set-dominant basic cell, an S, $R = 1, 1$ input condition eliminates mixed-rail output response and invalidates the circuit symbol in Figure 5.18d. The two Q's in the logic symbol assume the existence of mixed-rail output response.

5.5   Introduction to Flip-Flops

The basic cell, to which the last section was devoted, is not by itself an adequate memory element for a synchronous sequential machine. It lacks versatility and, more importantly, its operation cannot be synchronized with other parts of a logic circuit or system. Actually, basic cells are asynchronous FSMs without a timing control input but which are essential to the design of *flip-flops*, the memory elements that are used in the design synchronous state machines. A flip-flop may be defined as follows:

A flip-flop is an asynchronous one-bit memory element (device) that exhibits sequential behavior controlled exclusively by an enabling input called CLOCK.

A flip-flop samples a data input of one bit by means of a clock signal, issues an output response, and stores that one bit until it is replaced by another. One flip-flop is required for each state variable in a given state diagram. For example, FSMs that are capable of generating the 3-bit binary sequences shown in Figure 5.2 each require three flip-flops for their design.

The enabling input, clock, can be applied to the flip-flops as either a regular or irregular waveform. Both types of clock waveforms are represented in Figure 5.19. The regular clock waveform in Figure 5.19a is a periodic signal characterized by a clock period T_{CK} and frequency f_{CK} given by:

$$f_{CK} = \frac{1}{T_{CK}}, \qquad (5.8)$$

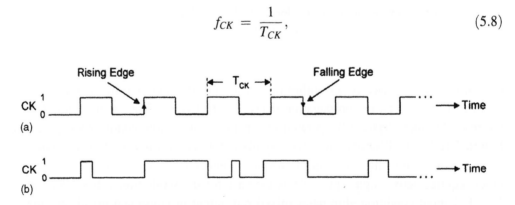

Figure 5.19: Clock logic waveforms (a) Regular clock waveform showing rising and falling edges and a fixed clock period T_{CK} (b) Irregular clock waveform having no fixed clock period

where f_{CK} is given in units of Hz (hertz) when the clock period is specified in seconds. The irregular clock waveform in Figure 5.19b has no fixed clock period associated with it. However, both regular and irregular clock waveforms must have rising $(0 \rightarrow 1)$ and falling $(1 \rightarrow 0)$ edges associated with them, as indicated in Figure 5.19.

5.5.1 Triggering Mechanisms

In synchronous sequential machines, state-to-state transitions occur as a result of a *triggering mechanism* that is either a rising or falling edge of the enabling clock waveform. Flip-flops and latches that trigger on the rising edge of the clock waveform are said to be *rising edge triggered* (*RET*), and those that trigger on the falling edge of the clock waveform are referred to as *falling edge triggered* (*FET*). These two triggering mechanisms are illustrated in Figure 5.20, together with the logic symbols used to represent them. The distinction between flip-flops and latches will be made in Section 5.7.

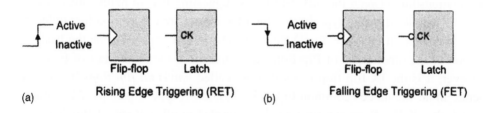

Figure 5.20: Flip-flop and latch logic circuit symbology (a) Rising-edge triggering (b) Falling-edge triggering

Mechanisms involving a two-stage triggering combination of RET and FET flip-flops are classified as *master–slave* (MS) triggering mechanisms and flip-flops that employ this two-stage triggering are called, accordingly, *master–slave (MS) flip-flops*. MS flip-flops will be dealt with together with edge-triggered flip-flops in subsequent sections.

5.5.2 Types of Flip-Flops

The designer has a variety of flip-flops and triggering mechanisms from which to choose for a given FSM design. The mechanisms are classified as either edge triggered (ET), meaning RET or FET, or master–slave (MS). The types of flip-flops and the mechanisms by which they operate are normally chosen from the following list:

D flip-flops (ET or MS triggered)

T flip-flops (ET or MS triggered)

JK flip-flops (ET or MS triggered)

The generalized definitions of the flip-flop types D, T, and JK are internationally accepted and will be discussed in turn in the sections that follow. There are other flip-flop types (e.g., SR flip-flops) and other triggering mechanism interpretations, and these will be noted where appropriate. It is the intent of this text to concentrate on the major types of flip-flop memory elements.

5.5.3   *Hierarchical Flow Chart and Model for Flip-Flop Design*

In checking the data books on flip-flops it becomes clear that there exists an interrelationship between the different types suggesting that in many cases there exists a "parent" flip-flop type from which the others are created—a hierarchy for flip-flop design. In fact, it is the D flip-flop (D-FF) that appears to be the basis for the creation of the other types of flip-flops, as indicated in the flow chart of Figure 5.21. However, it is the JK flip-flop types that are called *universal flip-flops* because they operate in all the modes common to the D, T, and SR type flip-flops. Also, once created, the JK flip-flops are most easily converted to other types of flip-flops (e.g., JKs converted to Ts), as suggested by the flow chart. With few exceptions,

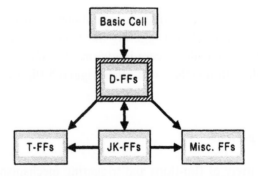

Figure 5.21: Flow chart for flip-flop design hierarchy, showing D type flip-flops as central to the design of other flip-flops

flip-flops other than D, JK, and SR types are rarely available commercially. Of the latches, one finds that only the D and SR latches are available commercially.

There are, of course, exceptions to this hierarchy for flip-flop design, but it holds true for most of the flip-flops. The miscellaneous category of flip-flops includes those with special properties for specific applications. The SR flip-flop types fall into this category.

The model that is used for flip-flop design is the basic model given in Figure 5.3c but adapted specifically to flip-flops. This model, presented in Figure 5.22a, is applied to a generalized, fictitious RET XY type flip-flop and features one or more basic memory cells as the memory, the next state (NS) forming logic, external inputs including clock (CK), the *S* and *R* next state functions, and the present state (PS) feedback paths. Had the fictitious XY-FF been given an FET mechanism, a bubble would appear on the outside of the clock triggering symbol (the triangle). Note that the *S* and *R* next state functions would each be represented by dual lines if two basic cells are used as the memory for the XY flip-flop. The logic circuit symbol for the RET XY flip-flop (XY-FF) is given in Figure 5.22b.

Not all flip-flops to be discussed in the sections that follow have two data inputs and not all have PS feedback paths as in Figure 5.22. And not all flip-flops will be rising edge triggered as in this fictitious flip-flop. Furthermore, flip-flops classified as master–slave flip-flops do not adhere to the model of Figure 5.22, since they are two-stage memory elements composed of two memory elements of one type or another. Nevertheless, the model of Figure 5.22 presents a basis for flip-flop design and will be used in the discussions that follow.

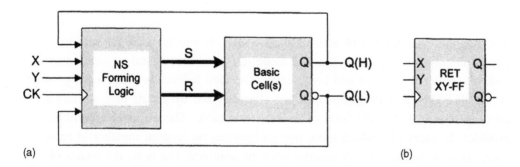

(a) (b)

Figure 5.22: (a) The basic model adapted to a fictitious RET XY type flip-flop showing the basic cell(s) as memory, the NS forming logic, the *S* and *R* next state functions, the external data inputs *X* and *Y*, the clock (CK) input, and the present state (PS) feedback lines from the mixed-rail outputs *Q* (b) Circuit symbol for the RET XY flip-flop

5.6   Procedure for FSM (Flip-Flop) Design and the Mapping Algorithm

The following three-step procedure will be used in the design of FSMs including flip-flops:

1. Select the FSM (e.g., a flip-flop type) to be designed and represent this FSM in the form of a state diagram. The output-forming logic can be mapped and obtained at this point.

2. Select the memory element (e.g., a basic cell or flip-flop) to be used in the design of the FSM (e.g., in the design of another flip-flop) and represent this memory element in the form of an excitation table.

3. Obtain the NS function(s) for the FSM in the form of NS K-maps by combining the information represented in the state diagram with that represented in the excitation table for the memory. To accomplish this, apply the following *mapping algorithm*:

> **Mapping Algorithm for FSM Design**
> AND the memory input logic value in the excitation table with the corresponding branching condition (BC) in the state diagram for the FSM to be designed, and enter the result in the appropriate cell of the NS K-map.

The mapping algorithm is of general applicability. It will be used not only to design and convert flip-flops, but also to design synchronous and asynchronous state machines of any size and complexity. The idea behind the mapping algorithm is that all FSMs, including flip-flops, are characterized by a state diagram and a memory represented in the form of an excitation table. The mapping algorithm provides the means by which these two entities can be brought together in some useful fashion so that the NS functions can be obtained. For now, the means of doing this centers around the NS K-maps. But the procedure is general enough to be computerized for CAD purposes by using a state table in place of the state diagram.

5.7   The D Flip-Flops: General

Every properly designed and operated D-FF behaves according to a *single* internationally accepted definition that is expressed in any one or all of the three ways. Presented in Figure 5.23 are the three means of defining the D flip-flop of any type. The first is the operation table for any D-FF given in Figure 5.23a. It specifies that when D is active Q must be active (Set condition), and when D is inactive Q must be inactive (Reset condition). The state diagram for any D-FF, given in Figure 5.23b, is best derived from the operation table and expresses the same information about the operation of the D-FF. Thus, state 0 is the Reset state ($Q_t = 0$) when $D = 0$, and state 1 is the Set state ($Q_t = 1$) when $D = 1$.

The excitation table for any D-FF, given in Figure 5.23c, is the third means of expressing the definition of a D-FF. It is best derived directly from the state diagram in Figure 5.23b. In this table the $Q_t \rightarrow Q_{t+1}$ column represents the state variable change from PS to NS, and the D column gives the input logic value for the corresponding branching path in the state diagram. For example, the Reset hold branching path $0 \rightarrow 0$ is assigned $D = 0$ (for $\overline{D}$), and the Set branching path $0 \rightarrow 1$ is assigned the $D = 1$ for branching condition D. The excitation table for the D-FF is extremely important to the design of other state machines, including other flip-flops, as will be demonstrated in later sections.

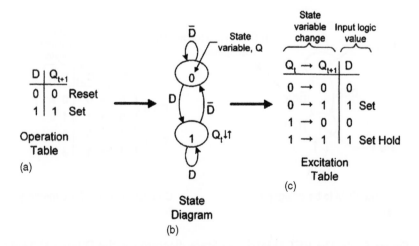

Figure 5.23: Generalized D flip-flop definition expressed in terms of the (a) operation table (b) the state diagram and (c) the excitation table

Now that the foundation for flip-flop design has been established, it is necessary to consider specific types of D flip-flops. There are three types to be considered: the D-latch, the edge triggered (ET) D flip-flop, and the master–slave (MS) D flip-flop, all of which adhere to the generalized definition of a D flip-flop expressed in Figure 5.23. Each of these D-type flip-flops is represented by a unique state diagram containing the enabling input clock (CK) in such a way as to identify the triggering mechanism and character of the D flip-flop type. In each case the memory element used for the design of the D flip-flop is the basic cell (set-dominant or reset-dominant) that is characterized by the combined excitation table given in Figure 5.15c. The design procedure follows that given in Section 5.6 where use is made of the important *mapping algorithm*.

5.7.1 The D Latch

A flip-flop whose sequential behavior conforms to the state diagram presented in Figure 5.24a is called an RET *transparent (high) D latch or simply D latch*. Under normal flip-flop action the RET D latch behaves according to the operation table in Figure 5.23a, but only when enabled by *CK*. The transparency effect occurs when *CK* is active (*CK* = 1). During this time *Q* goes active when *D* is active, and Q goes inactive when *D* is inactive—that is, *Q* tracks D when *CK* = 1. Under this transparency condition, data (or noise) on the *D* input is passed directly to the output and normal

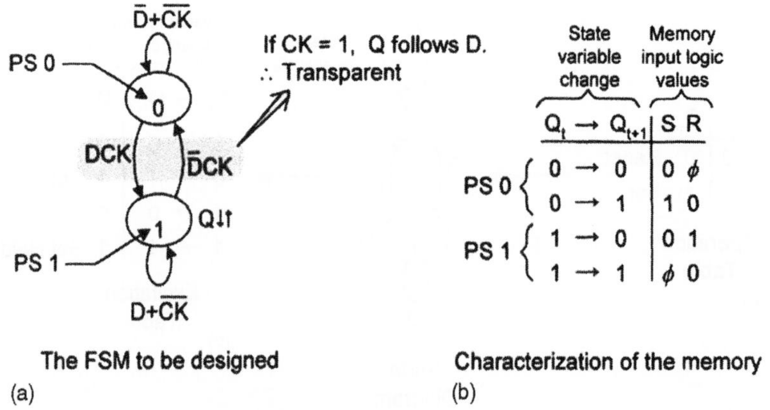

The FSM to be designed

(a)

Characterization of the memory

(b)

Figure 5.24: The RET D latch (a) State diagram for the D latch showing transparency effect when CK = 1 (b) Excitation table for the basic cell and characterization of the memory

flip-flop action (regulated by *CK*) does not occur. If the D latch is itself to be used as a memory element in the design of a synchronous FSM, the transparent effect must be avoided. This can be accomplished by using a pulse narrowing circuit of the type discussed later. The idea here is that minimizing the active portions of the clock waveform also minimizes the probability that the transparent effect can occur. Although this is likely to be true, it is generally recommended that the *D* latch not be considered as a viable option when selecting a memory element for FSM design. Of course, if *D* can never go active when *CK* is active, the D latch can be considered as a viable option for memory in FSM design.

The memory element to be used in the design of the D latch is one or the other of two basic cells (Figure 5.18) characterized by the combined excitation table given in Figure 5.24b. The plan for design of the D latch is simply to take the information contained in the state diagram of Figure 5.24a and in the excitation table in Figure 5.24b, and bring the two kinds of information together in the form of next-state K-maps by using the mapping algorithm given in Section 5.6. When this is done the following information is used for the K-map entries:

For $0 \rightarrow 0$
$$\begin{cases} \text{place } 0 \cdot \left(\overline{D} + \overline{CK}\right) = 0 \text{ in Cell 0 of the } S \text{ } K\text{-}map \\ \text{place } \phi \cdot \left(\overline{D} + \overline{CK}\right) = \phi\left(\overline{D} + \overline{CK}\right) \text{ in Cell 0 of the } R \text{ } K\text{-}map \end{cases}$$

For $0 \rightarrow 1$
$$\begin{cases} \text{place } 1 \cdot (DCK) = DCK \text{ of Cell 0 of the } S \text{ } K\text{-}map \\ \text{place } 0 \cdot (DCK) = 0 \text{ in Cell 0 of the } R \text{ } K\text{-}map \end{cases}$$

For $1 \rightarrow 0$
$$\begin{cases} \text{place } 0 \cdot \left(\overline{D}CK\right) = 0 \text{ in Cell 1 of the } S \text{ } K\text{-}map \\ \text{place } 0 \cdot \left(\overline{D}CK\right) = \overline{D}CK \text{ in Cell 1 of the } R \text{ } K\text{-}map \end{cases}$$

For $1 \rightarrow 1$
$$\begin{cases} \text{place } \phi \cdot \left(D + \overline{CK}\right) = \phi\left(D + \overline{CK}\right) \text{ in Cell 1 of the } S \text{ } K\text{-}map \\ \text{place } 0 \cdot \left(\overline{D}CK\right) = 0 \text{ in Cell 1 in the } R \text{ } K\text{-}map \end{cases}.$$

This results in the next state EV K-maps, minimum next state functions for *S* and *R*, and the logic circuit and symbol all shown in Figure 5.25. The four null (zero) entries are omitted in the EV K-maps, leaving only the two essential and two nonessential (don't care) entries for use in extracting minimum cover. Note that DCK (in the *S* K-map) is contained in ϕD, that $\overline{D}CK$ (in the *R* K-map) is contained in $\phi\overline{D}$, and that the logic

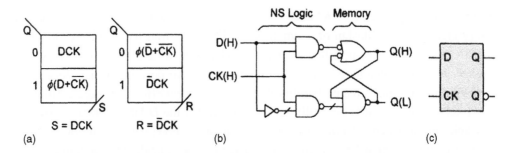

Figure 5.25: Design of the RET D latch by using a basic cell as the memory (a) EV K-maps and minimum Boolean expressions for the *S* and *R* next-state functions (b) Logic circuit showing the NS logic from part (a) and the set-dominant basic cell as the memory (c) Logic circuit symbol for the RET D latch

circuit conforms to the model in Figure 5.22 exclusive of PS feedback. The *CK* input to the circuit symbol in Figure 5.25c is consistent with that for a latch as indicated in Figure 5.20a.

The behavior of the RET D latch is best demonstrated by the timing diagram shown in Figure 5.26. Here, normal D flip-flop (D-FF) action is indicated for *D* pulse durations much longer than a *CK* period. For normal D-FF behavior, Q goes active when *CK* samples (senses) *D* active, and *Q* goes inactive when *CK* samples *D* inactive. However, when *CK* is active and *D* changes activation level, the transparency effect occurs. This is demonstrated in the timing diagram of Figure 5.26.

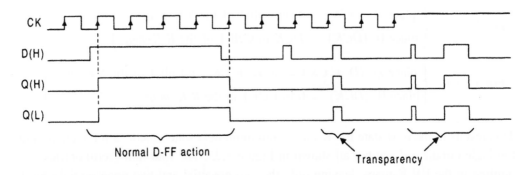

Figure 5.26: Timing diagram for an RET D latch showing normal D-FF action and the transparency effect that can occur when CK is active, where no account is taken of gate path delays

The FET (*transparent low*) D latch is designed in a similar manner to the RET D latch just described. All that is required is to complement *CK* throughout in the state diagram of Figure 5.24a, as shown in Figure 5.27a. Now, the transparency effect occurs when *CK* is inactive ($CK = 0$). If a set-dominant basic cell is again used as the memory, there results the logic circuit of Figure 5.27b, where an inverter is the only added feature to the logic circuit shown in Figure 5.25b. The logic circuit symbol for the FET D latch is given in Figure 5.27c. Here, the active low indicator bubble on the clock input identifies this as a falling edge-triggered device consistent with Figure 5.20(b). A *CK*(H) or *CK*(L) simply means RET or FET, respectively.

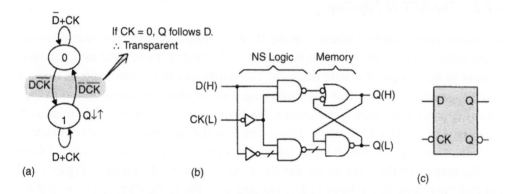

(a)   (b)   (c)

Figure 5.27: The FET D latch (a) State diagram showing condition for transparency (b) Logic circuit assuming the use of a set-dominant basic cell as the memory for design (c) Logic circuit symbol

If either the RET D latch or the FET D latch is to be used as the memory element in the design of a synchronous FSM, extreme care must be taken to ensure that the transparency effect does not occur. Transparency effects in flip-flops result in unrecoverable errors and must be avoided. This can be accomplished by using a pulse narrowing circuit of the type shown in Figure 5.28a. Here, an inverting delay element of duration Δt is used to produce narrow pulses of the same duration in the output logic waveform as indicated in Figure 5.28b. The delay element can be one or any odd number of inverters, an inverting buffer, or an inverting Schmitt trigger. In any case, the delay element must be long enough to allow the narrow pulses to reliably cross the switching threshold. If the delay is too long, the possibility of transparency exists; if it is too short, flip-flop triggering will not occur.

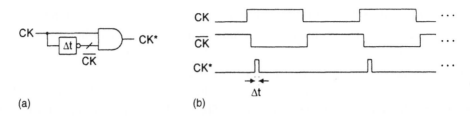

Figure 5.28: Pulse narrowing circuit (a) Logic circuit showing an inverting delay element Δt used to produce narrow pulses from long input pulses (b) Positive logic timing diagram showing the resulting narrow pulses of duration Δt on the output waveform

5.7.2   The RET D Flip-Flop

The transparency problem inherent in the D latch, discussed in the previous subsection, places rather severe constraints on the inputs if the latch is to be used as a memory element in the design of a state machine. This problem can be overcome by using an edge triggered D flip-flop that possesses *data lockout character* as discussed in the following paragraph. Shown in Figure 5.29a is the *resolver* FSM that functions as the input stage of an RET D flip-flop. Here, state *a* is the sampling (unresolved) state, *CK* is the sampling (enabling) input, and states *b* and c are the resolved states. Observe that the outputs of the resolver are the inputs to the basic cell shown in Figure 5.29b, and that the output of the basic cell is the output of the D flip-flop. Thus, an input FSM (the resolver) drives an output FSM (the basic cell) to produce the D flip-flop which conforms to the general D flip-flop definitions given in Figure 5.23. Note that both the resolver and basic cell are classified as asynchronous state machines, yet they combine to produce a state machine (flip-flop) that is designed to operate in a synchronous (clock-driven) environment. But the flip-flop itself is an asynchronous FSM!

To understand the function of the RET D flip-flop, it is necessary to move stepwise through the operation of the two FSMs in Figure 5.29: Initially, let Q be inactive in state a of the resolver. Then, if CK samples D active in state *a*, the resolver transits $a \rightarrow c$ and issues the output S, which drives the basic cell in Figure 5.29b to the set state 1 where Q is issued. In state *c*, the resolver holds on CK, during which time Q remains active; and the data input D can change at any time without altering the logic status of the flip-flop—this is the *data lockout* feature. When CK goes inactive ($\overline{CK}$), the resolver transits back to state a, where the sampling process begins all over again, but where Q remains active. Now, if CK samples D inactive ($\overline{D}$) in state *a*, the resolver transits $a \rightarrow b$, at which time R is issued. Since the branching condition $\overline{S}\,R$ is now satisfied, the basic cell is forced to transit to the reset state 0, where Q is deactivated.

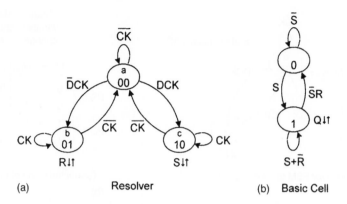

(a)   Resolver   (b)   Basic Cell

Figure 5.29: The RET D flip-flop as represented by state diagrams (a) Resolver FSM input stage (b) Set-dominant basic cell output stage

The resolver holds in state b on active CK. Then when CK goes inactive $(\overline{CK})$, the resolver transits back to the unresolved state a, at which time the sampling process begins all over again, but with Q remaining inactive.

The design of the RET D flip-flop follows the design procedure and mapping algorithm given in Section 5.6. Since the logic circuit for the set-dominant basic cell is known and given in Figure 5.18a, all that is necessary is to design the resolver circuit. This is done by using what is called the *nested cell model*, which uses the basic cells as the memory elements. Shown in Figure 5.30 are the state diagram for the resolver (the FSM to be designed), the characterization of the memory (combined excitation table for the basic cell), and the EV K-maps for the next state and output functions.

The mapping algorithm requires that the information contained in the state diagram of Figure 5.30a be combined with the excitation table of Figure 5.30b to produce the next state EV K-maps. This has been done in Figure 5.30c by introducing the following information obtained by a step-by-step application of the mapping algorithm:

State 00 (K-map cell 0)

$$\text{Bit A} \begin{cases} 0 \to 1, \text{ place } 1 \cdot (DCK) = DCK \text{ in the } S_A \text{ K-map} \\ 0 \to 0, \text{ place } \phi \cdot \left(\overline{D} + \overline{CK}\right) \text{ in the } R_A \text{ K-map} \end{cases}$$

$$\text{Bit B} \begin{cases} 0 \to 1, \text{ place } 1 \cdot \left(\overline{D}CK\right) = \overline{D}CK \text{ in the } S_B \text{ K-map} \\ 0 \to 0, \text{ place } \phi \cdot \left(D + \overline{CK}\right) \text{ in the } R_B \text{ K-map} \end{cases}$$

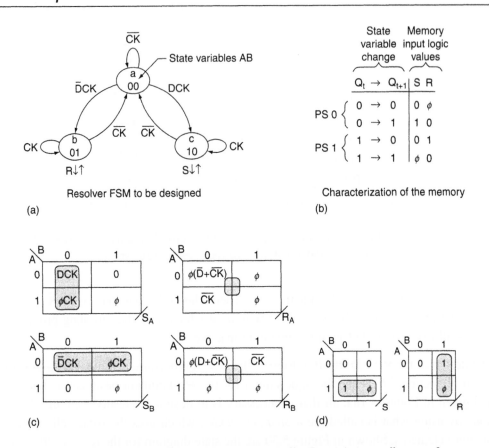

Figure 5.30: Resolver design for the RET D flip-flop (a) State diagram for the resolver (b) Characterization of the memory (c) EV K-maps for the next state functions required to drive the two basic cells (d) Output K-maps for the resolver

State 01 (K-map cell 1)

$$\text{Bit A} \begin{cases} 0 \to 0, \text{ place } 0 \text{ in the } S_A \text{ K-map} \\ \text{place } \phi \text{ in the } R_A \text{ K-map} \end{cases}$$

$$\text{Bit B} \begin{cases} 1 \to 0, \text{ place } 1 \cdot (\overline{CK}) = \overline{CK} \text{ in the } R_B \text{ K-map} \\ 1 \to 1, \text{ place } \phi \cdot (CK) = \phi CK \text{ in the } S_B \text{ K-map} \end{cases}$$

State 10 (K-map cell 2)

$$\text{Bit A} \begin{cases} 1 \to 0, & \text{place } 1 \cdot (\overline{CK}) = \overline{CK} \text{ in the } R_A \text{ K-map} \\ 1 \to 1, & \text{place } \phi \cdot (CK) = \phi CK \text{ in the } S_A \text{ K-map} \end{cases}$$

$$\text{Bit B} \begin{cases} 0 \to 0, & \text{place } 0 \text{ in the } S_B \text{ K-map,} \\ & \text{place } \phi \text{ in the } R_A \text{ K-map} \end{cases}.$$

Notice that for every essential EV entry in a given K-map cell there exists the complement of that entry ANDed with ϕ in the same cell of the other K-map. This leads to the following modification of the mapping algorithm in Section 5.6 as it pertains to S/R mapping:

1. Look for Sets $(0 \to 1)$ and Resets $(1 \to 0)$ and make the entry $1 \cdot ($ Appropriate BC$)$ in the proper S_i or R_i K-map, respectively, according to the combined excitation table for the basic cell. (Note: BC = branching condition.)

2. For each Set entry (from [1]) in a given cell of the S_i K-map, enter $\phi \cdot \overline{(Appropriate\ BC)}$ in the *same cell* of the corresponding R_i K-map.
   For each Reset entry (from [1]) in a given cell of the R_i K-map, enter $\phi \cdot \overline{(Appropriate\ BC)}$ in the *same cell* of the corresponding S_i K-map.

3. For Hold Resets $(0 \to 0)$ and Hold Sets $(1 \to 1)$, enter $(0,\phi)$ and $(\phi,1)$, respectively, in the (S_i, R_i) K-maps in accordance with the combined excitation table for basic cell given in Figure 5.15c.

Continuing with the design of the RET D flip-flop, the minimum NS and output functions extracted from the EV K-maps in Figs. 5.30c and 5.30d are

$$\begin{cases} S_A = \overline{B}DCK & R_A = \overline{CK} \\ S_B = \overline{A}\overline{D}CK & R_B = \overline{CK} \\ & S = A \quad R = B \end{cases} \tag{5.9}$$

which are implemented in Figure 5.31a. Here, the basic cells for bits A and B are highlighted by the shaded areas within the resolver section of the RET D flip-flop. Notice that the requirement of active low inputs to the three set-dominant basic cells is satisfied. For example, in the resolver FSM this requirement is satisfied by

$R_A(L) = R_B(L) = \overline{CK}(L) = CK(H)$. The circuit symbol for the RET D flip-flop is given in Figure 5.31b, where the triangle on the CK input is indicative of an edge triggered flip-flop with data-lockout character and is consistent with Figure 5.20a.

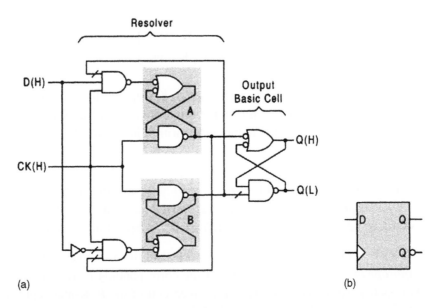

(a)   (b)

Figure 5.31: (a) Logic circuit for the RET D flip-flop as constructed from equations (5.9) showing the resolver and output basic cell stage (b) Logic circuit symbol

The operation of the RET D flip-flop is best represented by the timing diagram in Figure 5.32, where arrows on the rising edge of the clock waveform provide a reminder that this is an RET flip-flop. The edge-triggering feature is made evident by the vertical dashed lines, and the data lockout character is indicated by the absence of a flip-flop output response to narrow data pulses during the active and inactive portions of the clock waveform. For the sake of simplicity, no account is taken of gate propagation delay in Figure 5.32.

5.7.3   *The Master-Slave D Flip-Flop*

Another useful type of D flip-flop is the master-slave (MS) D flip-flop defined by the two state diagrams in Figure 5.33a and that conforms to the general definitions for a D flip-flop given in Figure 5.23. The MS D flip-flop is a two-stage device consisting of an RET D latch as the master stage and an FET D latch as the slave stage. The output

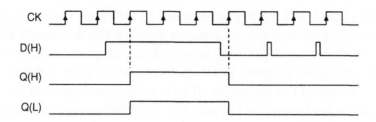

Figure 5.32: Timing diagram showing proper operation of the RET D flip-flop

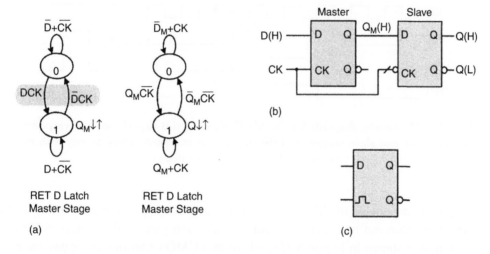

Figure 5.33: The master-slave (MS) D flip-flop (a) State diagram for the master and slave stages (b) Logic circuit (c) Circuit symbol

of the master stage is the input to the slave stage. Thus, the transparency problem of the D latch in Figure 5.24a has been eliminated by the addition of the slave stage that is triggered *antiphase* to the master. Thus, should signals pass through the master stage when *CK* is active, they would be held up at the slave stage input until *CK* goes inactive.

The design of the MS D flip-flop can be carried out following the same procedure as given in Figs. 5.24, 5.25, and 5.27. However, this is really unnecessary, since the logic circuits for both stages are already known from these earlier designs. The result is the logic circuit given in Figure 5.33b, where the output of the master RET D latch symbol is the input to the slave FET D latch symbol. The logic circuit symbol is shown in Figure 5.33c and is identified by the pulse symbol on the clock input.

The operation of the MS D flip-flop is illustrated by the timing diagram in Figure 5.34, where no account is taken of gate propagation delay. Notice that signals that are passed through the master stage during active *CK* are not passed through the slave stage, which is triggered antiphase to the master. However, there is the possibility of noise transfer, though of low probability. If logic noise should appear at the input to the slave stage just at the instant that CK goes through a falling edge, that noise can be transferred to the output.

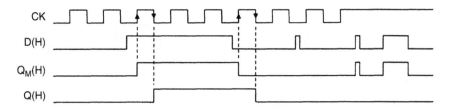

Figure 5.34: Timing diagram for the MS D flip-flop showing the output response from master and slave stages, and the absence of complete transparency with no account taken of gate path delays

One important advantage the MS D flip-flop has over the edge triggered variety is that the MS D flip-flop can be configured with transmission gates and inverters. Such a configuration is shown in Figure 5.35a, where two CMOS transmission gates are used together with two inverters. To achieve the two-stage effect required by the MS configuration, the CMOS transmission gates must be operated by using two-phase (2Φ) clocking such that the active portions of the clock phases are nonoverlapping. Shown in Figure 5.35b is a reset-dominant basic cell used to generate the two clock phases

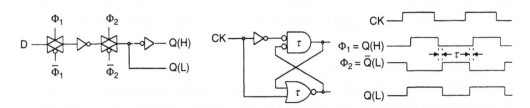

Figure 5.35: (a) The MS D flip-flop configured with CMOS transmission gates and inverters and requiring two-phase (2Φ) clocking (b) The reset-dominant basic cell used to generate 2Φ clocking as indicated by the output logic waveforms

(Φ_1 and Φ_2) whose active portions are separated in time by an amount Φ the path delay of a NOR gate. Notice that both phase waveforms (Φ_1 and Φ_2) are given in positive logic, similar to the physical voltage waveforms but without rise and fall times. These clock phase signals must each be supplied to the CMOS transmission gates in complementary form. This means that when Φ_1 is at LV, Φ_1 must be at HV and vice versa. The same must be true for Φ_2. Each complementary form is achieved by the use of an inverter with a buffer in the HV path for delay equalization, if necessary.

5.8   Flip-Flop Conversion: The T, JK Flip-Flops and Miscellaneous Flip-Flops

In Figure 5.21 a hierarchy for flip-flop design is given with the understanding that the D flip-flop is central to such a process. In this text, this is the case, as will be demonstrated by the design of the other important types of flip-flops. First, however, certain information must be understood.

To design one flip-flop from another, it is important to remember the following:

> The new flip-flop to be designed inherits the triggering mechanism of the old (memory) flip-flop.

This important fact can best be understood by considering the fictitious XY flip-flop shown in Figure 5.36. This fictitious flip-flop has been derived from a D flip-flop of some arbitrary triggering mechanism indicated by the question mark (?) on the clock input.

The model in Figure 5.36a can be compared with the basic model in Figure 5.22 for the same fictitious XY flip-flop, where now a D flip-flop is used as the memory instead of basic cells. In either case the XY flip-flop is designed according to the design procedure and mapping algorithm presented in Section 5.6, but the characterization of memory is different. As will be recalled from Section 5.7, flip-flops designed by using one or more basic cells require that the memory be characterized by the combined excitation table for the basic cell given in Figure 5.15c. Now, for flip-flop conversion by using a D flip-flop as the memory, the excitation table for the D flip-flop in Figure 5.23c must be used.

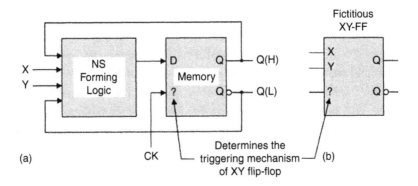

Figure 5.36: (a) Model and (b) logic symbol for a fictitious XY flip-flop derived from a D flip-flop having an unspecified triggering mechanism

5.8.1 The T Flip-Flops and Their Design from D Flip-Flops

All types of T flip-flops behave according to an internationally accepted definition that is expressed in one or all of three ways. Presented in Figure 5.37 are three ways of defining the T flip-flop, all expressed in positive logic as was true in the definition of the D flip-flops. Shown in Figure 5.37a is the operation table for any T flip-flop. It specifies

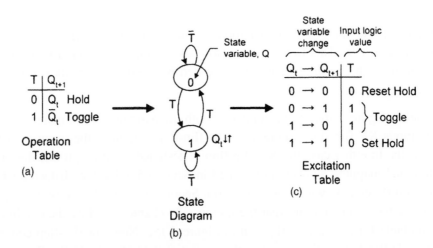

Figure 5.37: Generalized T flip-flop definition expressed in terms of the operation table (a), the state diagram (b), and the excitation table (c)

that when T is active, the device must toggle, meaning that $0 \to 1$ and $1 \to 0$ transitions occur as long as $T = 1$. When $T = 0$, the T flip-flop must hold in its present state. The state diagram for T flip-flops in Figure 5.37b is derived from the operation table and conveys the same information as the operation table. Here, the toggle character of the T flip-flop is easily shown to take place between Set and Reset states when T is active, but holding in these states when T is inactive.

The excitation table presented in Figure 5.37c is the third means of expressing the definition of T flip-flops. It is easily derived from the state diagram and hence conveys the same information regarding T flip-flop operation. This excitation table will be used to characterize the memory in the design of FSMs that require the use of T flip-flops as the memory elements.

Design of the T Flip-Flops from D Flip-Flops Since T flip-flops are to be designed (converted) from D flip-flops, the excitation table for the D flip-flop must be used to characterize the memory. This excitation table and the state diagram representing for the family of T flip-flops must be brought together by using the mapping algorithm set forth in Section 5.6. This is done in Figure 5.38, parts (a), (b), and (c), where the next state logic for flip-flop conversion is found to be

$$D = T \oplus Q. \tag{5.10}$$

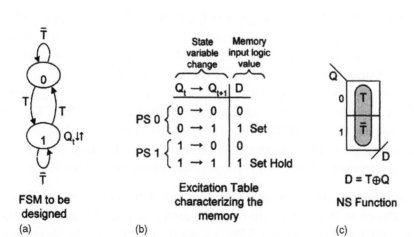

FSM to be
designed
(a)

Excitation Table
characterizing the
memory
(b)

$D = T \oplus Q$

NS Function
(c)

Figure 5.38: Design of the T flip-flops (a) The state diagram for any T flip-flop (b) Excitation table for the D flip-flop memory (c) NS K-map and NS function resulting from the mapping algorithm

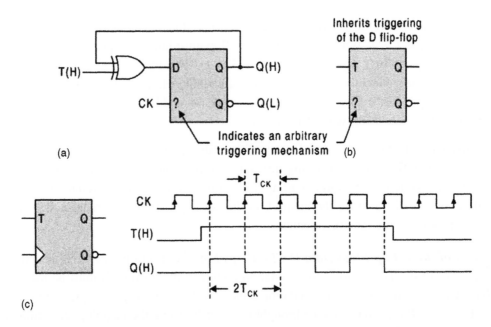

Figure 5.39: (a), (b) Implementation of Eq. (5.10) and logic circuit symbol for a T flip-flop of arbitrary triggering mechanism (c) Logic symbol and timing diagram for an RET T flip-flop showing toggle and hold modes of operation

Implementation of the NS function given in Eq. (5.10) is shown in Figure 5.39a together with the symbol for the T flip-flop in Figure 5.39b, which as yet has not been assigned a triggering mechanism—the designer's choice indicated by the question mark (?) on the clock input. Remember that the new FSM (in this case a T flip-flop) inherits the triggering mechanism of the memory flip-flop (in this case a D flip-flop). Shown in Figure 5.39c is the logic circuit symbol and timing diagram for an RET T flip-flop, the result of choosing an RET D flip-flop as the memory. The timing diagram clearly indicates the toggle and hold modes of operation of the T flip-flop. For the sake of simplicity no account is taken of the propagation delays through the logic.

Were it desirable to produce an MS T flip-flop, the memory element in Figure 5.39a would be chosen to be a MS D flip-flop. The timing diagram for an MS T flip-flop would be similar to that of Figure 5.39c, except the output from the slave stage would be delayed from the master stage by a time period $T_{CK}/2$. This is so because the slave stage picks up the output from the master stage only on the falling edge of CK, that is, the two stages are triggered antiphase to one another.

5.8.2 The JK Flip-Flops and Their Design from D Flip-Flops

The flip-flops considered previously are single data input flip-flops. Now, consideration centers on a type of flip-flop that has two data inputs, J and K. The members of the JK flip-flop family conform to the internationally accepted definition expressed in terms of an operation table, a state diagram, or an excitation table provided in Figure 5.40. The operation table in Figure 5.40a reveals the four modes of JK flip-flop operation: Hold, Reset, Set, and Toggle. Thus, it is seen that the JK type flip-flops operate in all the modes common to SR, T, and D type flip-flops, though SR flip-flops (clocked SR latches) are yet to be discussed. For this reason the JK flip-flops are sometimes referred to as the universal flip-flops. The state diagram in Figure 5.40b is best derived from the operation table. For example, the Set $(0 \to 1)$ branching condition follows from the Boolean sum (Set + Toggle) $= J\overline{K} + JK = J$, and the Reset $(1 \to 0)$ branching condition results from the sum (Reset + Toggle) $= \overline{J}K + JK = K$. The Set-Hold and Reset-Hold conditions result from the sums $J\overline{K} + \overline{J}\,\overline{K} = \overline{K}$ and $\overline{J}K + \overline{J}\,\overline{K} = \overline{J}$, respectively. However, given the set and reset branching conditions, the sum rule in equation (5.3) can and should be used to obtain the two hold conditions.

The excitation table for the JK flip-flops in Figure 5.40c is easily derived from the state diagram in (b). For example, the Reset–Hold branching path requires a branching condition $\overline{J}$ that places a 0 and a ϕ in the J and K columns of the excitation table.

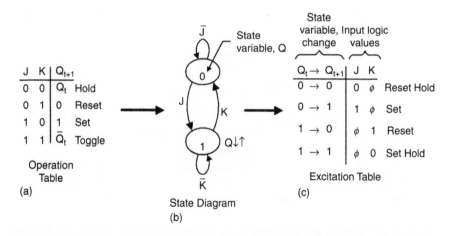

Figure 5.40: Generalized JK flip-flop definition expressed in terms of the operation table (a), the state diagram (b), and the excitation table (c)

A ϕ is used for unspecified inputs in branching conditions. Similarly, a 1 and ϕ are placed in the J and K columns for the Set branching condition J. Notice that this excitation table bears some resemblance to that of the combined excitation table for the basic cells in Figure 5.15c, but with two additional don't cares. The excitation table for the JK flip-flops will be used rather extensively to characterize the memory in the design of FSMs that require JK flip-flops as memory elements.

Design of the JK Flip-Flops from the D Flip-Flops The process used previously in the design of T flip-flops from D flip-flops is now repeated for the case of the JK flip-flops defined in Figure 5.40 in terms of the operation table, state diagram, and excitation table. Shown in Figure 5.41a is the state diagram representing the family of JK flip-flops, the FSMs to be designed. Since a D flip-flop is to be used as the memory element in the design, its excitation table must be used to characterize the memory and is provided in Figure 5.41b. By using the mapping algorithm in Section 5.6 together with the state diagram for a JK flip-flop and the excitation table for the memory D flip-flop, there results the NS logic K-map and NS forming logic shown in Figure 5.41c. Notice that only the Set and Set Hold branching paths produce non-null entries in the NS K-map for D, a fact that is always true when applying the mapping algorithm to D flip-flop memory elements.

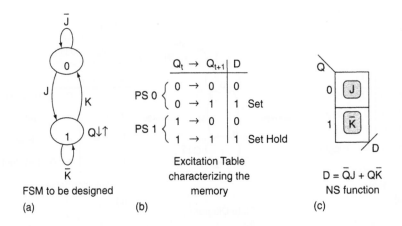

Figure 5.41: Design of the JK flip-flops (a) State diagram for any JK flip-flop (b) Excitation table for the D flip-flop memory (c) NS K-map and NS function required for flip-flop conversion

The minimum NS logic function extracted from the K-map is:

$$D = \overline{Q}J + Q\overline{K} \qquad (5.11)$$

and is shown implemented in Figure 5.42a with a D flip-flop of an arbitrary triggering mechanism as the memory. Its circuit symbol is given in Figure 5.42b, also with a question mark (?) indicating an arbitrary triggering mechanism determined from the D flip-flop memory element. In Figure 5.42c is shown the circuit symbol and timing diagram for an FET JK flip-flop that has been derived from an FET D flip-flop. The timing diagram illustrates the four modes of JK flip-flop operation: Hold (Reset or Set), Reset, Set, and Toggle. Notice that once a set condition is sampled by clock, that condition is maintained by the flip-flop until either a reset or toggle condition is sampled by the falling edge of the clock waveform. Similarly, once a reset condition is executed by clock, that condition is maintained until either a set or toggle condition is initiated. As always, the toggle mode results in a divide-by-two of the clock frequency.

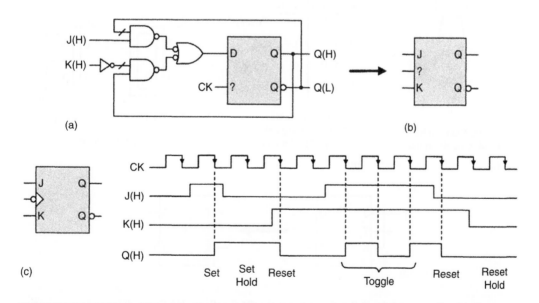

Figure 5.42: (a) Implementation of Eq. (5.11), and (b) logic circuit symbol for a JK flip-flop of arbitrary triggering mechanism (c) Logic symbol and timing diagram for an FET JK flip-flop designed from an FET D flip-flop showing all four modes of operation indicated by the operation table in Figure 5.40a

Eq. (5.11) has application beyond that of converting a D flip-flop to a JK flip-flop. It is also the basis for converting D K-maps to JK K-maps and vice versa. K-map conversion is very useful in FSM design and analysis since it can save time and reduce the probability for error. The subject of K-map conversion will be explored in detail later in this chapter.

5.8.3   Design of T and D Flip-Flops from JK Flip-Flops

The procedures for converting D flip-flops to T and JK flip-flops, used in the preceding subsections, will now be used for other flip-flop conversions. The conversions JK-to-T and JK-to-D are important because they emphasize the universality of the JK flip-flop types. Presented in Figure 5.43, for JK-to-T flip-flop conversion, are the state diagram for the T flip-flops (a), the excitation table characterizing the JK memory (b), and the NS K-maps and NS functions for J and K (c). Plotting the NS K-maps follows directly from application of the mapping algorithm given earlier in

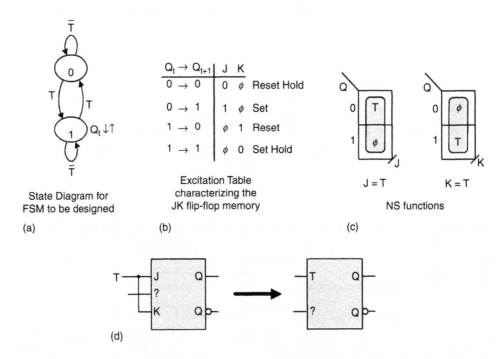

Figure 5.43: Design of the T flip-flops from the JK flip-flops (a) State diagram representing the family of T flip-flops (b) Excitation table characterizing the JK memory element (c) NS K-maps and NS functions for the flip-flop conversion (d) Logic circuit and symbol for a T flip-flop of arbitrary triggering mechanism

Section 5.6. Notice that the ϕ's in the NS K-maps result from summing of the branching condition values relative to the branching paths of a particular present state (PS). For example, in PS state 1, a ϕ is placed in cell 1 of the J K-map, since $\phi T + \phi \overline{T} = \phi$ as required by the $1 \rightarrow 0$ and $1 \rightarrow 1$ branching paths, respectively. By using the don't cares in this manner, the minimum cover for the NS functions is:

$$J = K = T. \tag{5.12}$$

Thus, to convert any JK flip-flop to a T flip-flop of the same triggering character, all that is necessary is to connect the J and K input terminals together to become the T input, as indicated by the logic circuit symbols in Figure 5.43d. Equation (5.12) will also be useful for converting JK K-maps to T K-maps and vice versa.

The conversion of JK flip-flops to D flip-flops follows in a similar manner to that just described for converting JK to T flip-flops. Presented in Figure 5.44 are the state diagram for the family of D flip-flops (a), the excitation table for the memory JK

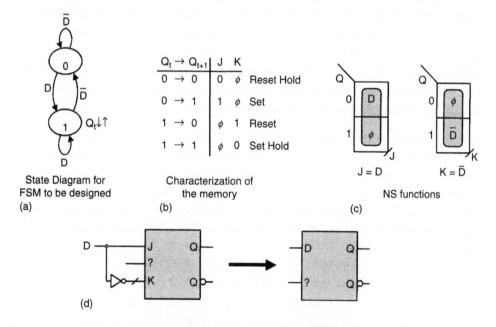

Figure 5.44: Design of the D flip-flops from the JK flip-flops (a) State diagram representing the family of D flip-flops (b) Excitation table characterizing the JK memory element (c) NS K-maps and NS functions for the flip-flop conversion (d) The logic circuit and symbol for a D flip-flop of arbitrary triggering mechanism

flip-flop (b), and the NS K-maps and conversion logic extracted from the K-maps (c). The minimum NS functions, as extracted from the NS K-maps, are given by

$$J = D \text{ and } K = \overline{D}. \tag{5.13}$$

Shown in Figure 5.44d is the logic circuit and its circuit symbol for D flip-flop conversion from a JK flip-flop of arbitrary triggering mechanism. Clearly, all that is necessary to convert a JK flip-flop to a D flip-flop is to connect D to J and D to K via an inverter.

5.8.4 Review of Excitation Tables

For reference purposes, the excitation tables for the families of D, T, and JK flip-flops, discussed previously, are provided in the table of Figure 5.45. Also shown in the table is the excitation table for the family of SR flip-flops and all related SR devices which include the basic cells. Notice the similarity between the JK and SR excitation tables, which leads to the conclusion that *J is like S and K is like R*, but not exactly. The only difference is that there are two more don't cares in the JK excitation table than in the SR excitation table. Also observe that the D values are active for Set and Set Hold conditions, and that the T values are active only under toggle $1 \rightarrow 0$ and $0 \rightarrow 1$ conditions. These facts should serve as a mnemonic means for the reader in remembering these important tables. Eventually, construction of the NS K-maps will become so commonplace that specific mention of either the mapping algorithm or the particular excitation table in use will not be necessary.

Any of the excitation tables given in Figure 5.45 can be used to characterize the flip-flop memory for the purpose of applying the mapping algorithm in Section 5.6 to obtain the NS forming logic for an FSM. In fact, that is their only purpose. For example, if D flip-flops are required as the memory in the design of an FSM, the excitation table for the family of D flip-flops is used. Or if JK flip-flops are to be used as the memory, the excitation table for the JK flip-flops is used for the same purpose, etc.

	$Q_t \rightarrow Q_{t+1}$	D	T	J	K	S	R
Reset Hold	$0 \rightarrow 0$	0	0	0	ϕ	0	ϕ
Set	$0 \rightarrow 1$	1	1	1	ϕ	1	0
Reset	$1 \rightarrow 0$	0	1	ϕ	1	0	1
Set Hold	$1 \rightarrow 1$	1	0	ϕ	0	ϕ	0

Figure 5.45: Summary of the excitation tables for the families of D, T, JK, and SR flip-flops

5.8.5 Design of Special-Purpose Flip-Flops and Latches

To emphasize the applicability and versatility of the design procedure and mapping algorithm given in Section 5.6, other less common or even "nonsense" flip-flops will now be designed. These design examples are intended to further extend the reader's experience in design procedures.

An Unusual (Nonsense) Flip-Flop Suppose it is desirable to design an FET ST (Set/Toggle) flip-flop that is defined according to the operation table in Figure 5.46a. The state diagram for the family of ST flip-flops, derived from the operation table, is shown in Figure 5.46b. Also, suppose it is required that this flip-flop is to be designed from an RET D flip-flop. Therefore, the memory must be characterized by the

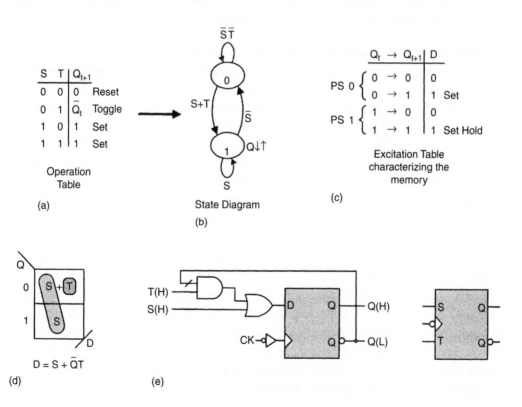

Figure 5.46: Design of a special-purpose FET ST flip-flop (a) Operation table for the family of ST flip-flops (b) State diagram derived from (a) (c) Characterization of the D flip-flop memory (d) NS K-map and NS function for flip-flop conversion (e) Logic circuit and circuit symbol for the FET ST flip-flop

excitation table for the D flip-flop presented in Figure 5.46c. By using the mapping algorithm, the NS K-map and NS forming logic are obtained and are given in Figure 5.46d. Implementation of the NS logic with the RET D flip-flop to obtain the FET ST flip-flop is shown in part (e) of the figure. Notice that the external Q feedback is necessary to produce the toggle character required by the operation table and state diagram for the family of ST flip-flops. If it had been required to design an MS ST flip-flop, then an MS D flip-flop would have been used as the memory element while retaining the same NS forming logic.

The external hardware requirements in the design of the FET ST flip-flop can be minimized by using an RET JK flip-flop as the memory in place of a D flip-flop. If the D excitation table in Figure 5.46c is replaced by that for the JK flip-flops in Figure 5.40c, the NS functions become $J = S + T$ and $K = \bar{S}$, a reduction of one gate. It is left to the reader to show the mapping details.

A Special-Purpose Clocked SR Latch As used in this text, the term *latch* refers to gated or clocked memory elements that do not have data lockout character and that exhibit transparency, or that lose their mixed-rail output character under certain input conditions. The D latch in Figure 5.24 is an example, since it exhibits the transparency effect under the condition $CK(H) = 1(H)$. The family of SR latches also fall into this category. One such SR latch is defined by the operation table in Figure 5.47a from which the state diagram in Figure 5.47b is derived. This latch is observed to Set under the $S\bar{R}$ branching condition, Reset under condition $\bar{S}R$, and hold if S, R is either 0,0 or 1,1. Notice that CK is part of the input branching conditions, and that the basic cell is to be used as the memory characterized by the excitation table in Figure 5.47c. Applying the mapping algorithm yields the NS K-maps and NS-forming logic given in part (d) of the figure. Implementing with a reset-dominant basic cell yields the logic circuit and circuit symbol shown in Figure 5.47e. Clearly, an $S, R = 1,1$ condition cannot be delivered to the basic cell output stage. But there is a partial transparency effect. For example, a change $\bar{S}\,\bar{R} \rightarrow S\,\bar{R}$ while in state 0 with CK active ($CK = 1$) will cause a transition to state 1 where Q is issued. Thus, Q follows S in this case, which is a transparency effect. Similarly, a change $\bar{S}\,\bar{R} \rightarrow \bar{S}\,R$ while in state 1 when CK is active causes a transition $1 \rightarrow 0$ with an accompanying deactivation of Q. Again, this is a transparency effect, since Q tracks R when $CK = 1$.

The Data Lockout MS Flip-Flop The *data lockout MS flip-flop* is a type of master–slave flip-flop whose two stages are composed of edge-triggered flip-flops or are an edge-triggered/latch combination. Only the master stage must have the data lockout

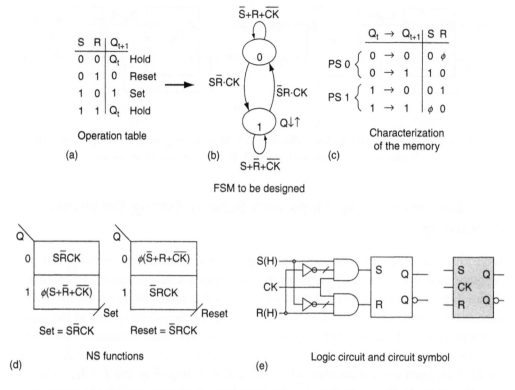

Figure 5.47: Design of a special-purpose SR latch (a) Operation table for this family of SR flip-flops and latches (b) State diagram for the special SR latch derived from the operation table in (a) (c) Characterization of the basic cell memory (d) NS K-maps and NS-forming logic (e) Logic circuit and circuit symbol

character (hence must be edge triggered). Shown in Figure 5.48a is a D data lockout flip-flop composed of an RET D flip-flop master stage and an FET D flip-flop slave stage, and in (b) an RET D flip-flop master with an FET D latch as the slave stage. The design in Figure 5.48b needs less hardware than that in (a) because of the reduced logic requirements of the D latch. Another possibility is to use JK flip-flops in place of the D flip-flops in Figure 5.48a, thus creating a JK data lockout flip-flop. But the JK flip-flops require more logic than do the D flip-flops, making the JK data lockout flip-flop less attractive. In any case, there is little advantage to using a data lockout flip-flop except when it is necessary to operate peripherals antiphase off of the two stage outputs, Q_M and Q, in Figure 5.48.

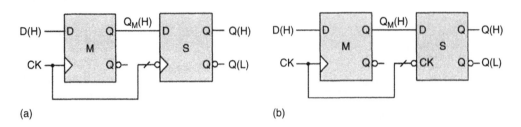

Figure 5.48: The D data lockout flip-flop (a) All edge triggered flip-flop variety (b) Same as (a) except with an FET D latch as the slave stage

5.9 Latches and Flip-Flops with Serious Timing Problems: A Warning

With very few exceptions, two-state flip-flops have serious timing problems that preclude their use as memory elements in synchronous state machines. Presented in Figure 5.49 are four examples of two-state latches that have timing problems—none have the data lockout feature. The RET D latch (a) becomes transparent to the input data when $CK = 1$, causing flip-flop action to cease. The three remaining exhibit even more severe problems. For example, the FET T latch (b) will oscillate when $T \cdot \overline{CK} = 1$, and the RET JK latch (c) will oscillate when $JK \cdot CK = 1$, requiring that $J = K = CK = 1$, as indicated in the figure. Notice that the branching conditions required to cause any of

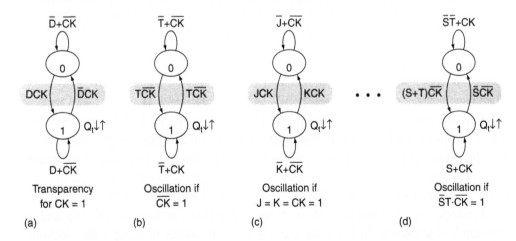

Figure 5.49: Timing problems in latches (a) RET D latch (b) FET T latch (c) RET JK latch (d) FET ST latch

the latches to oscillate is found simply by ANDing the $0 \rightarrow 1$ and $1 \rightarrow 0$ branching conditions. Any nonzero result is the branching condition that will cause oscillation. Thus, the FET ST latch in Figure 5.49d will oscillate under the condition $(S + T)\overline{S} \cdot \overline{CK} = \overline{S}T \cdot \overline{CK} = 1$, that is if $S = CK = 0$ and $T = 1$. The reason for the oscillation in these latches is simply that CK no longer controls the transition between states since the branching condition between the two states is logic 1. These FSMs are asynchronous, as are all flip-flops and latches, and if the transitions are unrestricted by CK, they will oscillate. Thus, none of these two-state latches should ever be considered for use as memory elements in the design of synchronous FSMs. The one exception is the JK latch, which can be used as a memory element providing that J and K are never active at the same time—thus, operating as an SR latch.

There is an MS flip-flop that is particularly susceptible to timing problems. It is the MS JK flip-flop defined by the two state diagrams shown in Figure 5.50a and implemented in (b). Here, a handshake configuration exists between the master and slave stages. A handshake configuration occurs when the output of one FSM is the input to another and vice versa. This FSM is susceptible to a serious *error catching* problem: In the reset state, if CK is active and a glitch or pulse occurs on the J input to the master stage, the master stage is irreversibly set, passing that set condition on to the slave stage input. Then when CK goes inactive, the output is updated to the set state. This is called *1's catching* and is an unrecoverable error, since the final set state was not regulated by CK. Similarly, in the set state, if CK is active and a glitch or pulse

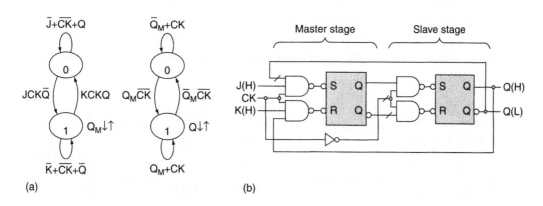

(a) (b)

Figure 5.50: An MS JK flip-flop that exhibits the error catching problem (a) State diagrams for the MS JK flip-flop which exhibit a handshake configuration (b) Logic circuit derived from the state diagrams in (a)

occurs on the K input, the master stage is irreversibly reset, passing that reset condition on to the slave stage input. Then when CK goes inactive, the output is updated to the reset state. This is called $0's$ *catching* and is also an unrecoverable error.

Because of the error catching problem just described, the MS JK flip-flop in Figure 5.50b, derived from the "handshake" state diagrams in Figure 5.50a, should never be considered for application as a memory element in a synchronous state machine. If an MS JK flip-flop is needed as the memory element, it is best designed by using equation (5.11) and Figure 5.42a for conversion from an MS D flip-flop that has no error catching problem. Also, because the MS D flip-flop can be implemented by using transmission gates and inverters, as in Figure 5.35, the conversion to a MS JK can be accomplished with a minimum amount of hardware.

5.10   Asynchronous Preset and Clear Overrides

There are times when the flip-flops in a synchronous FSM must be initialized to a logic 0 or logic 1 state. This is done by using the asynchronous preset and clear override inputs to the flip-flops. To illustrate, a D latch is shown in Figs. 5.51a and 5.51b with both preset and clear overrides. If the flip-flop is to be initialized a logic 0, then a $CL(L) = 1(L)$ is presented to NAND gates 1 and 4, which produces a mixed-rail reset condition, $Q(H) = 0(H)$ and $Q(L) = 0(L)$ while holding $PR(L) = 0(L)$. Or to initialize a logic 1, a $PR(L) = 1(L)$ is presented to NAND gates 2 and 3, which produces a mixed-rail set condition, $Q(H) = 1(H)$ and $Q(L) = 1(L)$, but with CL (L) held at $0(L)$. Remember from subsection 5.4.4 that $S(L)$ and $R(L)$ cannot both be $1(L)$ at the same time or else there will be loss of mixed-rail output. Thus, the $CL,PR = 1,1$ input condition is forbidden for this reason. The following relations summarize the various possible preset and clear override input conditions applicable to any flip-flop:

$$\left.\begin{cases} \begin{array}{ll} CL(L) = 1(L) & \textit{Initialize 0} \\ PR(L) = 0(L) & \\ CL(L) = 0(L) & \textit{Normal Operarion} \\ PR(L) = 0(L) & \end{array} \qquad \begin{array}{ll} CL(L) = 0(L) & \textit{Initialize 1} \\ PR(L) = 1(L) & \\ CL(L) = 1(L) & \textit{Forbidden} \\ PR(L) = 1(L) & \end{array} \end{cases}\right\}. \qquad (5.14)$$

The timing diagram in Figure 5.51c best illustrates the effect of the asynchronous preset and clear overrides. In each case of a PR(L) or CL(L) pulse, normal operation of the latch is interrupted until that pulse disappears and a clock triggering (rising) edge occurs. This asynchronous override behavior is valid for any flip-flop regardless of its type or triggering mechanism, as indicated in Figure 5.52. For all flip-flops, these asynchronous overrides act directly on the output stage, which is a basic cell.

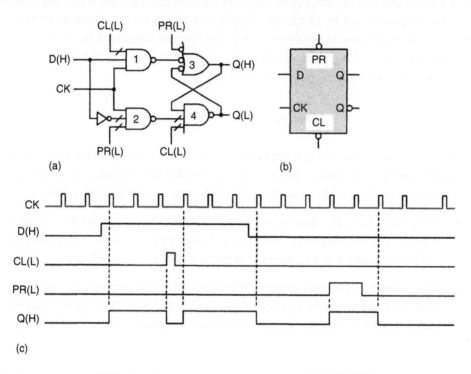

(a)

(b)

(c)

Figure 5.51: Asynchronous preset and clear overrides applied to the D latch (a) Logic circuit for the D latch showing the active low preset and clear connections (b) Logic circuit symbol with active low preset and clear inputs indicated (c) Timing diagram showing effects of the asynchronous overrides on the flip-flop output

Figure 5.52: Examples of flip-flops with asynchronous preset and/or clear overrides

5.11   Setup and Hold Time Requirements of Flip-Flops

Flip-flops will operate reliably only if the data inputs remain stable at their proper logic levels just before, during, and just after the triggering edge of the clock waveform. To put this in perspective, the data inputs must meet the *setup and hold-time* requirements established by clock, the *sampling variable* for synchronous FSMs. The setup and hold-time requirements for a flip-flop are illustrated by voltage waveforms in Figure 5.53, where both rising and falling edges of the clock signal are shown. The sampling interval is defined as:

$$Sampling\ interval = (t_{su} + t_h), \qquad (5.15)$$

where t_{su} is the *setup time* and t_h is the *hold time*. It is during the sampling interval that the data inputs must remain fixed at their proper logic level if the outcome is to be predictable. This fact is best understood by considering the definitions of setup and hold times:

- Setup time t_{su} is the time interval preceding the active (or inactive) transition point (t_{tr}) of the triggering edge of *CK* during which all data inputs must remain stable to ensure that the intended transition will be initiated.

- Hold time t_h is the time interval following the active (or inactive) transition point (t_{tr}) of the triggering edge of CK during which all data inputs must remain stable to ensure that the intended transition is successfully completed.

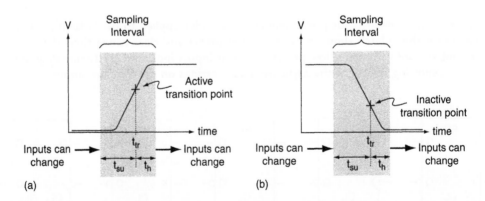

Figure 5.53: Clock voltage waveforms showing sampling interval ($t_{su} + t_h$) during which time the data inputs must remain stable at their proper logic levels (a) Rising edge of the clock waveform (b) Falling edge of the clock waveform

Failure to meet the setup and hold-time requirements of the memory flip-flops in an FSM can cause improper sampling of the data that could, in turn, produce erroneous transitions, or even metastability. A change of the data input at the time *CK* is in its sampling interval can produce a *runt pulse*, a pulse that barely reaches the switching threshold. An incompletely sampled runt pulse may cause erroneous FSM behavior. As an example of proper and improper sampling of an input, consider a portion of the resolver state diagram for an RET D flip-flop shown in Figure 5.54a. Assuming that the FSM is in state *a* and that the rising edge of *CK* is to sample the *D* input waveform, two sampling possibilities are illustrated by the voltage waveforms for *CK* and *D* in Figure 5.54b. Proper sampling occurs when the data input *D* is stable at logic level 1 in advance of the rising edge of CK and maintained during the sampling interval. Improper sampling results when *D* changes during the sampling interval.

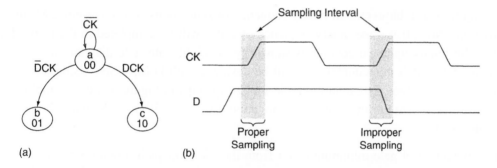

Figure 5.54: Examples of proper and improper sampling of the data input (a) Portion of the resolver state diagram for an RET D flip-flop (b) Voltage waveforms showing proper and improper sampling of the D waveform during the sampling interval of CK

The setup and hold-time intervals are important design parameters for which manufacturers will normally provide worst-case data for their flip-flops. Awareness and proper use of this data is vital to good state machine design practice. Ignoring this data may lead to state machine unpredictability or even failure.

5.12 Design of Simple Synchronous State Machines with Edge-Triggered Flip-Flops: Map Conversion

Where nearly ideal, high-speed sampling is required, and economic considerations are not a factor, edge-triggered flip-flops may be the memory elements of choice. The setup and hold-time requirements for these flip-flops are the least stringent of all, and they possess none of the problems associated with either the latches or MS flip-flops discussed earlier. In this section, two relatively simple FSMs will be designed to demonstrate the methodology to be used. The emphasis will be on the procedure required to obtain the next state and output functions of the FSM. This procedure will involve nothing new. Rather, it will be the continued application of the design procedure and mapping algorithm discussed in Section 5.6, and an extension of the flip-flop conversion examples covered in Section 5.8 but now applied to K-map conversion and FSM design.

5.12.1 Design of a Three-Bit Binary Up/Down Counter: D-to-T K-Map Conversion

In Figure 5.2d a bidirectional binary sequence of states is used to represent a Mealy machine. Now, that same binary sequence of states will be completed in the form of a three-bit binary up/down counter as shown by the eight-state state diagram in Figure 5.55a. It is this counter that will be designed with D flip-flops. Using the mapping algorithm, the excitation table for D flip-flops in Figure 5.55b is combined with the state diagram in (a) to yield the entered variable (EV) NS K-maps shown in Figure 5.55c.

The extraction of gate-minimum cover from the EV K-maps in Figure 5.55c is sufficiently complex as to warrant some explanation. Shown in Figure 5.56 are the compressed EV Kmaps for NS functions D_A and D_B, which are appropriate for use by the CRMT method, discussed at length in Section 5.7, to extract multilevel gate minimum forms. The secondorder K-maps in Figure 5.56 are obtained by entering the BC subfunction forms shown by the shaded loops in Figure 5.55c. For D_A, the CRMT coefficients g_i are easily seen to be $g_0 = A \oplus \overline{C}\,\overline{X}$ and $g_1 = (A \oplus CX) = (A \oplus \overline{C}\,\overline{X}) = CX \oplus \overline{C}\,\overline{X}$, as obtained from the first-order K-maps in Figure 5.56b. Similarly, for D_B the CRMT coefficients are $g_0 = \overline{B} \oplus X$ and $g_1 = 1$. When combined with the f coefficients, the gate minimum becomes,

$$\left\{ \begin{array}{l} D_A = A \oplus \overline{B}\,\overline{C}\,\overline{X} \oplus BCX \\ D_B = \overline{B} \oplus C \oplus X \\ D_C = \overline{C} \\ Z = ABCX \end{array} \right\}, \qquad (5.16)$$

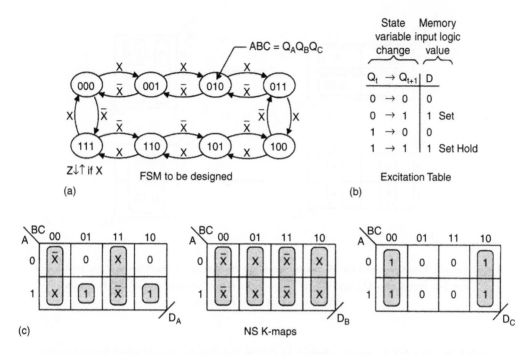

(a)

FSM to be designed

$Z\downarrow\uparrow$ if X

State variable change		Memory input logic value	
Q_t	$\rightarrow Q_{t+1}$	D	
0	$\rightarrow$ 0	0	
0	$\rightarrow$ 1	1	Set
1	$\rightarrow$ 0	0	
1	$\rightarrow$ 1	1	Set Hold

Excitation Table

(b)

(c)

NS K-maps

Figure 5.55: Design of a three-bit up/down binary counter by using D flip-flops (a) State diagram for the three-bit up/down counter with a conditional (Mealy) output, Z (b) Excitation table characterizing the D flip-flop memory (c) NS K-maps plotted by using the mapping algorithm showing BC domain subfunctions indicated with shaded loops

which is a three-level result (due to D_A) with an overall gate/input tally of 7/18, excluding inverters. The next state function for D_C is obtained by inspection of the third-order K-map in Figure 5.55c, and the output Z is read directly off of the state diagram. Note that the expressions for D_A and D_B in equations (5.16) can be obtained directly from the first-order K-maps in Figure 5.56b by applying the mapping methods discussed in Section 5.2. The minimum cover is indicated by the shaded loops.

Toggle character is inherent in the binary code. This is evident from an inspection of the state diagram in Figure 5.55a. State variable C toggles with each transition, state variable B toggles in pairs of states, and state variable A toggles in groups of four states. Thus, it is expected that the T flip-flop design of a binary counter will lead to a logic minimum, and this is the case. Shown in Figure 5.57 is the design of the binary up/down counter by using T flip-flops as the memory represented by the excitation table in

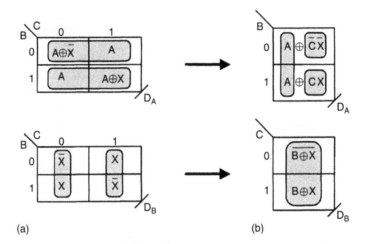

Figure 5.56: Compressed EV K-maps required to extract a multilevel logic minimum for NS functions D_A and D_B of Figure 5.54 (a) Second-order EV K-maps (b) First-order EV K-maps

Figure 5.57b. The NS K-maps, shown in (c) of the figure, are plotted by using the mapping algorithm. Extracting minimum cover from these K-maps (see shaded loops) yields the two-level results

$$\left\{ \begin{array}{l} T_A = \overline{B}\,\overline{C}\,\overline{X} + BCX \\ T_B = \overline{C}\overline{X} + CX \\ T_C = \overline{X} + X = 1 \\ Z = ABCX \end{array} \right\} \tag{5.17}$$

with an overall gate input tally 7/18 excluding inverters. Although the gate/input tally is the same as that produced by the three-level result given by equations (5.16), the two-level result is expected to be faster and, of course, amenable to implementation by two-level programmable logic devices (e.g., PLAs).

Implementation of equation (5.17) is shown in Figure 5.58a, where the NS forming logic, memory and output forming logic are indicated. The present state is read from the flip-flop outputs A(H), B(H), and C(H), where $Q_A = A$, $Q_B = B$, and $Q_C = C$, and the Mealy output Z is issued from the AND gate in state 111 but only when input X is active, i.e., only when the counter is in an up-count mode. The block symbol for this counter is shown in Figure 5.58b.

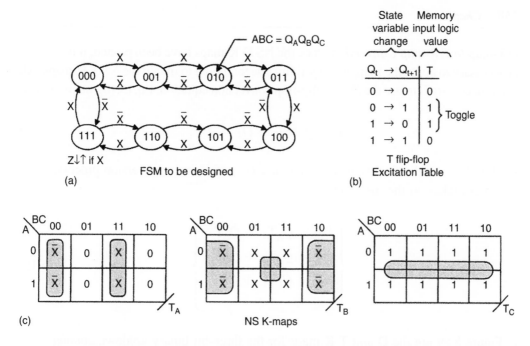

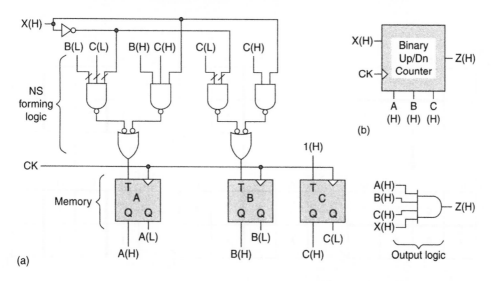

Figure 5.57: Design of the three-bit up/down binary counter by using T flip-flops (a) State diagram for the three-bit up/down counter with a conditional (Mealy) output, Z (b) Excitation table characterizing the T flip-flop memory (c) NS K-maps plotted by using the mapping algorithm and showing minimum cover

Figure 5.58: Implementation of the up/down binary counter represented by equations (5.17) (a) NS-forming logic, T flip-flop memory, and output-forming logic stages (b) block diagram for the counter

D K-map to T K-map Conversion Once the NS D K-maps have been plotted, it is unnecessary to apply the mapping algorithm a second time to obtain the NS T K-maps. All that is necessary is to use the $D \rightarrow T$ flip-flop conversion equation, Eq. (5.10), but written as

$$D = Q \oplus T = \overline{Q}T + Q\overline{T}. \tag{5.18}$$

Applied to the individual state variables in a $D \rightarrow T$ K-map conversion process, Eq. (5.18) takes on the meaning

$$\begin{cases} D_A = \overline{Q}_A T_A + Q_A \overline{T}_A \\ D_B = \overline{Q}_B T_B + Q_B \overline{T}_B \\ D_C = \overline{Q}_C T_C + Q_C \overline{T}_C \\ \quad \vdots \end{cases} \tag{5.19}$$

In Figure 5.59 are the D and T K-maps for the three-bit binary up/down counter reproduced from Figs. 5.55 and 5.57. The heavy lines indicate the domain boundaries for the three state variables A, B, and C. An inspection of the K-maps together with equations (5.19) results in the following algorithm for D-to-T K-map conversion:

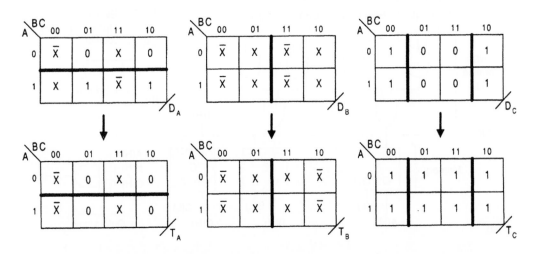

Figure 5.59: D and T K-maps for the three-bit binary up/down counter showing the domain boundaries for state variable bits A, B, and C

Algorithm 5.1: $D \rightarrow T$ K-map Conversion (Refer to Eq. 5.19)

(1) For all that is NOT A in the D_A K-map, transfer it to the T_A K-map directly $(\overline{A}T_A)$.

(2) For all that is A in the D_A K-map, transfer it to the T_A K-map complemented $(A\overline{T}_A)$.

(3) Repeat steps (1) and (2) for the $D_B \rightarrow T_B$ and $D_C \rightarrow T_C$, etc., K-map conversions.

Notice that the word "complemented," as used in the map conversion algorithm, refers to the complementation of the contents of each cell in the domain indicated.

5.12.2 Design of a Sequence Recognizer: D-to-JK K-Map Conversion

It is required to design a sequence recognizer that will issue an output any time an overlapping sequence ... 01101 ... is detected as indicated in Figure 5.60a. To do this a choice is made between the Moore or Mealy constructions shown in Figs. 5.60b and 5.60c, respectively. For the purpose of this example, the Mealy construction is chosen. Let the external input X be synchronized *antiphase* to clock, meaning, for example, that X is synchronized to the rising edge of clock when the memory is FET flip-flops. An *overlapping sequence* is one for which a given sequence can borrow from the latter portions of an immediately preceding sequence as indicated in Figure 5.60a. The loop $\cdots d \rightarrow e \rightarrow f \rightarrow d \cdots$ in the Moore construction or the loop $\cdots c \rightarrow d \rightarrow e \rightarrow c \cdots$ in the Mealy construction illustrates the overlapping sequence. A *nonoverlapping sequence* requires that each sequence of pulses be separate, i.e., independent, of any immediately preceding sequence. Note that the Mealy state diagram is constructed from the Moore version by merging states e and f in Figure 5.60b, and by changing the unconditional output to a conditional output.

The timing diagram showing the sequence of states leading to the conditional (Mealy) output is presented in Figure 5.61, where the state identifiers and state code assignments are indicated below the Z waveform. Notice that input X is permitted to change only on the rising edge of the clock waveform and that the arrows indicate a FET flip-flop memory. Thus, when the FSM enters state e on the falling edge of clock, an output is issued when X goes active, and is deactivated when the FSM leaves state e. Any deviation from the sequence ...01101... would prevent the sequence recognizer from entering state e and no output would be issued. Also, once in state e the overlapping loop $\cdots e \rightarrow c \rightarrow d \rightarrow e \cdots$ would result in repeated issuance of the output Z.

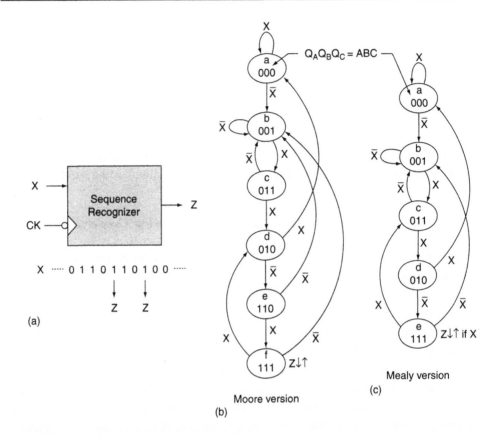

(a)

Moore version

(b)

Mealy version

(c)

Figure 5.60: A simple sequence recognizer for an overlapping sequence $\cdots$ 01101 $\cdots$.
(a) Block diagram and sample overlapping sequence (b) Moore FSM representation
(c) Mealy FSM representation

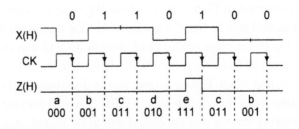

Figure 5.61: Timing diagram for the Mealy version of the sequence recognizer in
Figure 5.60c

Consider that the Mealy version of the sequence recognizer is to be designed by using D flip-flops. Shown in Figure 5.62 are the excitation table and the resulting D K-maps obtained by applying the mapping algorithm. The shaded loops reveal the minimum covers for the output and NS functions, which are easily read as

$$\left\{ \begin{array}{l} D_A = B\overline{C}\overline{X} \\ D_B = B\overline{C}\overline{X} + CX \\ D_C = \overline{X} + \overline{B}C + A \\ Z = AX \end{array} \right\} \tag{5.20}$$

Notice that the term $B\overline{C}\overline{X}$ is a shared PI since it appears in two of the three NS functions.

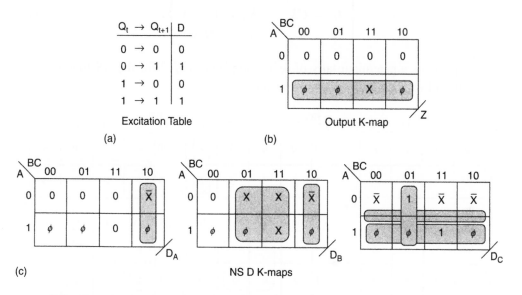

Figure 5.62: D K-map construction for the Mealy version of the sequence recognizer in Figure 5.60c (a) Excitation table for D flip-flops (b), (c) Output K-map and NS D K-maps showing minimum cover

D K-map to JK K-map Conversion Assuming it is desirable to design the sequence recognizer of Figure 5.60c by using JK flip-flops instead of D flip-flops, the process of obtaining the NS JK K-maps can be expedited by K-map conversion. It will be recalled from Eq. (5.11) that D → JK flip-flop conversion logic is given by

$$D = \overline{Q}J + Q\overline{K}.$$

When this equation is applied to the individual state variables in a D $\rightarrow$ JK K-map conversion, equation (5.11) takes the meaning

$$\begin{cases} D_A = \overline{Q}_A J_A + Q_A \overline{K}_A B \\ D_B = \overline{Q}_B J_B + Q_B \overline{K}_B \\ D_C = \overline{Q}_C J_C + Q_C \overline{K}_C \\ \quad \vdots \end{cases}. \tag{5.21}$$

Shown in Figure 5.63 are the JK K-maps converted from the D K-maps. From these K-maps the minimum cover is easily observed to be,

$$\begin{cases} J_A = B\overline{C}\overline{X} & K_A = 1 \\ J_B = CX & K_B = C \oplus X \\ J_C = \overline{X} & K_C = \overline{A}BX \end{cases}, \tag{5.22}$$

which represents a gate/input tally of 4/10 compared to 5/12 for the NS functions in Eq. (5.20), all exclusive of possible inverters.

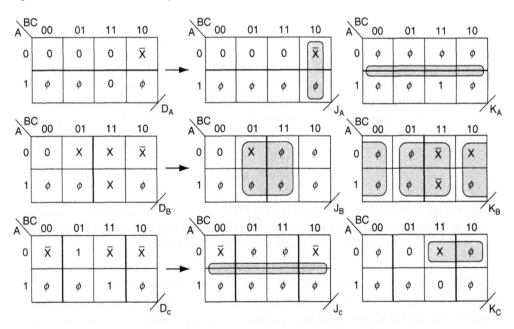

Figure 5.63: D-to-JK K-map conversion for the sequence recognizer of Figure 5.60c, showing domain boundaries for state variables A, B, and C, and minimum cover for the JK K-maps

Implementation of equations (5.22) is provided in Figure 5.64, together with the output-forming logic given in equations (5.20). Notice that NOR/XOR/INV logic is used for this purpose and that notation for the present state follows established practice in this text, namely $Q_A = A$, $Q_B = B$, and $Q_C = C$. The clock symbol $CK(L)$ simply indicates FET memory elements.

An inspection of the $D \rightarrow JK$ K-map conversion in Figure 5.63 together with equations (5.21) evokes the following algorithm:

> Algorithm 5.2:$D \rightarrow JK$ K-map Conversion [Refer to Eq. (5.21)]
>
> (1) For all that is NOT A in the D_A K-map, transfer it to the J_A K-map directly ($A\overline{J}_A$).
>
> (2) For all that is A in the D_A K-map, transfer it to the K_A K-map complemented ($A\overline{K}_A$).
>
> (3) Fill in empty cells with don't cares.
>
> (4) Repeat steps (1), (2), and (3) for the $D_B \rightarrow J_B$, K_B and $D_C \rightarrow J_C$, K_C, etc., K-map conversions.

It is important to note that the "fill-in" of the empty cells with don't cares is a result of the don't cares that exist in the excitation table for JK flip-flops. The reader should verify that the JK K-map results in Figure 5.63 are also obtained by directly applying the JK excitation table and the mapping algorithm to the Mealy form of the sequence recognizer given in Figure 5.60c. In doing so, it will become apparent that the $D \rightarrow JK$ K-map conversion method is quicker and easier than the direct method by using the excitation table for JK flip-flops. Furthermore, the K-map conversion approach permits a comparison between, say, a D flip-flop design and a JK K-map design, one often producing a more optimum result than the other. For these reasons the K-map conversion approach to design will be emphasized in this text.

Missing-State Analysis To this point no mention has been made of the missing (don't care) states in Figure 5.60c. Missing are the states 100, 101, and 110, which do exist but are not part of the primary routine expressed by the state diagram in Figure 5.60c. Each don't care state goes to ($\rightarrow$) a state in the state diagram as indicated in Figure 5.65. For example, 100 $\rightarrow$ 001 unconditionally, but 110 $\rightarrow$ 111 if $\overline{X}$ or 110 $\rightarrow$ 001 if X, etc. The NS values are determined by substituting the present state values A, B, and C into the NS functions given in equation (5.20).

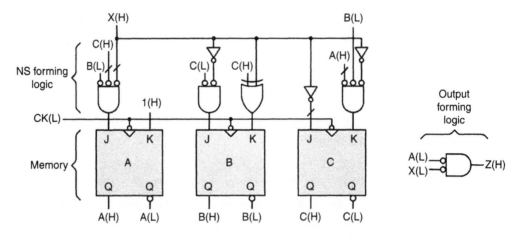

Figure 5.64: Implementation of equations (5.22) for the sequence recognizer of Figure 5.60c showing the NS-forming logic, memory, and output-forming logic

The missing state analysis gives emphasis to the fact that FSMs, such as the sequence recognizer in Figure 5.60, must be initialized into a specific state. On power-up, the sequence recognizer of Figure 5.64 could initialize into any state, including a don't care state. For example, if the FSM should power up into don't care state 110 with X inactive $(\bar{X})$, it would transit to state 111 on the next clock triggering edge and would falsely issue an output Z if X goes active. Ideally, on power-up, this FSM should be initialized into state 000 to properly begin the sequence.

Present State A B C	Next State $D_A D_B D_C$	Conclusion
1 0 0	0 0 1	$100 \longrightarrow 001$
1 0 1	0 X 1	$101 \xrightarrow{\bar{X}} 001$ or $101 \xrightarrow{X} 011$
1 1 0	X̄ X̄ 1	$110 \xrightarrow{\bar{X}} 111$ or $110 \xrightarrow{X} 001$

Figure 5.65: Missing state analysis for the Mealy version of the sequence recognizer given in Figure 5.60c

5.13   Analysis of Simple State Machines

The purpose of analyzing an FSM is to determine its sequential behavior and to identify any problems it may have. The procedure for FSM analysis is roughly the reverse of the procedure for FSM design given in Section 5.6. Thus, in a general sense, one begins with a logic circuit and ends with a state diagram. There are six principal steps in the analysis process:

1. Given the logic circuit for the FSM to be analyzed, carefully examine it for any potential problems it may have and note the number and character of its flip-flops and its outputs (Mealy or Moore).

2. Obtain the NS and output logic functions by carefully reading the logic circuit.

3. Map the output logic expressions into K-maps, and map the NS logic expressions into K-maps appropriate for the flip-flops used. If the memory elements are other than D flip-flops, use K-map conversion to obtain D K-maps.

4. From the D K-maps, construct the Present State/Inputs/Next State (PS/NS) table. To do this, observe which inputs control the branching, as indicated in each cell, and list these in ascending canonical word form together with the corresponding NS logic values. Ascending canonical word form means the use of minterm code such as $\overline{X}\overline{Y}\overline{Z}$, $\overline{X}\overline{Y}Z$, $\overline{X}Y\overline{Z}$ for branching dependency on inputs X, Y, and Z relative to a given K-map cell.

5. Use the PS/NS table in step 4 and the output K-maps to construct the fully documented state diagram for the FSM.

6. Analyze the state diagram for any obvious problems the FSM may have. These problems may include possible hang (isolated) states, subroutines from which there are no exits, and timing defects. Thus, a redesign of the FSM may be necessary.

A Simple Example To illustrate the analysis procedure, consider the logic circuit given in Figure 5.66a, which is seen to have one input X and one output, Z, and to be triggered on the falling edge of the clock waveform. Also, the external input arrives from a negative logic source. Reading the logic circuit yields the NS and output logic expressions,

$$J_A = B \odot X, \qquad J_B = AX$$

$$K_A = X, \qquad K_B = A \tag{5.23}$$

$$Z = \overline{A}\,\overline{B}X.$$

where A and B are the state variables. These expressions are mapped into JK K-maps and converted to D K-maps as shown in Figure 5.66b. Here, use is made of Algorithm 5.2 for the reverse conversion process, that is, for the JK-to-D K-map conversion. Notice that the domain boundaries are indicated by heavy lines as was done in Figure 5.63.

Step 4 in the analysis procedure, given previously, requires the construction of the PS/NS table from the D K-maps that are provided in Figure 5.66b. This is done in Figure 5.67a, from which the state diagram follows directly as shown in Figure 5.67b.

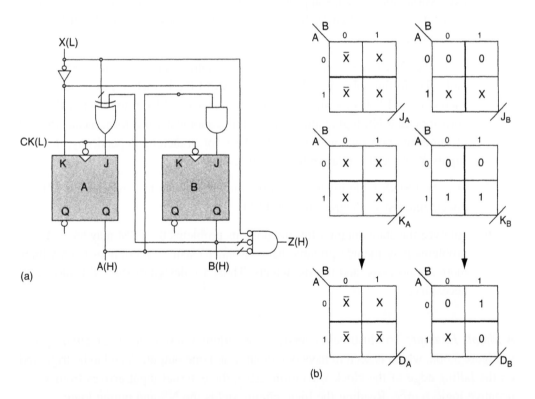

(a)

(b)

Figure 5.66: (a) The logic circuit to be analyzed (b) The NS JK K-maps for equations (5.23) and their conversion to the D K-maps needed to construct the PS/NS table

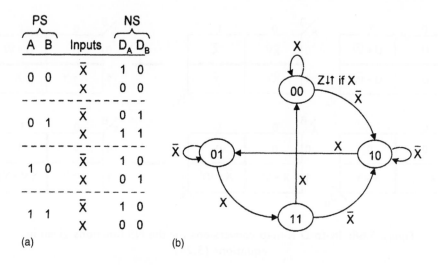

Figure 5.67: (a) PS/NS table constructed from the D K-maps in Figure 5.66, and (b) the resulting state diagram for the FSM of Figure 5.66a

There are no serious problems with this FSM other than the potential to produce an output race glitch (ORG) as a result of the transition $10 \rightarrow 01$ under branching condition X. The problem arises because two state variables are required to change during this transition, but do not do so simultaneously. The result is that the FSM must transit from state 10 to 01 via one of two race states, 00 or 11. If the transition is by way of state 00, Z will be issued as a glitch that could cross the switching threshold.

A More Complex Example The following NS and output expressions are read from a logic circuit that has five inputs, U, V, W, X, and Y, and two outputs, *LOAD* (*LD*) and *COUNT* (*CNT*):

$$J_A = U + B\overline{W} \qquad J_B = AX + AY$$
$$K_A = B\overline{X} + \overline{X}Y \qquad K_B = A(\overline{X} + VY) \qquad (5.24)$$
$$LD = \overline{A}\,\overline{B}X \qquad CNT = \overline{A}B\overline{X}Y.$$

Presented in Figure 5.68 are the JK-to-D K-map conversions for the NS functions given in equations (5.24). As in the previous example, Algorithm 5.2 is used for the reverse conversion from JK to D K-maps. The domain boundaries are again indicated by heavy lines.

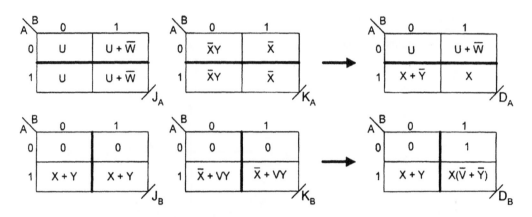

Figure 5.68: JK-to-D K-map conversions for the NS functions given in equations (5.24)

The PS/NS table for the NS functions, shown in Figure 5.69a, is constructed from the D K-maps in Figure 5.68. Notice that only the input literals indicated in a given cell of the D K-maps are represented in the PS/NS table as required by step 4 of the analysis procedure given previously in this section. Representation of these literals in canonical form ensures that the sum rule is obeyed—all possible branching conditions relative to a given state are taken into account.

The state diagram for the Mealy FSM represented by equations (5.24) is derived from the PS/NS table in Figure 5.69a and is shown in Figure 5.69b. Both Mealy outputs are deduced directly from the output expressions in equations (5.24). This FSM has the potential to form an output race glitch (ORG) during the transition from state 11 to state 00 under branching condition $\overline{X}Y$. Thus, if state variable A changes first while in state 11, the FSM could transit to state 00 via race state 01, causing a positive glitch in the output CNT, which is issued conditional on the input condition $\overline{X}Y$. No other potential ORGs exist.

5.14   VHDL Description of Simple State Machines

In this section, the behavioral descriptions of two FSMs (a flip-flop and a simple synchronous state machine) are presented by using the IEEE standard package *std_logic_1164*.

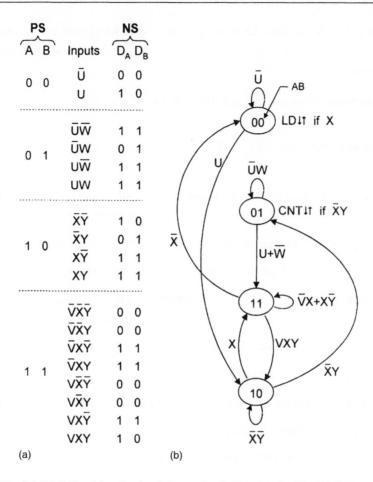

PS			NS	
A B		Inputs	D_A	D_B
0 0		$\bar{U}$	0	0
		U	1	0
0 1		$\bar{U}\bar{W}$	1	1
		$\bar{U}W$	0	1
		$U\bar{W}$	1	1
		UW	1	1
1 0		$\bar{X}\bar{Y}$	1	0
		$\bar{X}Y$	0	1
		$X\bar{Y}$	1	1
		XY	1	1
1 1		$\bar{V}\bar{X}\bar{Y}$	0	0
		$\bar{V}\bar{X}Y$	0	0
		$\bar{V}X\bar{Y}$	1	1
		$\bar{V}XY$	1	1
		$V\bar{X}\bar{Y}$	0	0
		$V\bar{X}Y$	0	0
		$VX\bar{Y}$	1	1
		VXY	1	0

(a)   (b)

Figure 5.69: (a) PS/NS table obtained from the D K-maps in Figure 5.68, and (b) the resulting state diagram for the FSM represented by equations (5.24)

5.14.1 The VHDL Behavioral Description of the RET D Flip-flop

(Note: Figure 5.51a provides the symbol for the RET D flip-flop that is being described here.)

```
library IEEE

use IEEE.std_logic_1164.all;

entity RETDFF is

    generic (SRDEL, CKDEL: Time);
```

```
        port (PR, CL, D, CK: in bit; Q, Qbar: out bit) −PR and CL are active
                                                          low inputs

    end RETDFF

    architecture behavioral of RETDFF is
begin
        process (PR, CL, CK)
begin
        if PR = '1' and CL = '0' then         −PR = '1' and CL = '0' is a
                                                  clear condition

            Q <= '0' after SRDEL;             −'0' represents LV

            Qbar <= '1' after SRDEL;          −'1' represents HV

        elseif PR = '0' and CL = '1' then     −PR = '0' and CL = '1' is a
                                                  preset condition

            Q <= '1' after SRDEL;

            Qbar <= '0' after SRDEL;

        elseif CK' event and CK = '1' and PR = '1' and CL = '1' then

            Q <= D after CKDEL;

            Qbar <= (not D) after CKDEL;

            end if;

        end process;

    end behavioral;
```

In the example just completed, the reader is reminded that the asynchronous overrides are active low inputs as indicated in Figure 5.51a. However, VHDL descriptions treat the "1" and "0" as HV and LV, respectively.

5.14.2   The VHDL Behavioral Description of a Simple FSM

Shown in Figure 5.70 is the state diagram for a two-state FSM having one input, X, and one output, Z. It also has a Sanity input for reset purposes.

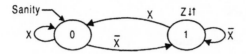

Figure 5.70: A simple FSM that is used for a VHDL description

The following is a VHDL behavioral description of the FSM in Figure 5.70:

```
library IEEE;

use IEEE.std_logic_1164.all;

entity FSM is

    port (Sanity, CK, X: in bit; Z: out bit); —Sanity is an active
                                              low reset input

end FSM;

architecture behavorial of FSM is

    type statetype is (state0, state1);

    signal state, NS : statetype := state0;

begin

sequence_process: process (state, X);

begin

    case state is

    when state0 =>

            if X = '0' then

                NS <= state1;

                Z <= '1';

            else NS <= state0;

                Z <= '0';

        end if;
```

```
    when state1=>

            if X = '1' then

                NS <= state0;

                Z <= '0';

            else NS <= state1;

                Z <= '1';

            end if;

    end case;

end process sequence_process;

CK_process: process;

begin

wait until (CK' event and CK = '1');

            if Sanity = '0' then —'0' represents LV

                state <= state0;

            else state <= NS;

            end if;

    end process CK_process;

end behavorial;
```

In this example the effect of the Sanity input is presented at the end of the behavioral description. But it could have been placed in front of the sequence_process. Also, a keyword not encountered in all previous examples is **type**. This keyword is used to declare a name and a corresponding set of declared values of the type. Usages include scalar types, composite types, file types, and access types.

Nearly all texts on the subject of digital design offer coverage, to one extent or another, of flip-flops and synchronous state machines. However, only a few texts approach these subjects by using fully documented state (FDS) diagrams, sometimes called *mnemonic state diagrams*. The FDS diagram approach is the simplest, most versatile, and most powerful pencil-and-paper means of representing the sequential behavior of an FSM in graphical form. The text by Fletcher is believed to be the first to use the FDS diagram approach to FSM design. Other texts that use FDS diagrams to one degree or another are those of Comer and Shaw. The text by Tinder is the only text to use the FDS diagram approach in the design and analysis of latches, flip-flops, and state machines (synchronous and asynchronous). Also, the text by Tinder appears to be the only one that covers the subject of K-map conversion as it is presented in the present text.

References

1. D. L. Comer, *Digital Logic and State Machine Design*, 3rd ed., Saunders College Publishing, Fort Worth, TX, 1995.
2. W. I. Fletcher, *An Engineering Approach to Digital Design*, Prentice Hall, Englewood Cliffs, NJ, 1980.
3. A. W. Shaw, *Logic Circuit Design*, Saunders College Publishing, Fort Worth, TX, 1993.
4. R. F. Tinder, *Digital Engineering Design: A Modern Approach*, Prentice Hall, Englewood Cliffs, NJ, 1991.
   The subjects of setup and hold times for flip-flops are adequately treated in the texts by Fletcher (previously cited), Katz, Taub, Tinder (previously cited), Wakerly, and Yarbrough.
5. R. H. Katz, *Contemporary Logic Design*, Benjamin/Cummings Publishing,Redwood City, CA, 1994.
6. H. Taub, *Digital Circuits and Microprocessors*, McGraw-Hill, New York, 1982.
7. J. F.Wakerly, *Digital Design Principles and Practices*, 2nd ed., Prentice-Hall, Englewood Cliffs, NJ, 1994.
8. J. M. Yarbrough, *Digital Logic Applications and Design*, West Publishing Co., Minneapolis/ St. Paul, MN, 1997.
   With the exception of texts by Katz and Taub, all of the previously cited references cover adequately the subject of synchronous machine analysis. The texts by Fletcher, Shaw, and Tinder in particular present the subject in a fashion similar to that of the present text. Other texts that can be recommended for further reading on this subject are those by Dietmeyer and by Nelson *et al.*, the former being more for the mathematically inclined.
9. D. L. Dietmeyer, *Logic Design of Digital Systems*, 2nd ed., Allyn and Bacon, Inc., Boston, MA, 1978.
10. V. P. Nelson, H. T. Nagle, B. D. Carroll, and J. D. Irwin, *Digital Logic Circuit Analysis and Design*, Prentice Hall, Englewood Cliffs, NJ, 1995.

CHAPTER 6

Embedded Processors

Peter R. Wilson

This chapter comes from the book Design Recipes for FPGAs *by Peter Wilson, ISBN: 9780750668453. As we discussed in Chapter 3, Peter's book is designed to provide the methods and understanding to make the reader able to develop practical, operational VHDL that will run correctly on FPGAs. Of particular interest to me is the chapter on embedded processors. This application example chapter concentrates on the key topic of integrating processors into FPGA designs. This ranges from simple 8-bit microprocessors up to large IP processor cores that require an element of hardware-software codesign.*

This chapter takes the reader through the basics of implementing a behavioral-based microprocessor for evaluation of algorithms, through to the practicalities of structurally correct models that can be synthesized and implemented on an FPGA. As part of this chapter, Peter describes the design and implementation of a very simple 8-bit microprocessor core. After describing the major functional blocks, Peter walks us through the process of defining a simple instruction set, designing a simple "fetch-execute" cycle, and then creating the VHDL corresponding to the various functional elements. This is a lot of fun that should inspire you to start playing with your own microprocessor architecture and/or instruction set.

"One of the major challenges facing hardware designers in the 21st century is the problem of hardware-software codesign. This has moved on from a basic partitioning mechanism based on standard hardware architectures to the current situation where the algorithm itself can be optimized at a compilation level for performance or power by implementing appropriately at different levels with hardware or software as required.

> *"This aspect suits FPGAs perfectly, as they can handle a fixed hardware architecture that runs software compiled onto memory, they can implement optimal hardware running at much faster rates than a software equivalent could, and there is now the option of configurable hardware that can adapt to the changing requirements of a modified environment."*
>
> **—Clive "Max" Maxfield**

6.1   Introduction

This application example chapter concentrates on the key topic of Integrating Processors onto field programmable gate array (FPGA) designs. This ranges from simple 8-bit microprocessors up to large IP processor cores that require an element of hardware–software codesign involved. This chapter will take the reader through the basics of implementing a behavioral-based microprocessor for evaluation of algorithms, through to the practicalities of structurally correct models that can be synthesized and implemented on an FPGA.

One of the major challenges facing hardware designers in the 21st century is the problem of hardware-software codesign. This has moved on from a basic partitioning mechanism based on standard hardware architectures to the current situation where the algorithm itself can be optimized at a compilation level for performance or power by implementing appropriately at different levels with hardware or software as required.

This aspect suits FPGAs perfectly, as they can handle fixed hardware architecture that runs software compiled onto memory, they can implement optimal hardware running at much faster rates than a software equivalent could, and there is now the option of configurable hardware that can adapt to the changing requirements of a modified environment.

6.2   A Simple Embedded Processor

6.2.1   Embedded Processor Architecture

A useful example of an embedded processor is to consider a generic microcontroller in the context of an FPGA platform. Take a simple example of a generic 8-bit microcontroller shown in Figure 6.1.

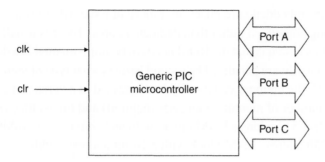

Figure 6.1: Simple microcontroller

As can be seen from Figure 6.1, the microcontroller is a "general-purpose microprocessor", with a simple clock (clk) and reset (clr), and three 8-bit ports (A, B and C). Within the microcontroller itself, there needs to be the following basic elements:

1. A control unit: This is required to manage the clock and reset of the processor, manage the data flow and instruction set flow, and control the port interfaces. There will also need to be a program counter (PC).

2. An arithmetic logic unit (ALU): a PIC will need to be able to carry out at least some rudimentary processing—carried out in the ALU.

3. An address bus.

4. A data bus.

5. Internal registers.

6. An instruction decoder.

7. A read only memory (ROM) to hold the program.

While each of these individual elements (1–6) can be implemented simply enough using a standard FPGA, the ROM presents a specific difficulty. If we implement a ROM as a set of registers, then obviously this will be hugely inefficient in an FPGA architecture. However, in most modern FPGA platforms, there are blocks of random access memory (RAM) on the FPGA that can be accessed and it makes a lot of sense to design a RAM block for use as a ROM by initializing it with the ROM values on reset and then using that to run the program.

This aspect of the embedded core raises an important issue, which is the reduction in efficiency of using embedded rather than dedicated cores. There is usually a compromise involved and in this case it is that the ROM needs to be implemented in a different manner, in this case with a hardware penalty. The second issue is what type of memory core to use? In an FPGA RAM, the memory can usually be organized in a variety of configurations to vary the depth (number of memory addresses required) and the width (width of the data bus). For example a 512 address RAM block, with an 8-bit address width would be equivalent to a 256 address RAM block with a 16-bit address width.

If the equivalent ROM is, say 12 bits wide and 256, then we can use the 256×16 RAM block and ignore the top four bits. The resulting embedded processor architecture could be of the form shown in Figure 6.2.

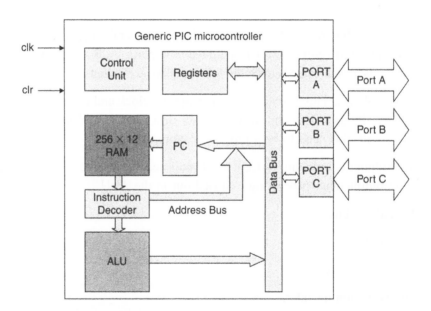

Figure 6.2: Embedded microcontroller architecture

6.2.2 Basic Instructions

When we program a microprocessor of any type, there are three different ways of representing the code that will run on the processor. These are machine code (1's and 0's), assembler (low-level instructions such as LOAD, STORE, ...) and high-level code (such as C, Fortran or Pascal). Regardless of the language used, the code will always be compiled or assembled into machine code at the lowest level for programming into memory. High-level code (e.g., C) is compiled and assembler code is assembled (as the name suggests) into machine code for the specific platform.

Clearly a detailed explanation of a compiler is beyond the scope of this book, but the same basic process can be seen in an assembler and this is useful to discuss in this context.

Every processor has a basic "Instruction Set" which is simply the list of functions that can be run in a program on the processor. Take the simple example of the following pseudocode expression:

$$b = a + 2$$

In this example, we are taking the variable a and adding the integer value 2 to it, and then storing the result in the variable b. In a processor, the use of a variable is simply a memory location that stores the value, and so to load a variable we use an assembler command as follows:

LOAD a

What is actually going on here? Whenever we retrieve a variable value from memory, the implication is that we are going to put the value of the variable in the register called the accumulator (ACC). The command "LOAD a" could be expressed in natural language as "LOAD the value of the memory location denoted by a into the accumulator register ACC".

The next stage of the process is to add the integer value 2 to the accumulator. This is a simple matter, as instead of an address, the value is simply added to the current value stored in the accumulator. The assembly language command would be something like:

ADD #x02

Notice that we have used the x to denote a hexadecimal number. If we wished to add a variable, say called c, then the command would be the same, except that it would use the address c instead of the absolute number. The command would therefore be:

ADD c

Now we have the value of a + 2 stored in the accumulator register (ACC). This could be stored in a memory location, or put onto a port (e.g., PORT A). It is useful to notice that for a number we use the key character # to indicate that we are adding the value and not using the argument as the address.

In the pseudocode example, we are storing the result of the addition in the variable called b, so the command would be something like this:

STORE b

While this is superficially a complete definition of the instruction set requirements, there is one specific design detail that has to be decided on for any processor. This is the number of instructions and the data bus size. If we have a set of instructions with the number of instructions denoted by I, then the number of bits in the opcode (n) must conform to the following rule:

$$2^n \leq I$$

In other words, the number of bits provides the number of unique different codes that can be defined, and this defines the size of the instruction set possible. For example, if $n = 3$, then with three bits there are eight possible unique opcodes, and so the maximum size of the instruction set is eight.

6.2.3   Fetch Execute Cycle

The standard method of executing a program in a processor is to store the program in memory and then follow a strict sequence of events to carry out the instructions. The first stage is to use the PC to increment the program line, this then calls up the next command from memory in the correct order, and then the instruction can be loaded into the appropriate register for execution. This is called the "fetch execute cycle".

What is happening at this point? First the contents of the PC is loaded into the memory address register (MAR). The data in the memory location are then retrieved and loaded into the memory data register (MDR). The contents of the MDR can then be transferred into the Instruction Register (IR). In a basic processor, the PC can then be incremented by one (or in fact this could take place immediately after the PC has been loaded into the MDR).

Once the opcode (and arguments if appropriate) are loaded, then the instruction can be executed. Essentially, each instruction has its own state machine and control path, which is linked to the IR and a sequencer that defines all the control signals required to move the data correctly around the memory and registers for that instruction. We will discuss registers in the next section, but in addition to the PC, IR and accumulator (ACC) mentioned already, we require two memory registers as a minimum, the MDR and MAR.

For example, consider the simple command LOAD a, from the previous example. What is required to actually execute this instruction? First, the opcode is decoded and this defines that the command is a "LOAD" command. The next stage is to identify the address. As the command has not used the # symbol to denote an absolute address, this is stored in the variable "a". The next stage, therefore is to load the value in location "a" into the MDR, by setting MAR a and then retrieving the value of a from the RAM. This value is then transferred to the accumulator (ACC).

6.2.4   Embedded Processor Register Allocation

The design of the registers partly depends on whether we wish to "clone" a PIC device or create a modified version that has more custom behavior. In either case, there are some mandatory registers that must be defined as part of the design. We can assume that we need an accumulator (ACC), a program counter (PC), and the three input/output ports (PORTA, PORTB, PORTC). Also, we can define the IR, MAR, and MDR.

In addition to the data for the ports, we need to have a definition of the port direction and this requires three more registers for managing the tristate buffers into the data bus to and from the ports (DIRA, DIRB, DIRC). In addition to this, we can define a number (essentially arbitrary) of registers for general-purpose usage. In the general case the naming, order and numbering of registers does not matter, however, if we intend to use a specific device as a template, and perhaps use the same bit code, then it is vital that the registers are configured in exactly the same way as the original device and in the same order.

In this example, we do not have a base device to worry about, and so we can define the general-purpose registers (24 in all) with the names REG0 to REG23. In conjunction with the general-purpose registers, we need to have a small decoder to select the correct register and put the contents onto the data bus (F).

6.2.5   A Basic Instruction Set

In order for the device to operate as a processor, we must define some basic instructions in the form of an instruction set. For this simple example we can define some very basic instructions that will carry out basic program elements, ALU functions, memory functions. These are summarized in Table 6.1.

In this simple instruction set, there are 10 separate instructions. This implies, from the rule given in equation $2^n \leq I$ previously in this chapter that we need at least 4 bits to

Table 6.1: Basic instructions

Command	Description
LOAD arg	This command loads an argument into the accumulator. If the argument has the prefix # then it is the absolute number, otherwise it is the address and this is taken from the relevant memory address. Examples: LOAD #01 LOAD abc
STORE arg	This command stores an argument from the accumulator into memory. If the argument has the prefix # then it is the absolute address, otherwise it is the address and this is taken from the relevant memory address. Examples: STORE #01 STORE abc
ADD arg	This command adds an argument to the accumulator. If the argument has the prefix # then it is the absolute number, otherwise it is the address and this is taken from the relevant memory address. Examples: ADD #01 ADD abc
NOT	This command carries out the NOT function on the accumulator.
AND arg	This command ands an argument with the accumulator. If the argument has the prefix # then it is the absolute number, otherwise it is the address and this is taken from the relevant memory address. Examples: AND #01 AND abc
OR arg	This command ors an argument with the accumulator. If the argument has the prefix # then it is the absolute number, otherwise it is the address and this is taken from the relevant memory address. Examples: OR #01 OR abc
XOR arg	This command xors an argument with the accumulator. If the argument has the prefix # then it is the absolute number, otherwise it is the address and this is taken from the relevant memory address. Examples: XOR #01 XOR abc
INC	This command carries out an increment by one on the accumulator.

Table 6.1: Basic instructions (Cont'd)

Command	Description

SUB arg	This command subtracts an argument from the accumulator. If the argument has the prefix # then it is the absolute number, otherwise it is the address and this is taken from the relevant memory address.
	  Examples:
	  SUB #01
	  SUB abc

BRANCH arg	This command allows the program to branch to a specific point in the program. This may be very useful for looping and program flow. If the argument has the prefix # then it is the absolute number, otherwise it is the address and this is taken from the relevant memory address.
	  Examples:
	  BRANCH #01
	  BRANCH abc

describe each of the instructions given in the table above. Given that we wish to have 8 bits for each data word, we need to have the ability to store the program memory in a ROM that has words of at least 12 bits wide. In order to cater for a greater number of instructions, and also to handle the situation for specification of different addressing modes (such as the difference between absolute numbers and variables), we can therefore suggest a 16-bit system for the program memory.

Notice that at this stage there are no definitions for port interfaces or registers. We can extend the model to handle this behavior later.

6.2.6 *Structural or Behavioral?*

So far in the design of the simple microprocessor, we have not specified details beyond a fairly abstract structural description of the processor in terms of registers and buses. At this stage we have a decision about the implementation of the design with regard to the program and architecture.

One option is to take a program (written in assembly language) and simply convert this into a state machine that can easily be implemented in a VHDL model for testing out the algorithm. Using this approach, the program can be very simply modified and recompiled based on simple rules that restrict the code to the use of registers and techniques applicable to the processor in question. This can be useful for investigating and developing algorithms, but is more ideal than the final implementation as there will be control signals and delays due to memory access in a processor plus memory configuration, that will be better in a dedicated hardware design.

Another option is to develop a simple model of the processor that does have some of the features of the final implementation of the processor, but still uses an assembly language description of the model to test. This has advantages in that no compilation to machine code is required, but there are still not the detailed hardware characteristics of the final processor architecture that may cause practical issues on final implementation.

The third option is to develop the model of the processor structurally and then the machine code can be read in directly from the ROM. This is an excellent approach that is very useful for checking both the program and the possible quirks of the hardware/software combination as the architecture of the model reflects directly the structure of the model to be implemented on the FPGA.

6.2.7   Machine Code Instruction Set

In order to create a suitable instruction set for decoding instructions for our processor, the assembly language instruction set needs to have an equivalent machine code instruction set that can be decoded by the sequencer in the processor. The resulting opcode/instruction table is given in Table 6.2.

Table 6.2: Opcode/instruction table

Command	Opcode (Binary)
LOAD arg	0000
STORE arg	0001
ADD arg	0010
NOT	0011
AND arg	0100
OR arg	0101
XOR arg	0110
INC	0111
SUB arg	1000
BRANCH arg	1001

6.2.8 Structural Elements of the Microprocessor

Taking the abstract design of the microprocessor given in Figure 6.3 we can redraw with the exact registers and bus configuration as shown in the structural diagram in Figure 6.3. Using this model we can create separate VHDL models for each of the blocks that are connected to the internal bus and then design the control block to handle all the relevant sequencing and control flags to each of the blocks in turn.

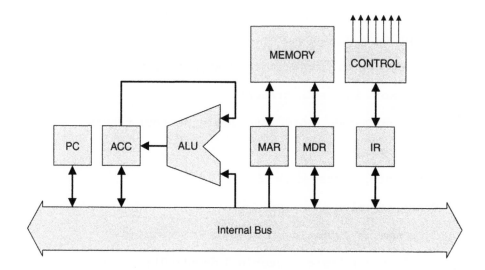

Figure 6.3: Structural model of the microprocessor

Before this can be started, however, it makes sense to define the basic criteria of the models and the first is to define the basic type. In any digital model (as we have seen elsewhere in this book) it is sensible to ensure that data can be passed between standard models and so in this case we shall use the std_logic_1164 library that is the standard for digital models.

In order to use this library, each signal shall be defined as of the basic type std_logic and also the library ieee.std_logic_1164.all shall be declared in the header of each of the models in the processor.

Finally, each block in the processor shall be defined as a separate block for implementation in VHDL.

6.2.9   Processor Functions Package

In order to simplify the VHDL for each of the individual blocks, a set of standard functions have been defined in a package call processor_functions. This is used to define useful types and functions for this set of models. The VHDL for the package is given below:

```
Library ieee;

Use ieee.std_logic_1164.all;


Package processor_functions is

    Type opcode is (load, store, add, not, and, or,

        xor, inc, sub, branch);

    Function Decode (word : std_logic_vector) return

        opcode;

    Constant n : integer := 16;

    Constant oplen : integer := 4;

    Type memory_array is array (0 to 2**(n-oplen-1)

        of Std_logic_vector(n-1 downto 0);

    Constant reg_zero : unsigned (n-1 downto 0) :=

        (others => '0');

End package processor_functions;


Package body processor_functions is

    Function Decode (word : std_logic_vector) return

        opcode is

            Variable opcode_out : opcode;
```

```
    Begin

        Case word(n-1 downto n-oplen-1) is

                When "0000" => opcode_out : = load;

                When "0001" => opcode_out : = store;

                When "0010" => opcode_out : = add;

                When "0011" => opcode_out : = not;

                When "0100" => opcode_out : = and;

                When "0101" => opcode_out : = or;

                When "0110" => opcode_out : = xor;

                When "0111" => opcode_out : = inc;

                When "1000" => opcode_out : = sub;

                When "1001" => opcode_out : = branch;

                When others => null;

        End case;

        Return opcode_out;

    End function decode;

  End package body processor_functions;
```

6.2.10   The PC

The PC needs to have the system clock and reset connections, the system bus (defined
as inout so as to be readable and writable by the PC register block). In addition,
there are several control signals required for correct operation. The first is the signal
to increment the PC (PC_inc), the second is the control signal load the PC with a
specified value (PC_load) and the final is the signal to make the register contents visible
on the internal bus (PC_valid). This signal ensures that the value of the PC register
will appear to be high impedance ("Z") when the register is not required on the

processor bus. The system bus (PC_bus) is defined as a std_logic_vector, with direction inout to ensure the ability to read and write. The resulting VHDL entity is given below:

```
library ieee;

use ieee.std_logic_1164.all;

entity pc is

    Port (

            Clk : IN std_logic;

            Nrst : IN std_logic;

            PC_inc : IN std_logic;

            PC_load : IN std_logic;

            PC_valid : IN std_logic;

            PC_bus : INOUT std_logic_vector(n-1 downto 0)

    );
End entity PC;
```

The architecture for the PC must handle all of the various configurations of the PC control signals and also the communication of the data into and from the internal bus correctly. The PC model has an asynchronous part and a synchronous section. If the PC_valid goes low at any time, the value of the PC_bus signal should be set to "Z" across all of its bits. Also, if the reset signal goes low, then the PC should reset to zero.

The synchronous part of the model is the increment and load functionality. When the clk rising edge occurs, then the two signals PC_load and PC_inc are used to define the function of the counter. The precedence is that if the increment function is high, then regardless of the load function, then the counter will increment. If the increment function (PC_inc) is low, then the PC will load the current value on the bus, if and only if the PC_load signal is also high.

The resulting VHDL is given below:

```
architecture RTL of PC is

        signal counter : unsigned (n-1 downto 0);
```

```
begin

    PC_bus <= std_logic_vector(counter)

            when PC_valid='1' else (others =>'Z');

    process (clk, nrst) is

    begin

            if nrst='0' then

                    count <= 0;

            elsif rising_edge(clk) then

                    if PC_inc='1' then

                            count <= count + 1;

                    else

                            if PC_load='1' then

                                    count <= unsigned(PC_bus);

                            end if;

                    end if;

            end if;

    end process;

end architecture RTL;
```

6.2.11 The IR

The IR has the same clock and reset signals as the PC, and also the same interface to the bus (IR_bus) defined as a std_logic_vector of type INOUT. The IR also has two further control signals, the first being the command to load the IR (IR_load), and the second being to load the required address onto the system bus (IR_address). The final connection is the decoded opcode that is to be sent to the system controller. This is defined as a simple unsigned integer value with the same size as the basic system bus. The basic VHDL for the entity of the IR is given below:

```
library ieee;

use ieee.std_logic_1164.all;

use work.processor_functions.all;

entity ir is

    Port (

                Clk : IN std_logic;

                Nrst : IN std_logic;

                IR_load : IN std_logic;

                IR_valid : IN std_logic;

                IR_address : IN std_logic;

                IR_opcode : OUT opcode;

                IR_bus : INOUT std_logic_vector(n-1 downto 0)

        );

    End entity IR;
```

The function of the IR is to decode the opcode in binary form and then pass to the control block. If the IR_valid is low, the bus value should be set to "Z" for all bits. If the reset signal (nsrt) is low, then the register value internally should be set to all 0's.

On the rising edge of the clock, the value on the bus shall be sent to the internal register and the output opcode shall be decoded asynchronously when the value in the IR changes.

The resulting VHDL architecture is given below:

```
architecture RTL of IR is

    signal IR_internal : std_logic_vector (n-1 downto 0);

begin

    IR_bus <= IR_internal

            when IR_valid='1' else (others => 'Z');
```

```
IR_opcode <= Decode(IR_internal);

process (clk, nrst) is

begin

    if nrst = '0' then

            IR_internal <= (others => '0');

    elsif rising_edge(clk) then

            if IR_load = '1' then

                    IR_internal <= IR_bus;

            end if;

    end if;

end process;

end architecture RTL;
```

In this VHDL, notice that we have used the predefined function Decode from the processor_functions package previously defined. This will look at the top four bits of the address given to the IR and decode the relevant opcode for passing to the controller.

6.2.12 The Arithmetic and Logic Unit

The arithmetic and logic unit (ALU) has the same clock and reset signals as the PC, and also the same interface to the bus (ALU_bus) defined as a std_logic_vector of type INOUT. The ALU also has three further control signals, which can be decoded to map to the eight individual functions required of the ALU. The ALU also contains the accumulator (ACC) which is a std_logic_vector of the size defined for the system bus width. There is also a single-bit output ALU_zero, which goes high when all the bits in the accumulator are zero.

The basic VHDL for the entity of the ALU is given below:

```
library ieee;

use ieee.std_logic_1164.all;
```

```
use work.processor_functions.all;

entity alu is

    Port (

            Clk : IN std_logic;

            Nrst : IN std_logic;

            ALU_cmd : IN std_logic_vector(2 downto 0);

            ALU_zero : OUT std_logic;

            ALU_valid : IN std_logic;

            ALU_bus : INOUT std_logic_vector(n-1 downto 0)

    );

    End entity alu;
```

The function of the ALU is to decode the ALU_cmd in binary form and then carry out the relevant function on the data on the bus, and the current data in the accumulator. If the ALU_valid is low, the bus value should be set to "Z" for all bits. If the reset signal (nsrt) is low, then the register value internally should be set to all 0's.

On the rising edge of the clock, the value on the bus shall be sent to the internal register and the command shall be decoded.

The resulting VHDL architecture is given below:

```
architecture RTL of ALU is

        signal ACC : std_logic_vector (n-1 downto 0);
begin

    ALU_bus <= ACC

        when ACC_valid'1' else (others => 'Z');

    ALU_zero <= '1' when acc reg_zero else '0';

    process (clk, nrst) is
```

```vhdl
    begin

      if nrst = '0' then

          ACC <= (others => '0');

      elsif rising_edge(clk) then

          case ACC_cmd is

          -- Load the Bus value into the accumulator

          when "000" => ACC <= ALU_bus;

          -- Add the ACC to the Bus value

          When "001" => ACC <= add(ACC,ALU_bus);

          -- NOT the Bus value

          When "010" => ACC <= NOT ALU_bus;

          -- OR the ACC to the Bus value

          When "011" => ACC <= ACC or ALU_bus;

          -- AND the ACC to the Bus value

          When "100" => ACC <= ACC and ALU_bus;

          -- XOR the ACC to the Bus value

          When "101" => ACC <= ACC xor ALU_bus;

          -- Increment ACC

          When "110" => ACC <= ACC + 1;

          -- Store the ACC value

          When "111" => ALU_bus <= ACC;

      end if;

    end process;

  end architecture RTL;
```

6.2.13 The Memory

The processor requires a RAM memory, with an address register (MAR) and a data register (MDR). There therefore needs to be a load signal for each of these registers: MDR_load and MAR_load. As it is a memory, there also needs to be an enable signal (M_en), and also a signal denote Read or Write modes (M_rw). Finally, the connection to the system bus is a standard inout vector as has been defined for the other registers in the microprocessor.

The basic VHDL for the entity of the memory block is given below:

```
library ieee;

use ieee.std_logic_1164.all;

use work.processor_functions.all;

entity memory is

    Port (

            Clk : IN std_logic;

            Nrst : IN std_logic;

            MDR_load : IN std_logic;

            MAR_load : IN std_logic;

            MAR_valid : IN std_logic;

            M_en : IN std_logic;

            M_rw : IN std_logic;

            MEM_bus : INOUT std_logic_vector(n-1 downto 0)

        );

    End entity memory;
```

The memory block has three aspects. The first is the function that the memory address is loaded into the MAR. The second function is either reading from or writing to the memory using the MDR. The final function, or aspect of the memory is to store the actual program that the processor will run. In the VHDL model, we will achieve this by using a constant array to store the program values.

The resulting basic VHDL architecture is given below:

```
architecture RTL of memory is

        signal mdr : std_logic_vector(wordlen-1 downto 0);

        signal mar : unsigned(wordlen-oplen-1 downto 0);

                    begin

MEM_bus <= mdr

    when MEM_valid = '1' else (others => 'Z');

process (clk, nrst) is

    variable contents : memory_array;

    constant program : contents :=

    (

                0 => "0000000000000011",

                1 => "0010000000000100",

                2 => "0001000000000101",

                3 => "0000000000001100",

                4 => "0000000000000011",

                5 => "0000000000000000" ,

                Others => (others => '0')

        );

begin

        if nrst='0' then

                mdr <= (others => '0');

                mdr <= (others => '0');
```

```
                    contents := program;

        elsif rising_edge(clk) then

            if MAR_load='1' then

                    mar <= unsigned(MEM_bus(n-oplen-1

                            downto 0));

            elsif MDR_load='1' then

                mdr <= MEM_bus;

            elsif MEM_en='1' then

                if MEM_rw='0' then

                        mdr <= contents(to_integer

                                (mar));

                else

                        mem(to_integer(mar))

                            := mdr;

                end if;

            end if;

        end if;

    end process;

end architecture RTL;
```

We can look at some of the VHDL in a bit more detail and explain what is going on at this stage. There are two internal signals to the block, mdr and mar (the data and address, respectively). The first aspect to notice is that we have defined the MAR as an unsigned rather than as a std_logic_vector. We have done this to make indexing direct. The MDR remains as a std_logic_vector. We can use an integer directly, but an unsigned translates easily into a std_logic_vector.

```
signal mdr : std_logic_vector(wordlen-1 downto 0);
```

```
signal mar : unsigned(wordlen-oplen-1 downto 0);
```

The second aspect is to look at the actual program itself. We clearly have the possibility of a large array of addresses, but in this case we are defining a simple three line program:

$$c = a + b$$

The binary code is shown below:

```
0 => "0000000000000011",

1 => "0010000000000100",

2 => "0001000000000101",

3 => "0000000000001100",

4 => "0000000000000011",

5 => "0000000000000000" ,

Others => (others => '0')
```

For example, consider the line of the declared value for address 0. The 16 bits are defined as 0000000000000011. If we split this into the opcode and data parts we get the following:

Opcode   0000

Data   000000000011 (3)

In other words this means LOAD the variable from address 3. Similarly, the second line is ADD from 4, finally the third command is STORE in 5. In addresses 3, 4 and 5, the three data variables are stored.

6.2.14 Microcontroller: Controller

The operation of the processor is controlled in detail by the sequencer, or controller block. The function of this part of the processor is to take the current PC address, look up the relevant instruction from memory, move the data around as required, setting up all the relevant control signals at the right time, with the right values.

As a result, the controller must have the clock and reset signals (as for the other blocks in the design), a connection to the global bus and finally all the relevant control signals must be output. An example entity of a controller is given below:

```vhdl
library ieee;

use ieee.std_logic_1164.all;

use work.processor_functions.all;

entity controller is

    generic (

        n : integer :=16

    );

    Port (

        Clk : IN std_logic;

        Nrst : IN std_logic;

        IR_load : OUT std_logic;

        IR_valid : OUT std_logic;

        IR_address : OUT std_logic;

        PC_inc : OUT std_logic;

        PC_load : OUT std_logic;

        PC_valid : OUT std_logic;

        MDR_load : OUT std_logic;

        MAR_load : OUT std_logic;

        MAR_valid : OUT std_logic;

        M_en : OUT std_logic;

        M_rw : OUT std_logic;

        ALU_cmd : OUT std_logic_vector(2 downto 0);

        CONTROL_bus : INOUT std_logic_vector(n-1 downto 0)

    );

End entity controller;
```

Using this entity, the control signals for each separate block are then defined, and these can be used to carry out the functionality requested by the program. The architecture for the controller is then defined as a basic state machine to drive the correct signals. The basic state machine for the processor is defined in Figure 6.4.

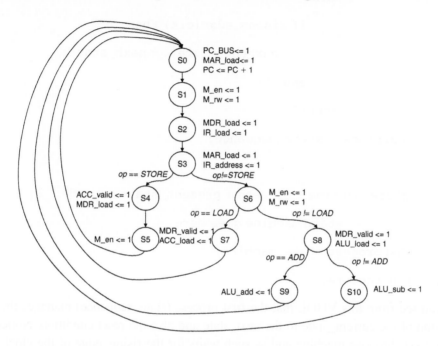

Figure 6.4: Basic processor controller state machine

We can implement this using a basic VHDL architecture that implements each state using a new state type and a case statement to manage the flow of the state machine. The basic VHDL architecture is shown below and it includes the basic synchronous machine control section (reset and clock) the management of the next stage logic:

```
architecture RTL of controller is

    type states is

        (s0,s1,s2,s3,s4,s5,s6,s7,s8,s9,s10);

    signal current_state, next_state : states;

begin

        state_sequence: process (clk, nrst) is
```

```
        if nrst = '0' then

            current_state < = s0;

        else

            if rising_edge(clk) then

                current_state < = next_state;

            end if;

        end if;

    end process state_sequence;



    state_machine : process ( present_state, opcode ) is

        -- state machine goes here

    End process state_machine;

end architecture;
```

You can see from this VHDL that the first process (state_sequence) manages the transition of the current_state to the next_state and also the reset condition. Notice that this is a synchronous machine and as such waits for the rising_edge of the clock, and that the reset is asynchronous. The second process (state_machine) waits for a change in the state or the opcode and this is used to manage the transition to the next state, although the actual transition itself is managed by the state_sequence process. This process is given in the VHDL below:

```
state_machine : process ( present_state, opcode ) is

begin

    -- Reset all the control signals

    IR_load <= '0';

    IR_valid <= '0';

    IR_address <= '0';
```

```
PC_inc <= '0';

PC_load <= '0';

PC_valid <= '0';

MDR_load <= '0';

MAR_load <= '0';

MAR_valid <= '0';

M_en <= '0';

M_rw <= '0';

Case current_state is

When s0 =>

    PC_valid <= '1'; MAR_load <= '1';

    PC_inc <= '1'; PC_load <= '1';

    Next_state <= s1;

When s1 =>

    M_en <='1'; M_rw <= '1';

    Next_state <= s2;

When s2 =>

    MDR_valid <= '1'; IR_load <= '1';

    Next_state <= s3;

When s3 =>

    MAR_load <= '1'; IR_address <= '1';

    If opcode=STORE then

        Next_state <= s4;

    else

        Next_state <=s6;
```

```
        End if;
    When s4 =>
        MDR_load <= '1'; ACC_valid <= '1';
        Next_state <= s5;
    When s5 =>
        M_en <= '1';
        Next_state <= s0;
    When s6 =>
        M_en <= '1'; M_rw <= '1';
        If opcode = LOAD then
            Next_state <= s7;
        else
            Next_state <= s8;
        End if;
    When s7 =>
        MDR_valid <= '1'; ACC_load <= '1';
        Next_state <= s0;
    When s8 =>
        M_en<='1'; M_rw <= '1';
        If opcode = ADD then
            Next_state <= s9;
        Else
            Next_state <= s10;
        End if;
```

```
    When s9 =>

        ALU_add <= '1';

        Next_state <= s0;

    When s10 =>

        ALU_sub <= '1';

        Next_state <= s0;

    End case;

  End process state_machine;
```

6.2.15 Summary of a Simple Microprocessor

Now that the important elements of the processor have been defined, it is a simple matter to instantiate them in a basic VHDL netlist and create a microprocessor using these building blocks. It is also a simple matter to modify the functionality of the processor by changing the address/data bus widths or extend the instruction set.

6.3 Soft Core Processors on an FPGA

While the previous example of a simple microprocessor is useful as a design exercise and helpful to gain understanding about how microprocessors operate, in practice most FPGA vendors provide standard processor cores as part of an embedded development kit that includes compilers and other libraries. For example, this could be the Microblaze core from Xilinx or the NIOS core supplied by Altera. In all these cases the basic idea is the same, that a standard configurable core can be instantiated in the design and code compiled using a standard compiler and downloaded to the processor core in question.

Each soft core is different and rather than describe the details of a particular case, in this section the general principles will be covered and the reader is encouraged to experiment with the offerings from the FPGA vendors to see which suits their application the best.

In any soft core development system there are several key functions that are required to make the process easy to implement. The first is the system building function. This enables a core to be designed into a hardware system that includes memory modules, control functions, Direct Memory Access (DMA) functions, data interfaces and interrupts. The second is the choice of processor types to implement. A basic NIOS II or similar embedded core will typically have a performance in the region of 100–200 MIPS, and the processor design tools will allow the size of the core to be traded off with the hardware resources available and the performance required.

6.4 Summary

The topic of embedded processors on FPGAs would be suitable for a complete book in itself. In this chapter the basic techniques have been described for implementing a simple processor directly on the FPGA and the approach for implementing soft cores on FPGAs have been introduced.

CHAPTER 7

Digital Signal Processing

R.C. Cofer
Ben Harding

This chapter comes from the book *Rapid System Prototyping with FPGAs* by *R.C.Cofer and Ben Harding, ISBN: 9780750678667. It is important to develop and follow a disciplined and optimized design flow when implementing a rapid system prototyping effort. Effective design flow optimization requires addressing the trade-offs between additional project risk and associated schedule reduction.*

This balance is likely to vary between projects. In order to efficiently implement a project with an optimized design flow system, engineering decisions become even more important since these key decisions affect every subsequent design phase. It is important to understand the design phases, their order, their relationships, and the decisions that must be made in each of these phases. This book addresses all of these topics.

"For our purposes here, we are interested in the chapter on Digital Signal processing. The rapid growth of communication and multimedia technologies over the last decade has dramatically expanded the range of digital signal processing (DSP) applications. The ongoing need to implement an increasingly complex algorithm at higher speeds and lower price points is a result of increasing demand for advanced information services, increased bandwidth, and expanded media handling capability.

"Some of the evolving high performance applications include advanced wired and wireless voice, data, and video processing. The growth of communications and multimedia applications such as internet communications, secure wireless communications, and consumer entertainment devices has driven the need for devices and structures capable of efficiently implementing complex math and signal processing algorithms.

"FPGAs are ideal for performing signal processing tasks. However, implementing DSP functionality within an FPGA requires the right combination of algorithm, FPGA

architecture, tool set, and design flow. This chapter addresses the architectural features developed for implementing DSP functions within FPGAs, and provides an overview as to which algorithms can be more efficiently implemented within FPGA using these features. Design flow, critical design decisions, terminology, numeric representation, and arithmetic operations are considered. Performance and implementation cost trade-offs, available DSP IP and design verification and debug approaches are also discussed."

—Clive "Max" Maxfield

7.1　Overview

The rapid growth of communication and multimedia technologies over the last decade has dramatically expanded the range of digital signal processing (DSP) applications. The ongoing need to implement an increasingly complex algorithm at higher speeds and lower price points is a result of increasing demand for advanced information services, increased bandwidth and expanded media handling capability. Some of the evolving high performance applications include advanced wired and wireless voice, data and video processing.

The growth of communications and multimedia applications such as internet communications, secure wireless communications and consumer entertainment devices has driven the need for devices and structures capable of efficiently implementing complex math and signal processing algorithms.

Some typical DSP algorithms required by these applications include fast Fourier transform (FFT), discrete cosine transform (DCT), Wavelet Transform, and digital filters (finite impulse response (FIR), infinite impulse response (IIR) and adaptive filters), and digital up and down converter. Each of these algorithms have structural elements that may be implemented with parallel functionality. FPGA architectures are able to implement parallel architectures efficiently.

FPGA architectures include resources capable of more advanced, higher-performance signal processing with each new FPGA device family. FPGA technology supports an increasing range of complex math and signal processing intellectual property. Advances in tool integration now support simplified system-level design. With front-end tools such as MATLAB™, pushbutton conversion from block-level system design to HDL-level code is possible. Chip density and process technology advances also support larger, more capable signal processing implementations.

FPGA implementation provides the added benefits of reduced NRE costs along with design flexibility and future design modification options. However, implementing DSP functionality within an FPGA requires the right combination of algorithm, FPGA architecture, tool set and design flow. This chapter addresses the architectural features developed for implementing DSP functions within FPGAs, and an overview of which algorithms can be more efficiently implemented within FPGA using these features. Design flow, critical design decisions, terminology, numeric representation, and arithmetic operations are discussed. Performance and implementation cost trade-offs, available DSP IP and design verification and debug approaches are also discussed.

7.2 Basic DSP System

This section presents a high-level overview of a typical DSP system and its critical elements. Figure 7.1 shows a typical DSP system implementation. The digital portion of the system is from the output of the analog-to-digital converter (ADC) through the DSP system and into the digital-to-analog converter (DAC). The remainder of the system is in the analog domain.

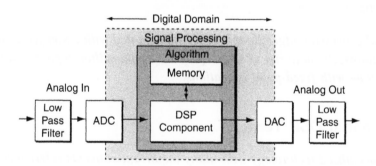

Figure 7.1: Basic DSP system

The ADC is responsible for converting the system input signal from the analog-to-digital domain. The ADC must be preceded by a low pass filter (LPF) based on the relationship between sampling speed and frequency described by the Nyquist sampling theorem. The LPF is required to limit the maximum frequency presented to the ADC to less than half of the ADC's sampling rate. This pre-filtering is known as anti-aliasing. Anti-aliasing prevents ambiguous data relationships known as aliasing from being translated into the digital domain.

The output of the ADC is a stream of sampled fixed-word-length values that represent the analog input signal at the discrete sample points determined by the ADC's sampling frequency. Each of these data samples is represented by a fixed-length binary word. The resolution of these samples is limited to the output data word width of the ADC and the data representation width internal to the DSP system. The ADC outputs are quantized representations of the input sampled analog values. This simply means that a value that has been translated from the analog domain (where the range of possible analog values occupies an infinite number of possible values with no word length limit) must be represented by one of a limited number of possible values in a finite word length system. The signal processing functionality that is most commonly implemented within FPGA components occurs in the digital domain.

The maximum number of values available to represent an individual data sample is 2^N, where N is the number of fixed bits of the word width. In an example where N equals 16, the full possible numeric range is 0 to 65,535. With more bits available in a system, the accuracy of the digital representation of the analog sample is improved. The difference between the original analog signal value and the quantized N-bit value is called *quantization error*.

While signal processing algorithms can be implemented with either fixed- or floating-point operations, the majority of signal processing algorithm implementations within FPGAs is done with fixed-point operations.

7.3 Essential DSP Terms

DSP is a specialized technology with many important concepts referenced by acronyms and specialized terms. Table 7.1 provides definitions for important DSP terms and abbreviations. Expanded definitions of these terms may be found in most DSP reference books. These terms will allow us to examine some elements of DSP design with FPGA components.

7.4 DSP Architectures

Many DSP algorithms require repetitive use of the operation group shown in Figure 7.2. This is clearly a multiply and addition operation group, also known as a multiply and accumulate (MAC) block.

Table 7.1: DSP terminology

Term	Definition
Accuracy	Magnitude of the difference between an element's real value and its represented value.
Complex math	Math performed on Complex numbers. Complex numbers have a real and imaginary part. Used in a wide range of DSP applications. How a DSP system performs complex arithmetic is a common benchmark for DSP.
CORDIC	(COordinate Rotation DIgital Computer) Algorithm to calculate trigonometric functions (sine, cosine, magnitude, and phase).
Decimation	The process of sample rate reduction. A digital low-pass filter may be used to remove samples.
DFT	(Discrete Fourier Transform) The digital form of the Fourier transform. The DFT result is a complex number.
DSP block (FPGA)	Term used to describe specialized circuitry within an FPGA optimized for implementing math intensive functions.
Dynamic range	Ratio of the maximum absolute value that can be represented and the minimum absolute value that can be represented.
FFT	(Fast Fourier Transform) An algorithm used to solve the DFT.
Filter coefficients	The set of constants (also called tap weights) are multiplied against filter data values within a filter structure.
Filter order	Equal to the number of delayed data values that must be stored in order to calculate a filter's output value.
Finite impulse response filter (FIR)	A class of nonrecursive digital filters with no internal data feedback paths. An FIR filter's output values will eventually return to zero after an input impulse. FIR filters are unconditionally stable.
Fixed-point	Architecture based on representing and operating on numbers represented in integer format.
Floating-point	Architecture based on representing and operating on numbers represented in floating-point format.
Floating-point format	Numerical values are represented by a combination of a mantissa (fractional part) and an exponent.
Format	Digital-system numeric representation style; fixed-point or floating-point.

(Continued)

Table 7.1: DSP terminology (Cont'd)

Term	Definition
Impulse response	A digital filter's output sequence after a single-cycle impulse (maximum value) input where the impulse is preceded and followed by an infinite number of zero-valued inputs.
Infinite impulse response (IIR) filter	A class of recursive digital filters with internal data feedback paths. An IIR filter's output values do not ever have to return to zero (theoretically) after an input impulse; however, in practice, output values do eventually reach negligibly small values. This filter form is prone to instability due to the feedback paths.
Interpolation	The process of increasing the sample rate. Up-sampling typically stuffs zero value samples between the original samples before digital filtering occurs.
Limit cycle effect	A filter's output will decay down to a specific range and then exhibit continuing oscillation within a limited amplitude range if the filter input is presented nonzero-value inputs (excited) followed by a long string of zero-value inputs.
Multirate	Data processing where the clock rate is not fixed. The clock rate may either be increased (interpolated), decreased (decimated) or re-sampled.
Overflow	A computation with a result number larger than the system's defined dynamic range or addition of numbers of like sign resulting in an output with an incorrect sum or sign; also called register overflow, large signal limit cycling or saturation.
Precision	Number of bits used to represent a value in the digital domain, also called bus width or fixed-word length.
Q-format	Format for representing fractional numbers within a fixed-length binary word. The designer assigns an implied binary point, which divides the fractional and integer numeric fields.
Quantization error	Difference in accuracy of representation of a signal's value in the analog domain and digital domain in a fixed-length binary word.
Radix point	Equivalent to a decimal point in base-10 math or a binary point in base-2 math; separates integer and fractional numeric fields.
Range	Difference between the most negative number and most positive number that can represent a value; ultimately determined by both numeric representation format and precision.

Table 7.1: DSP terminology (Cont'd)

Term	Definition
Recursive filter	A filter structure in which feedback takes place, and previous input and output samples are used in the calculation of the current filter output value.
Representation	Definition of how numbers are represented, including one's complement, two's complement, signed and unsigned.
Re-sampling	Re-sampling is the process of changing the sampling rate. May be achieved through a combination of decimation and interpolation.
Resolution	The smallest nonzero magnitude which can be represented.
Round-off error	Another term for truncation error.
Saturation level	The maximum value expected at the output.
Scaling	Adjusting the magnitude of a value; typically accomplished by multiplication or shifting the binary (radix) point. May be used to avoid over-flow and under-flow conditions.
Tap	An operation within a filter structure which multiplies a filter coefficient times a data value. The data value can be a current or delayed input, output or intermediate value.
Truncation error	Loss of numeric accuracy required when a value must be shortened or truncated to fit within a fixed-word length.
Word length effects	Errors and effects associated with reduced accuracy representation of numerical values within a fixed-word length.

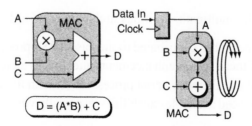

Figure 7.2: MAC block

MAC operations are heavily used in many DSP algorithms. The MAC operation is usually implemented within an iterative cycle. As the number of MAC operations that must be performed increases, system performance decreases.

7.5   Parallel Execution in DSP Components

Traditional sequential instruction DSP processors have evolved toward architectures that allow them to implement a broad range of DSP functions at an affordable price point. The majority of currently available popular DSP processors are inherently general-purpose devices by design. The development of DSP processors optimized for maximum performance of highly specialized functions has been accomplished by adding hardware accelerators such as Viterbi or Turbo coders to offload dedicated functions commonly seen in specific systems such as those used in high performance wired and wireless applications.

DSP processor suppliers have conventionally worked to improve performance by:

- Increasing clock cycle speeds

- Increasing the number of operations performed per clock cycle

- Adding optimized hardware coprocessing functionality (such as a Viterbi decoder)

- Implementing more complex (VLIW) instruction sets

- Minimizing sequential loop cycle counts

- Adding high-performance memory resources

- Implementing modifications, including deeper pipelines and superscalar architectural elements

Figure 7.3 illustrates the multipath, multibus architecture of a discrete DSP processor.

Each of these enhancements has contributed to increased DSP processor performance improvements. Each of these design enhancements attempts to increase the parallel processing capability of an inherently serial process. Even with these added features, DSP use has consistently outpaced available capabilities, especially when parallelism is required.

Algorithms that are inherently parallel process-oriented can stand to gain significant performance increases by migrating to a parallel process-oriented architecture.

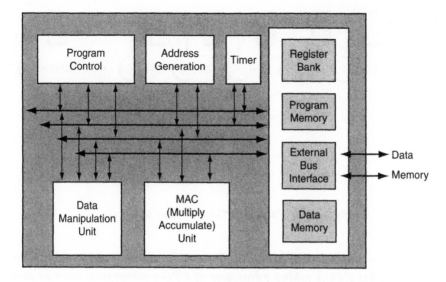

Figure 7.3: Discrete DSP implementing FIR filter

The traditional solution for increasing performance in discrete DSP processors has been to increase system clock speeds. However, even with high-clock rates, two MAC units and a modified Harvard bus architecture, there is a maximum level of performance that can be achieved. Higher levels of performance may potentially be achieved by implementing additional MAC units. ***FPGA components have been architected to support efficient parallel MAC functional implementations.***

7.6 Parallel Execution in FPGA

Higher-performance, resource-hungry, MAC-intensive DSP algorithms may benefit from implementation within FPGA components. FPGA architectural enhancements, development tool flow advances, speed increases and cost reductions are making implementation within FPGAs increasingly attractive. FPGA technology advances include increased clock speeds, specialized DSP blocks, tool enhancements and an increasing range of intellectual property solutions. Figure 7.4 illustrates an example parallel implementation of an FIR filter within an FPGA.

The MAC operational group may be implemented in one of several different configurations within an FPGA. ***Three popular implementation options for the MAC operational group within an FPGA are listed below.***

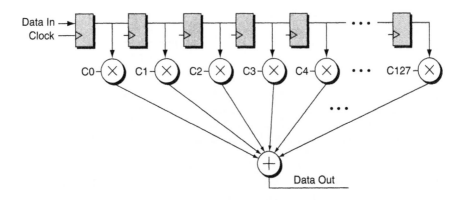

Figure 7.4: Parallel FPGA FIR filter structure

- *Both the multiplier and the accumulator may be implemented within the logic fabric of the FPGA taking advantage of FPGA structures, such as dedicated high-speed carry chains*

- *The multiplier may be implemented in an optimized multiplier block, avoiding use of FPGA fabric logic with the accumulator implemented within the logic fabric of the FPGA*

- *Both the multiplier and accumulator may be implemented within an advanced multiplier block requiring the use of no FPGA logic*

Figure 7.5 illustrates the three different MAC implementation options.

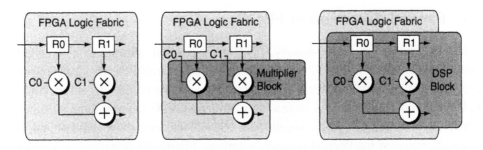

Figure 7.5: Three primary MAC implementation options

Each of these approaches has its own characteristics. The decision to use any of these approaches will be heavily dependent on the architecture of the FPGA fabric, the algorithms being implemented, the performance required and the amount of functionality being implemented on the FPGA component. For example, older device families may not support the integrated accumulator function within the DSP block. In this situation, the DSP block is actually just a multiplier. Likewise, if all the DSP blocks have been used for higher-performance algorithms, it may be possible to implement an algorithm with no DSP blocks within the FPGA logic fabric. FPGA DSP blocks are generally implemented in either a column or row structure within the FPGA fabric.

Different manufacturers have implemented significantly different DSP block architectures. Figure 7.6 illustrates simplified example DSP block architecture.

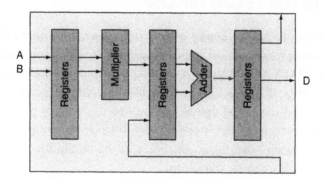

Figure 7.6: Simplified DSP block architecture

FPGA architectures are inherently oriented toward the implementation of parallel structures. FPGAs are also capable of implementing operations on wide and very wide data buses. They are also capable of implementing complex math operations in parallel.

Generally multiple options exist for implementing individual DSP-related operations within an FPGA. The structures that FPGA manufacturers have developed and continue to optimize for signal processing operations include:

- Advanced hardware multiplier blocks with associated accumulator functionality

- DSP blocks containing internal registers

- Common data widths natively supported ($\times 9$, $\times 18$, $\times 36$)

- Operational modes to perform various mathematical operations such as complex arithmetic (commonly used in DSP algorithms)

- Distributed and block memory within the FPGA

- Implementation of low-overhead shift registers

- Optimized clock management and distribution

- Access to memory external to the FPGA

DSP block implementations typically include dedicated control and carry logic circuitry and signal routing, allowing higher-performance, lower-overhead functional implementations of common signal processing functions.

7.7 When to Use FPGAs for DSP

Potential advantages to implementing a DSP function within an FPGA include performance improvements, design implementation flexibility, and higher system-level integration. FPGA-based signal processing performance may be improved through a combination of design adjustments. The operational speed or data path width may be increased, and sequential operations may be made more parallel in structure. Each of these will result in higher levels of performance. When an algorithm is implemented in a structure that takes advantage of the flexibility of the target FPGA architecture, the benefits can be significant.

Typically, there are a wider range of implementation options for signal processing algorithms within an FPGA than with fixed-function components. The design team must prioritize their design objectives. For example, it may be possible to implement an algorithm as a maximally parallelized architecture, or the algorithm could be implemented within a fully serial architecture. The serial structure would require the implementation of a loop counter function within an associated hardware counter. Another design option is a hybrid approach called a *semi-parallel structure*. The semi-parallel structure has elements of both the full-parallel and full-serial approaches. The algorithm would be separated into multiple parallel structures; however, multiple iterations would be required through each structure for each algorithm cycle. This contrasts with the single iteration required in the full-parallel approach and the maximum number of possible iterations with the full-serial approach. Each of these algorithm implementations will have a different level of performance, design effort and resource requirements. The design team has the flexibility to optimize for size, speed, cost or a targeted combination of these factors. Algorithms may also be reconfigured to dynamically meet changing operational requirements.

FPGA components also provide an opportunity to integrate multiple design functions into a single package. Functional integration may result in higher performance, and reduced real estate and power requirements. The resources integrated into the I/O blocks of FPGAs may improve system performance by allowing the design team control of device drive strength, signal slew rate, and on-chip signal termination. These options can optimize system performance and reduce board-level component count.

Another potential signal processing implementation advantage is the availability of preverified signal processing algorithms. IP cores and blocks can be used to efficiently implement common signal processing functions at the highest levels of performance. The ability to integrate multiple high-performance signal processing algorithms efficiently can potentially reduce project, cost, risk and schedule.

7.8   FPGA DSP Design Considerations

Some of the FPGA design issues that are important to signal processing algorithm implementation are presented below. These design factors must be carefully implemented in order to achieve the highest levels of performance and fastest design implementation.

- Synchronous design implementation

- Modular project structure

- Clock boundary transitions

- Clock architecture implementation

- Critical clock and control signal routing

- Pipeline depth and structure

- Effective design constraint

- Signal processing algorithm architecture decisions

- Incorporation of debug-friendly features

Many of these topics were covered in the first part of this book as standard FPGA design topics, but have particular impact on signal processing applications. A few topics that warrant additional discussion are covered in the following sections.

7.8.1 Clocking and Signal Routing

Many signal processing applications are performance limited. In other words, the faster they can run, the better. This makes the implementation of clocks and clock management critical to DSP functions. *Many of the most critical signal processing operations are directly affected by the clock architecture implementation of the design.* Important clock-related design factors that should be implemented with care include:

- Sufficient board-level device decoupling

- Clean low-jitter external clock sources (consider differential clock distribution for higher rate clocks)

- Careful clock source routing to the appropriate dedicated FPGA I/O pins

- Prioritized assignment (via constraints) of critical clocks to global resources

- Careful design analysis for clock function conflict

Signal processing functionality should be directed toward implementation within the optimized DSP blocks. If there are not enough DSP blocks to implement all of the desired signal processing functions within the available DSP blocks, then the algorithms with the highest level of required performance or largest amount of equivalent logic fabric to implement should be targeted toward the available DSP blocks. Design constraints can be used to guide the tools to place the desired functionality within the appropriate dedicated FPGA resources.

The design implementation layout or report file should be regularly checked to verify that the targeted functionality has been placed into the correct FPGA resources. This also applies to math function related signal routing such as carry logic. While the tools usually correctly identify and assign these signals it is possible for them to be assigned to regular priority logic fabric routing which can significantly reduce the level of performance which can be achieved.

7.8.2 Pipelining

Pipelining is an essential element of implementing high-speed signal processing algorithms. The register-rich nature of FPGA architectures naturally supports register-intensive algorithm implementations. The efficient implementation of signal processing algorithms within FPGA components is based on efficient implementation of the low-level algorithm arithmetic operations. These operations may be separated from each other by registers. The addition of registers in between math operations allows higher speeds of operation. Adding registers into the design is similar to higher-level architectural design partitioning. Adding registers to the design will result in a "deeper" pipeline through the design. The resource penalty of additional registers allows the highest level of performance possible.

7.8.3 Algorithm Implementation Choices

The wide range of potential algorithm implementation options with FPGA components will require the design team to run a number of design trade-off studies. *The most important design factors affecting DSP block resource allocation include the number of algorithms, which can benefit from DSP blocks, the number of available DSP blocks and associated block memories, the level of performance required for individual algorithms and the type of algorithm implemented.* Another design factor is how algorithm coefficients will be used and stored within the design. For fully-serial and semi-parallel algorithm implementations, if fixed coefficients are required, then shift registers may be used to store the coefficients saving valuable block memory resources for other functions. The design team will need to make architectural decisions regarding full-serial, semi-parallel or full-parallel for individual algorithm implementations since the tools may not be able to efficiently find an optimized implementation solution. *A final consideration is to ensure that all the available DSP blocks have been used. Implementing functionality within the DSP blocks results in higher performance and lower power consumption.*

7.8.4 DSP Intellectual Property (IP)

There are a wide range of potential signal processing algorithms. Some of the most popular functions have been implemented as intellectual property blocks. *IP blocks provide access to optimized, preverified DSP functionality.* Signal processing IP may be obtained from multiple sources including manufacturer, third-party, and open-access sources. IP designs that have been optimized for a particular FPGA architecture may often be found on an FPGA manufacturer's website.

There are several broad DSP IP categories. The categories divide into two groups: operational-level implementations and application-level implementations. Some of the most popular DSP IP categories and example algorithms are listed in Table 7.2.

Table 7.2: DSP IP categories

Group	Category	Example
Operational	Math Function	CORDIC, Parallel Multiplier, Pipelined Divider
Operational	base Function	Shift Register, Accumulator, Comparator, Adder
Application	DSP Function	Viterbi Decoder, FFT, MAC, FIR, Discrete Cosine Transform
Application	Memory Function	DDR-I/II Controller, ZbT Controller, Flash Controller
Application	Image Processing	Color Space Converter, JPEG Motion Encoder
Application	Communication	AES Encryption, Reed-Solomon Encoder, Turbo Decoder

7.9 FIR Filter Concept Example

In this section, we will consider the design and implementation of a high-performance FIR filter within an FPGA. The intent is to implement a low-pass FIR filter. The first step to implementing an FIR filter is the calculation of the number of taps. The calculation of the number of taps for the type of filter being implemented is shown below.

The number of filter taps is determined by the transition band and the desired stop band attenuation.

General Formulas
BW = Bandwidth
Transition_BW_Hz
Normalized_Transition_BW = Transition_BW_Hz / Sampling_Frequency_Hz
K(Attenuation_dB) = Attenuation_dB / 22 dB
Number_of_Taps = K(Attenuation_dB) / Normalized_Transition_BW

Filter Parameters
Sampling_Frequency = 16,000 Hz
Pass_Band = 0 Hz to 3,800 Hz
Transition_Band = 3,800 to 4,200 Hz
Stop_Band = 4,200 Hz to 5700 Hz
Stop_Band_Attenuation = 70 dB

Filter Calculations
Transition_BW_Hz = 400 Hz
Normalized_Transition_BW = 400 Hz / 16,000 Hz
Attenuation_dB = 70 db
K(Attenuation_dB) = 70 dB / 22 dB
Number_of_Taps = (70/22)/(400/16000)
Number_of_Taps = 128 (Rounded Up)

The first part of Figure 7.7 illustrates the implementation of the FIR filter in a fully-parallel DSP block implementation. With the implementation of the filter in a fully-parallel structure with fixed coefficients no block RAM elements will be required. If the design was implemented as a serial or semi-parallel structure, block RAM could be used to store filter coefficients. Depending on the operational speed, a distributed RAM implementation using LUT memory elements also could have been chosen for storing the filter coefficients.

The second part of Figure 7.7 illustrates an FIR filter implemented in a transpose filter structure.

As a further extension of this example, the implemented signal processing algorithm could be interfaced to an FPGA embedded processor. Xilinx application note XAPP717 presents an efficient method for interfacing an FPGA implemented signal processing algorithm with an embedded 405 hard processor IP through an APU block. This effectively allows the tight coupling of a DSP algorithm to a processor core as a DSP

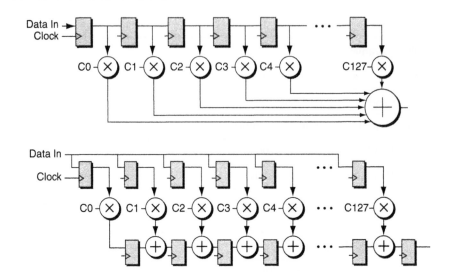

Figure 7.7: FIR filter example

coprocessor function. In an advanced application, an embedded processor core could update the filter coefficients efficiently if they are implemented in a block RAM element.

7.10   Summary

While traditional discrete DSP components provide a good balance of performance to cost, and have familiar development flows, advances in FPGA technology are providing an attractive alternative for implementing signal processing algorithms at an attractive price point. Traditional limitations to signal processing implementation on FPGAs are being addressed at the hardware and software design levels. FPGA hardware architectures are implementing enhanced DSP blocks with more functionality and higher performance. System-level design software is simplifying the process of translating designs from the block level to the HDL level code defining the hardware implementation. Integration with popular DSP algorithm development tools such as MATLAB™ continues to simplify the implementation of signal processing algorithms in FPGAs.

The MAC operational group may be implemented in three different configurations within an FPGA: both the multiplier and the accumulator in the logic fabric, the multiplier in a hard multiplier block with the adder in the logic fabric, or both the multiplier and the adder in the hard DSP block. The implementation chosen will be dependent on the algorithms implemented and the specialized DSP block resources available.

The following design factors should be given extra consideration when implementing signal processing algorithms. Some of these topics were discussed in the first part of the book and some are covered in more detail within this chapter.

- Synchronous design implementation

- Modular project structure

- Clock boundary transitions

- Clock architecture implementation

- Critical clock and control signal routing

- Pipeline depth and structure

- Effective design constraint

- Signal processing algorithm architecture decisions

- Incorporation of debug-friendly features

The chapter has presented a number of topics related to implementing DSP functionality within FPGAs, including when FPGA technology may be an attractive alternative to general-purpose DSP processors. Project design teams can benefit from using system-level design tools and implementing a hierarchical design block simulation flow, verifying elements before they are integrated into higher levels of functionality. by developing an understanding of the process for integrating signal processing algorithms, understanding the features of available DSP algorithm development tools, and making informed implementation decision, a design team can implement effective signal processing algorithms efficiently within FPGAs.

CHAPTER 8

Basics of Embedded Audio Processing

David Katz
Rick Gentile

This (and the following) chapter comes from the book Embedded Media Processing *by David Katz and Rick Gentile, ISBN: 9780750679121. When Intel Corporation and Analog Devices collaborated to introduce the Micro Signal Architecture (MSA) in 2000, the result was a processor uniting microcontroller (MCU) and DSP functionality into a single high-performance "convergent processing" device—a rather unique entry into the marketplace.*

What transpired with the MSA-based Blackfin® family was that a new class of application spaces opened up, from video-enabled handheld gadgets, to control-intensive industrial applications, to operating system/number-crunching combos running on a single processor core.

As a result, users needed to know how to partition memory intelligently, organize data flows, and divide processing tasks across multiple cores. From the authors' interactions with customers, it became clear that the same types of complex questions kept surfacing, so they said, "Why not encapsulate these basic ideas in a book?" They focused the text on embedded media processing (EMP) because this is the "sweet spot" today of high-performance, low-power, small-footprint processing—an area in which the Blackfin® processor sits squarely. EMP is a relatively new application area, because it requires performance levels only recently attainable.

This chapter, "Basics of Embedded Audio Processing," serves as a starting point for any embedded system that requires audio processing. The authors first provide a short introduction to the human auditory system and sampling theory. Next, they describe how to connect your embedded processor to audio converters or other inputs and outputs.

They devote a detailed section to the importance of dynamic range and data-type emulation on fixed-point processors. Subsequently, they discuss audio processing building blocks, basic programming techniques and fundamental algorithms. The chapter concludes with an overview of the most widely used audio and speech compression standards.

—**Clive "Max" Maxfield**

8.1   Introduction

Audio functionality plays a critical role in embedded media processing. While audio takes less processing power in general than video processing, it should be considered equally important.

In this chapter, we will begin with a discussion of sound and audio signals, and then explore how data is presented to the processor from a variety of audio converters. We will also describe the formats in which audio data is stored and processed.

Additionally, we'll discuss some software building blocks for embedded audio systems. Efficient data movement is essential, so we will examine data buffering as it applies to audio algorithms. Finally, we'll cover some fundamental algorithms and finish with a discussion of audio and speech compression.

8.1.1   What Is Sound?

Sound is a longitudinal displacement wave that propagates through air or some other medium. Sound waves are defined using two attributes: amplitude and frequency.

The amplitude of a sound wave is a gauge of pressure change, measured in decibels (dB). The lowest sound amplitude that the human ear can perceive is called the "threshold of hearing," denoted by 0 dBSPL. On this SPL (sound pressure level) scale, the reference pressure is defined as 20 microPascals (20 µPa). The general equation for dBSPL, given a pressure change x, is

$$\text{dBSPL} = 20 \times \log(x \text{ µPa}/20 \text{ µPa})$$

Table 8.1 shows decibel levels for typical sounds. These are all relative to the threshold of hearing (0 dBSPL).

Table 8.1: Decibel (dBSPL) values for various typical sounds

Source (Distance)	dBSPL
Threshold of hearing	0
Normal conversation (3-5 feet away)	60–70
Busy traffic	70–80
Loud factory	90
Power saw	110
Discomfort	120
Threshold of pain	130
Jet engine (100 feet away)	150

The main point to take away from Table 8.1 is that the range of tolerable audible sounds is about 120 dB (when used to describe ratios without reference to a specific value, the correct notation is dB without the SPL suffix). Therefore, all engineered audio systems can use 120 dB as the upper bound of dynamic range. In case you're wondering why all this is relevant to embedded systems, don't worry—we'll soon relate dynamic range to data formats for embedded media processing.

Frequency, the other key feature of sound, is denoted in hertz (Hz), or cycles per second. We can hear sounds in the frequency range between 20 and 20,000 Hz, but this ability degrades as we age.

Our ears can hear certain frequencies better than others. In fact, we are most sensitive to frequencies in the area of 2–4 kHz. There are other quirky features about the ear that engineers are quick to exploit. Two useful phenomena, employed in the lossy compression algorithms that we'll describe later, are *temporal masking* and *frequency masking*. In temporal masking (Figure 8.1a), loud tones can drown out softer tones that occur at almost the same time. Frequency masking (Figure 8.1b) occurs when a loud sound at a certain frequency renders softer sounds at nearby frequencies inaudible. The human ear is such a complex organ that only books dedicated to the subject can do it justice. For a more in-depth survey of ear physiology, consult Reference 1.

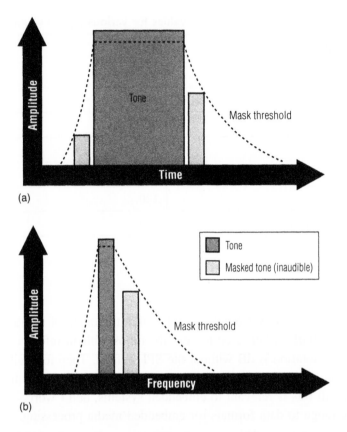

Figure 8.1: (a) Loud sounds at a specific time can mask out softer sounds in the temporal vicinity (b) Loud sounds at a specific frequency can mask softer sounds at nearby frequencies

8.1.2 Audio Signals

In order to create an analog signal that represents a sound wave, we must use a transducer to convert the mechanical pressure energy into electrical energy. A more common name for this audio source transducer is a microphone.

All transducers can be described with a sensitivity (or transduction) curve. In the case of a microphone, this curve dictates how well it translates pressure into an electrical signal. Ideal transducers have a linear sensitivity curve—that is, a voltage level is directly proportional to a sound wave's pressure.

Since a microphone converts a sound wave into voltage levels, we now need to use a new decibel scale to describe amplitude. This scale, called dBV, is based on a reference point of 1V. The equation describing the relationship between a voltage level x and dBV is

$$\text{dBV} = 20 \times \log(x \text{ volts}/1.0 \text{ volts})$$

An alternative analog decibel scale is based on a reference of 0.775V and uses dBu units. In order to create an audible mechanical sound wave from an analog electrical signal, we again need to use a transducer. In this case, the transducer is a speaker or headset.

8.1.3 Speech Processing

Speech processing is an important and complex class of audio processing, and we won't delve too deeply here. However, we'll discuss speech-processing techniques where they are analogous to the more general audio processing methods. The most common use for speech signal processing is in voice telecommunications for such algorithms as echo cancellation and compression.

Most of the energy in typical speech signals is stored within less than 4 kHz of bandwidth, thus making speech a subset of audio signals. However, many speech-processing techniques are based on modeling the human vocal tract, so these cannot be used for general audio processing.

8.2 Audio Sources and Sinks

8.2.1 Converting Between Analog and Digital Audio Signals

Assuming we've already taken care of converting sound energy into electrical energy, the next step is to digitize the analog signals. This is accomplished with an analog-to-digital converter (A/D converter or ADC). As you might expect, in order to create an analog signal from a digital one, a digital-to-analog converter (D/A converter or DAC) is used. Since many audio systems are really meant for a full-duplex media flow, the ADC and DAC are available in one package called an "audio codec." The term codec is used here to mean a discrete hardware chip. As we'll discuss later in the section on audio compression, this should not be confused with a software audio codec, which is a software algorithm.

All A/D and D/A conversions should obey the Shannon-Nyquist sampling theorem. In short, this theorem dictates that an analog signal must be sampled at a rate (Nyquist sampling rate) equal to or exceeding twice its highest-frequency component (Nyquist frequency) in order for it to be reconstructed in the eventual D/A conversion. Sampling below the Nyquist sampling rate will introduce aliases, which are low frequency "ghost" images of those frequencies that fall above the Nyquist frequency. If we take a sound signal that is band-limited to 0–20 kHz, and sample it at 2×20 kHz = 40 kHz, then the Nyquist Theorem assures us that the original signal can be reconstructed perfectly without any signal loss. However, sampling this 0–20 kHz band-limited signal at anything less than 40 kHz will introduce distortions due to aliasing. Figure 8.2 shows how sampling at less than the Nyquist sampling rate results in an incorrect representation of a signal. When sampled at 40 kHz, a 20 kHz signal is represented correctly (Figure 8.2a). However, the same 20 kHz sine wave that is sampled at a 30 kHz sampling rate actually looks like a lower frequency alias of the original sine wave (Figure 8.2b).

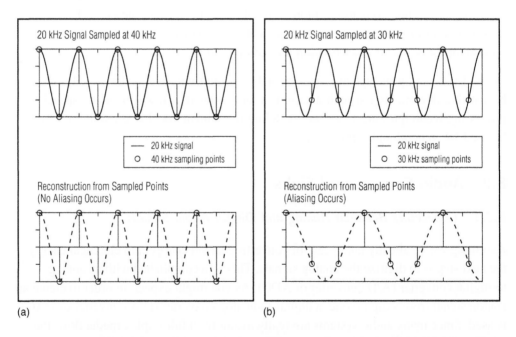

(a)   (b)

Figure 8.2: (a) Sampling a 20 kHz signal at 40 kHz captures the original signal correctly (b) Sampling the same 20 kHz signal at 30 kHz captures an aliased (low frequency ghost) signal

No practical system will sample at exactly twice the Nyquist frequency, however. For example, restricting a signal into a specific band requires an analog low-pass filter, but these filters are never ideal. Therefore, the lowest sampling rate used to reproduce music is 44.1 kHz, not 40 kHz, and many high-quality systems sample at 48 kHz in order to capture the 20 Hz–20 kHz range of hearing even more faithfully. As we mentioned earlier, speech signals are only a subset of the frequencies we can hear; the energy content below 4 kHz is enough to store an intelligible reproduction of the speech signal. For this reason, telephony applications usually use only 8 kHz sampling ($= 2 \times 4$ kHz). Table 8.2 summarizes some sampling rates used by common systems.

Table 8.2: Commonly used sampling rates

System	Sampling Frequency
Telephone	8000 Hz
Compact Disc	44100 Hz
Professional Audio	48000 Hz
DVD Audio	96000 Hz (for 6-channel audio)

The most common digital representation for audio is a pulse-code-modulated (PCM) signal. In this representation, an analog amplitude is encoded with a digital level for each sampling period. The resulting digital wave is a vector of snapshots taken to approximate the input analog wave. All A/D converters have finite resolution, so they introduce quantization noise that is inherent in digital audio systems. Figure 8.3 shows a PCM representation of an analog sine wave (Figure 8.3a) converted using an ideal A/D converter, in which the quantization manifests itself as the "staircase effect" (Figure 8.3b). You can see that lower resolution leads to a worse representation of the original wave (Figure 8.3c).

For a numerical example, let's assume that a 24-bit A/D converter is used to sample an analog signal whose range is –2.828V to 2.828V (5.656 Vpp). The 24 bits allow for 2^{24} (16,777,216) quantization levels. Therefore, the effective voltage resolution is 5.656V / 16,777,216 = 337.1 nV. Shortly, we'll see how codec resolution affects the dynamic range of audio systems.

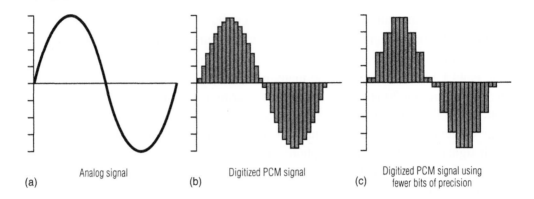

| | (a)   Analog signal | (b)   Digitized PCM signal | (c)   Digitized PCM signal using fewer bits of precision |

Figure 8.3: (a) An analog signal (b) Digitized PCM signal (c) Digitized PCM signal using fewer bits of precision

8.2.2   Background on Audio Converters

8.2.2.1   Audio ADCs

There are many ways to perform A/D conversion. One traditional approach is a successive approximation scheme, which uses a comparator to test the analog input signal against a number of interim D/A conversions to arrive at the final answer.

Most audio ADCs today, however, are sigma-delta converters. Instead of employing successive approximations to create wide resolutions, sigma-delta converters use 1-bit ADCs. In order to compensate for the reduced number of quantization steps, they are oversampled at a frequency much higher than the Nyquist frequency. Conversion from this super-sampled 1-bit stream into a slower, higher-resolution stream is performed using digital filtering blocks inside these converters, in order to accommodate the more traditional PCM stream processing. For example, a 16-bit 44.1 kHz sigma-delta ADC might oversample at 64x, yielding a 1-bit stream at a rate of 2.8224 MHz. A digital decimation filter (described in more detail later) converts this super-sampled stream to a 16-bit one at 44.1 kHz.

Because they oversample analog signals, sigma-delta ADCs relax the performance requirements of the analog low-pass filters that band-limit input signals. They also have the advantage of reducing peak noise by spreading it over a wider spectrum than traditional converters.

8.2.2.2 Audio DACs

Just as in the A/D case, sigma-delta designs rule the D/A conversion space. They can take a 16-bit 44.1 kHz signal and convert it into a 1-bit 2.8224 MHz stream using an interpolating filter (described later). The 1-bit DAC then converts the super-sampled stream to an analog signal.

A typical embedded digital audio system may employ a sigma-delta audio ADC and a sigma-delta DAC, and therefore the conversion between a PCM signal and an oversampled stream is done twice. For this reason, Sony and Philips have introduced an alternative to PCM, called Direct-Stream Digital (DSD), in their Super Audio CD (SACD) format. This format stores data using the 1-bit high-frequency (2.8224 MHz) sigma-delta stream, bypassing the PCM conversion. The disadvantage is that DSD streams are less intuitive to process than PCM, and they require a separate set of digital audio algorithms, so we will focus only on PCM in this chapter.

8.2.3 Connecting to Audio Converters

8.2.3.1 An ADC Example

OK, enough background information. Let's do some engineering now. One good choice for a low-cost audio ADC is the Analog Devices AD1871, which features 24-bit conversion at 96 kHz. The functional block diagram of the AD1871 is shown in Figure 8.4a. This converter has left (VINLx) and right (VINRx) input channels, which is really just another way of saying that it can handle stereo data. The digitized audio data is streamed out serially through the data port, usually to a corresponding serial port on a signal processor (like the SPORT interface on Blackfin processors). There is also an SPI (serial peripheral interface) port provided for the host processor to configure the AD1871 via software commands. These commands include ways to set the sampling rate, word width, and channel gain and muting, among other parameters.

As the block diagram in Figure 8.4b implies, interfacing the AD1871 ADC to a Blackfin processor is a glueless connection. The analog representation of the circuit is simplified, since only the digital signals are important in this discussion. The oversampling rate of the AD1871 is achieved with an external crystal. The Blackfin processor shown has two serial ports (SPORTs) and an SPI port used for connecting to the AD1871. The SPORT, configured in I^2S mode, is the data link to the AD1871, whereas the SPI port acts as the control link.

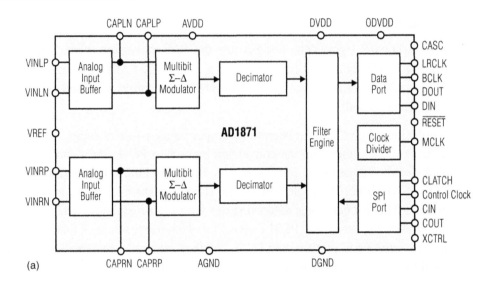

(a)

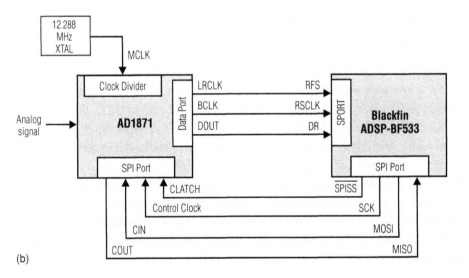

(b)

Figure 8.4: (a) Functional block diagram of the AD1871 audio ADC (b) Glueless connection of an ADSP-BF533 media processor to the AD1871

I²S (Inter-IC-Sound)   The I²S protocol is a standard developed by Philips for the digital transmission of audio signals. This standard allows for audio equipment manufacturers to create components that are compatible with each other.

In a nutshell, I^2S is simply a three-wire serial interface used to transfer stereo data. As shown in Figure 8.5a, it specifies a bit clock (middle), a data line (bottom), and a left/right synchronization line (top) that selects whether a left or right channel frame is currently being transmitted.

In essence, I^2S is a time-division-multiplexed (TDM) serial stream with two active channels. TDM is a method of transferring more than one channel (for example, left and right audio) over one physical link.

In the AD1871 setup of Figure 8.4b, the ADC can use a divided-down version of the 12.288 MHz sampling rate it receives from the external crystal to drive the SPORT clock (RSCLK) and frame synchronization (RFS) lines. This configuration insures that the sampling and data transmission are in sync.

SPI (Serial Peripheral Interface) The SPI interface, shown in Figure 8.5b, was designed by Motorola for connecting host processors to a variety of digital components. The entire interface between an SPI master and an SPI slave consists of a clock line (SCK), two data lines (MOSI and MISO), and a slave select (SPISELx) line. One of the data lines is driven by the master (MOSI), and the other is driven by the slave (MISO). In the example of Figure 8.4b, the Blackfin processor's SPI port interfaces gluelessly to the SPI block of the AD1871.

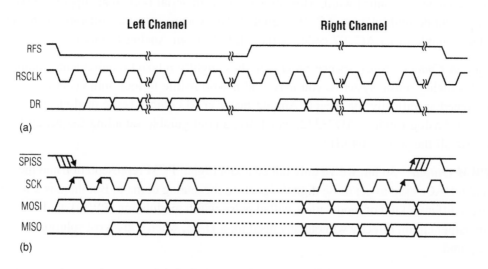

Figure 8.5: (a) The data signals transmitted by the AD1871 using the I^2S protocol
(b) The SPI interface used to control the AD1871

Audio codecs with a separate SPI control port allow a host processor to change the ADC settings on the fly. Besides muting and gain control, one of the really useful settings on ADCs like the AD1871 is the ability to place it in power-down mode. For battery-powered applications, this is often an essential function.

8.2.3.2 DACs and Codecs

Connecting an audio DAC to a host processor is an identical process to the ADC connection we just discussed. In a system that uses both an ADC and a DAC, the same serial port can hook up to both, if it supports bidirectional transfers.

But if you're tackling full-duplex audio, then you're better off using a single-chip audio codec that handles both the analog-to-digital and digital-to-analog conversions. A good example of such a codec is the Analog Devices AD1836, which features three stereo DACs and two stereo ADCs, and is able to communicate through a number of serial protocols, including I^2S.

AC '97 (Audio Codec '97) I^2S is only one audio specification. Another popular one is AC '97, which Intel Corporation created to standardize all PC audio and to separate the analog circuitry from the less-noise-susceptible digital chip. In its simplest form, an AC '97 codec uses a TDM scheme where control and data are interleaved in the same signal. Various timeslots in the serial transfer are reserved for a specific data channel or control word. Most processors with serial ports that support TDM mode can de-multiplex an AC '97 signal at the expense of some software overhead. One example of an AC '97 codec is the AD1847 from Analog Devices.

Speech Codecs   Since speech processing has slightly relaxed requirements compared to hi-fidelity music systems, you may find it worthwhile to look into codecs designed specifically for speech. Among many good choices is the dual-channel 16-bit Analog Devices AD73322, which has a configurable sampling frequency from 8 kHz all the way to 64 kHz.

PWM Output   So far, we've only talked about digital PCM representation and the audio DACs used to get those digital signals to the analog domain. But there is a way to use a different kind of modulation, called pulse-width modulation (PWM), to drive an output circuit directly without any need for a DAC, when a low-cost solution is required.

In PCM, amplitude is encoded for each sample period, whereas it is the duty cycle that describes amplitude in a PWM signal. PWM signals can be generated with general-purpose I/O pins, or they can be driven directly by specialized PWM timers, available on many processors.

To make PWM audio achieve decent quality, the PWM carrier frequency should be at least 12 times the bandwidth of the signal, and the resolution of the timer (i.e., granularity of the duty cycle) should be 16 bits. Because of the carrier frequency requirement, traditional PWM audio circuits were used for low-bandwidth audio, like subwoofers. However, with today's high-speed processors, it's possible to carry a larger audible spectrum.

The PWM stream must be low-pass-filtered to remove the high-frequency carrier. This is usually done in the amplifier circuit that drives a speaker. A class of amplifiers, called Class D, has been used successfully in such a configuration. When amplification is not required, then a low-pass filter is sufficient as the output stage. In some low-cost applications, where sound quality is not as important, the PWM streams can connect directly to a speaker. In such a system, the mechanical inertia of the speaker's cone acts as a low-pass filter to remove the carrier frequency.

8.3   Interconnections

Before we end this hardware-centric section, let's review some of the common connectors and interfaces you'll encounter when designing systems with embedded audio capabilities.

8.3.1   Connectors

Microphones, speakers, and other analog equipment connect to an embedded system through a variety of standard connectors (see Figure 8.6). Because of their small size, 1/8" connectors are quite common for portable systems. Many home stereo components support 1/4" connectors. Higher performance equipment usually uses RCA connectors, or even a coaxial cable connector, to preserve signal integrity.

8.3.2   Digital Connections

Some of the systems you'll design actually won't require any ADCs or DACs, because the input signals may already be digital and the output device may accept digital data. A few standards exist for transfer of digital data from one device to another.

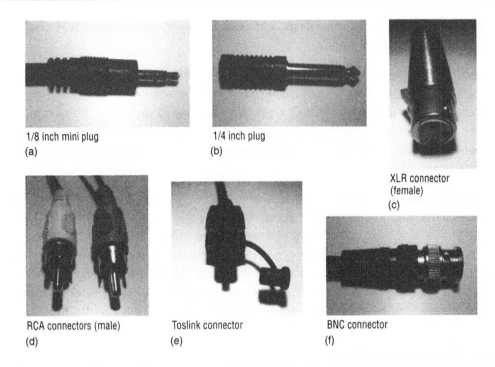

1/8 inch mini plug
(a)

1/4 inch plug
(b)

XLR connector
(female)
(c)

RCA connectors (male)
(d)

Toslink connector
(e)

BNC connector
(f)

Figure 8.6: Various audio connectors: (a) 1/8 inch mini plug (b) 1/4 inch plug (c) XLR connector (d) Male RCA connectors (e) Toslink connector (f) BNC connector

The *Sony Digital InterFace* (SDIF-2) protocol is used in some professional products. It requires an unbalanced BNC coaxial connection for each channel. The Audio Engineering Society (AES) introduced the AES3 standard for serial transmission of data; this one uses an XLR connector. A more ubiquitous standard, the S/PDIF (Sony/Philips Digital InterFace), is prevalent in consumer and professional audio devices. Two possible S/PDIF connectors are single-ended coaxial cable and the Toslink connector for fiber optic connections.

8.4 Dynamic Range and Precision

We promised earlier that we would get into a lot more detail on dynamic range of audio systems. You might have seen dB specs thrown around for various products available on the market today. Table 8.3 lists a few fairly established products along with their assigned signal quality, measured in dB.

Table 8.3: Dynamic range comparison of various audio systems

Audio Device	Typical Dynamic Range
AM Radio	48 dB
Analog TV	60 dB
FM Radio	70 dB
16-bit Audio Codecs	90–95 dB
CD Player	92–96 dB
Digital Audio Tape (DAT)	110 dB
20-bit Audio Codecs	110 dB
24-bit Audio Codecs	110–120 dB

So what exactly do those numbers represent? Let's start by getting some definitions down. Use Figure 8.7 as a reference diagram for the following discussion.

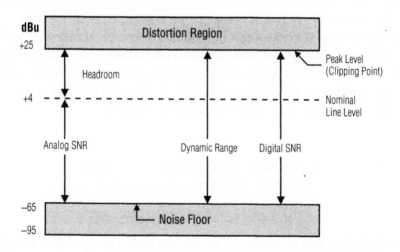

Figure 8.7: Relationship between some important terms in audio systems

As you might remember from the beginning of this chapter, the dynamic range for the human ear (the ratio of the loudest to the quietest signal level) is about 120 dB. In systems where noise is present, dynamic range is described as the ratio of the maximum signal level to the noise floor. In other words,

$$Dynamic\ Range\ (dB) = Peak\ Level\ (dB) - Noise\ Floor\ (dB)$$

The noise floor in a purely analog system comes from the electrical properties of the system itself. On top of that, audio signals also acquire noise from ADCs and DACs, including quantization errors due to the sampling of analog data.

Another important term is the signal-to-noise ratio (SNR). In analog systems, this means the ratio of the nominal signal to the noise floor, where "line level" is the nominal operating level. On professional equipment, the nominal level is usually 1.228 Vrms, which translates to +4 dBu. The headroom is the difference between nominal line level and the peak level where signal distortion starts to occur. The definition of SNR is a bit different in digital systems, where it is defined as the dynamic range.

Now, armed with an understanding of dynamic range, we can start to discuss how this is useful in practice. Without going into a long derivation, let's simply state what is known as the "6 dB rule." This rule holds the key to the relationship between dynamic range and computational word width. The complete formulation is described in equation (8.1), but it is used in shorthand to mean that the addition of one bit of precision will lead to a dynamic range increase of 6 dB. Note that the 6 dB rule does not take into account the analog subsystem of an audio design, so the imperfections of the transducers on both the input and the output must be considered separately. Those who want to see the statistical math behind the rule should consult Reference 1.

$$Dynamic\ Range\ (dB) = 6.02n + 1.76 \approx 6n\ dB$$
$$where\ n = the\ number\ of\ precision\ bits$$

The 6 dB rule dictates that the more bits we use, the higher the quality of the system we can attain. In practice, however, there are only a few realistic choices. Most devices suitable for embedded media processing come in three word-width flavors: 16-bit, 24-bit and 32-bit. Table 8.4 summarizes the dynamic ranges for these three types of processors.

Since we're talking about the 6 dB rule, it is worth noting something about nonlinear quantization methods typically used for speech signals. A telephone-quality linear PCM encoding requires 12 bits of precision. However, our ears are more sensitive to audio

Table 8.4: Dynamic range of various fixed-point architectures

Computation Word Width	Dynamic Range (Using 6 dB Rule)
16-bit fixed-point precision	96 dB
24-bit fixed-point precision	144 dB
32-bit fixed-point precision	192 dB

changes at small amplitudes than at high amplitudes. Therefore, the linear PCM sampling is overkill for telephone communications. The logarithmic quantization used by the A-law and μ–law companding standards achieves a 12-bit PCM level of quality using only 8 bits of precision. To make our lives easier, some processor vendors have implemented A-law and μ–law companding into the serial ports of their devices. This relieves the processor core from doing logarithmic calculations.

After reviewing Table 8.4, recall once again that the dynamic range for the human ear is around 120 dB. Because of this, 16-bit data representation doesn't quite cut it for high quality audio. This is why vendors introduced 24-bit processors that extended the dynamic range of 16-bit systems. The 24-bit systems are a bit nonstandard from a C compiler standpoint, so many audio designs these days use 32-bit processing.

Choosing the right processor is not the end of the story, because the total quality of an audio system is dictated by the level of the "lowest-achieving" component. Besides the processor, a complete system includes analog components like microphones and speakers, as well the converters to translate signals between the analog and digital domains. The analog domain is outside of the scope of this discussion, but the audio converters cross into the digital realm.

Let's say that you want to use the AD1871, the same ADC shown in Figure 8.4a, for sampling audio. The datasheet for this converter explains that it is a 24-bit converter, but its dynamic range is not 144 dB—it is 105 dB. The reason for this is that a converter is not a perfect system, and vendors publish only the useful dynamic range in their documentation.

If you were to hook up a 24-bit processor to the AD1871, then the SNR of your complete system would be 105 dB. The conversion error would amount to 144 dB – 105 dB = 39 dB. Figure 8.8 is a graphical representation of this situation. However, there is still another component of a digital audio system that we have not discussed yet: computation on the processor's core.

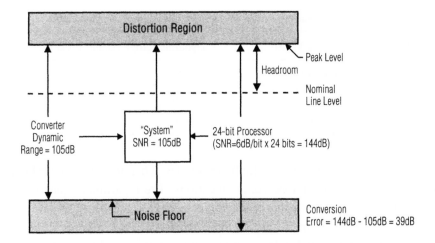

Figure 8.8: An audio system's SNR consists of the weakest component's SNR

Passing data through a processor's computation units can potentially introduce rounding and truncation errors. For example, a 16-bit processor may be able to add a vector of 16-bit data and store this in an extended-length accumulator. However, when the value in the accumulator is eventually written to a 16-bit data register, then some of the bits are truncated.

Take a look at Figure 8.9 to see how computation errors can affect a real system. If we take an ideal 16-bit A/D converter (Figure 8.9a), then its signal-to-noise ratio would be $16 \times 6 = 96$ dB. If arithmetic and storage errors did not exist, then 16-bit computation would suffice to keep the SNR at 96 dB. 24-bit and 32-bit systems would dedicate 8 and 16 bits, respectively, to the dynamic range below the noise floor. In essence, those extra bits would be wasted.

However, most digital audio systems do introduce some round-off and truncation errors. If we can quantify this error to take, for example, 18 dB (or 3 bits), then it becomes clear that 16-bit computations will not suffice in keeping the system's SNR at 96 dB (Figure 8.9b). Another way to interpret this is to say that the effective noise floor is raised by 18 dB, and the total SNR is decreased to 96 dB – 18 dB = 78 dB. This leads to the conclusion that having extra bits below the converter's noise floor helps to deal with the nuisance of quantization.

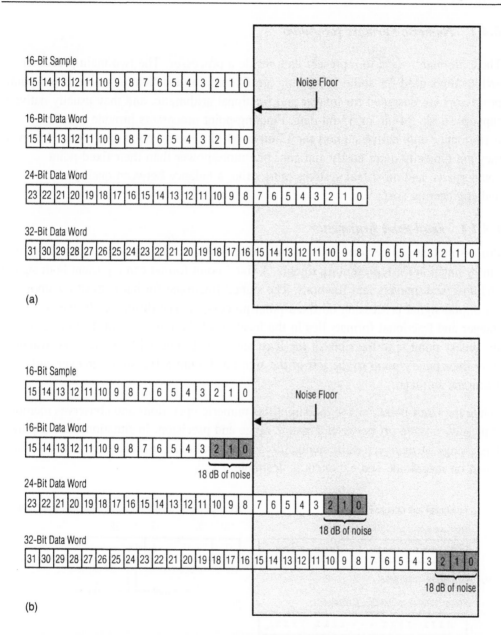

Figure 8.9: (a) Allocation of extra bits with various word-width computations for an ideal 16-bit, 96 dB SNR system, when quantization error is neglected (b) Allocation of extra bits with various word-width computations for an ideal 16-bit, 96 dB SNR system, when quantization noise is present

8.4.1 Numeric Formats for Audio

There are many ways to represent data inside a processor. The two main processor architectures used for audio processing are fixed-point and floating-point. Fixed-point processors are designed for integer and fractional arithmetic, and they usually natively support 16-bit, 24-bit, or 32-bit data. Floating-point processors provide excellent performance with native support for 32-bit or 64-bit floating-point data types. However, they are typically more costly and consume more power than their fixed-point counterparts, and most real systems must strike a balance between quality and engineering cost.

8.4.1.1 Fixed-Point Arithmetic

Processors that can perform fixed-point operations typically use a twos-complement binary notation for representing signals. A fixed-point format can represent both signed and unsigned integers and fractions. The signed fractional format is most common for digital signal processing on fixed-point processors. The difference between integer and fractional formats lies in the location of the binary point. For integers, the binary point is to the right of the least significant digit, whereas fractions usually have their binary point to the left of the sign bit. Figure 8.10a shows integer and fractional formats.

While the fixed-point convention simplifies numeric operations and conserves memory, it presents a trade-off between dynamic range and precision. In situations that require a large range of numbers while maintaining high resolution, a radix point that can shift based on magnitude and exponent is desirable.

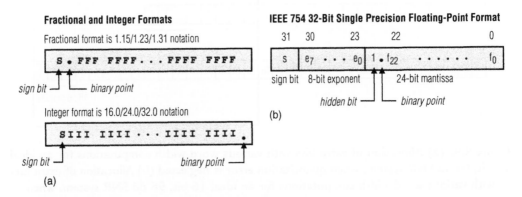

Figure 8.10: (a) Fractional and integer formats (b) IEEE 754 32-bit single-precision floating-point format

8.4.1.2 Floating-Point Arithmetic

Using floating-point format, very large and very small numbers can be represented in the same system. Floating-point numbers are quite similar to scientific notation of rational numbers. They are described with a mantissa and an exponent. The mantissa dictates precision, and the exponent controls dynamic range.

There is a standard that governs floating-point computations of digital machines. It is called IEEE 754 (Figure 8.10b) and can be summarized as follows for 32-bit floating-point numbers. Bit 31 (MSB) is the *sign bit*, where a 0 represents a positive sign and a 1 represents a negative sign. Bits 30 through 23 represent an exponent field (*exp_field*) as a power of 2, biased with an offset of 127. Finally, bits 22 through 0 represent a fractional mantissa (*mantissa*). The hidden bit is basically an implied value of 1 to the left of the radix point.

The value of a 32-bit IEEE 754 floating-point number can be represented with the following equation:

$$(-1)^{sign\_bit} \times (1.mantissa) \times 2^{(exp\_feild - 127)}$$

With an 8-bit exponent and a 23-bit mantissa, IEEE 754 reaches a balance between dynamic range and precision. In addition, IEEE floating-point libraries include support for additional features such as $\pm\infty$, 0 and NaN (not a number).

Table 8.5 shows the smallest and largest values attainable from the common floating-point and fixed-point types.

Table 8.5: Comparison of dynamic range for various data formats

Data Type	Smallest Positive Value	Largest Positive Value
IEEE 754 Floating-Point (single-precision)	$2^{-126} \approx 1.2 \times 10^{-38}$	$2^{128} \approx 3.4 \times 10^{38}$
1.15 16-bit fixed-point	$2^{-15} \approx 3.1 \times 10^{-5}$	$1 - 2^{-15} \approx 9.9 \times 10^{-1}$
1.23 24-bit fixed-point	$2^{-23} \approx 1.2 \times 10^{-7}$	$1 - 2^{-23} \approx 9.9 \times 10^{-1}$

8.4.1.3 Emulation on 16-Bit Architectures

As we explained earlier, 16-bit processing does not provide enough SNR for high quality audio, but this does not mean that you shouldn't choose a 16-bit processor for an audio system. For example, a 32-bit floating-point machine makes it easier to code an algorithm that preserves 32-bit data natively, but a 16-bit processor can also maintain 32-bit integrity through emulation at a much lower cost. Figure 8.11 illustrates some of the possibilities when it comes to choosing a data type for an embedded algorithm.

Floating-Point Emulation Techniques on a 16-bit Processor

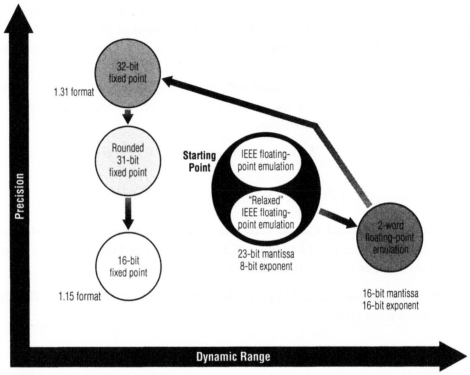

Path of arrows denotes decreasing core cycles required

Figure 8.11: Depending on the goals of an application, there are many data types that can satisfy system requirements

In the remainder of this section, we'll describe how to achieve floating-point and 32-bit extended-precision fixed-point functionality on a 16-bit fixed-point machine.

Floating-Point Emulation Techniques on a 16-bit Processor

Floating-Point Emulation on Fixed-Point Processors   On most 16-bit fixed-point processors, IEEE 754 floating-point functions are available as library calls from either C/C++ or assembly language. These libraries emulate the required floating-point processing using fixed-point multiply and ALU logic. This emulation requires additional cycles to complete. However, as fixed-point processor core-clock speeds venture into the 500 MHz–1 GHz range, the extra cycles required to emulate IEEE 754-compliant floating-point math become less significant.

It is sometimes advantageous to use a "relaxed" version of IEEE 754 in order to reduce computational complexity. This means that the floating-point arithmetic doesn't implement features such as ∞ and NaN.

A further optimization is to use a processor's native data register widths for the mantissa and exponent. Take, for example, the Blackfin architecture, which has a register file set that consists of sixteen 16-bit registers that can be used instead as eight 32-bit registers. In this configuration, on every core-clock cycle, two 32-bit registers can source operands for computation on all four register halves. To make optimized use of the Blackfin register file, a two-word format can be used. In this way, one word (16 bits) is reserved for the exponent and the other word (16 bits) is reserved for the fraction.

Double-Precision Fixed-Point Emulation   There are many applications where 16-bit fixed-point data is not sufficient, but where emulating floating-point arithmetic may be too computationally intensive. For these applications, extended-precision fixed-point emulation may be enough to satisfy system requirements. Using a high-speed fixed-point processor will insure a significant reduction in the amount of required processing. Two popular extended-precision formats for audio are 32-bit and 31-bit fixed-point representations.

32-Bit-Accurate Emulation   32-bit arithmetic is a natural software extension for 16-bit fixed-point processors. For processors whose 32-bit register files can be accessed as two 16-bit halves, the halves can be used together to represent a single 32-bit fixed-point number. The Blackfin processor's hardware implementation allows for single-cycle 32-bit addition and subtraction.

For instances where a 32-bit multiply will be iterated with accumulation (as is the case in some algorithms we'll talk about soon), we can achieve 32-bit accuracy with 16-bit multiplications in just three cycles. Each of the two 32-bit operands (R0 and R1) can be broken up into two 16-bit halves (R0.H I R0.L and R1.H I R1.L).

From Figure 8.12, it is easy to see that the following operations are required to emulate the 32-bit multiplication R0 × R1 with a combination of instructions using 16-bit multipliers:

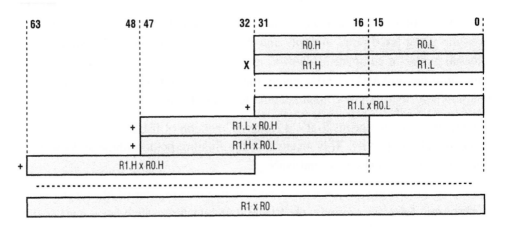

Figure 8.12: 32-bit multiplication with 16-bit operations

- Four 16-bit multiplications to yield four 32-bit results:

  1. R1.L × R0.L

  2. R1.L × R0.H

  3. R1.H × R0.L

  4. R1.H × R0.H

- Three operations to preserve bit place in the final answer (the $>>$ symbol denotes a right shift). Since we are performing fractional arithmetic, the result is 1.62 ($1.31 \times 1.31 = 2.62$ with a redundant sign bit). Most of the time, the result can be truncated to 1.31 in order to fit in a 32-bit data register. Therefore, the result of the multiplication should be in reference to the sign bit, or the most significant bit. This way the least significant bits can be safely discarded in a truncation.

  1. (R1.L × R0.L) $>>$ 32

  2. (R1.L × R0.H) $>>$ 16

  3. (R1.H × R0.L) $>>$ 16

The final expression for a 32-bit multiplication is:

$$((R1.L \times R0.L) >> 32 + (R1.L \times R0.H) >> 16)$$
$$+ ((R1.H \times R0.L) >> 16 + R1.H \times R0.H)$$

On the Blackfin architecture, these instructions can be issued in parallel to yield an effective rate of a 32-bit multiplication in three cycles.

31-Bit-Accurate Emulation   We can reduce a fixed-point multiplication requiring at most 31-bit accuracy to just two cycles. This technique is especially appealing for audio systems, which usually require at least 24-bit representation, but where 32-bit accuracy may be a bit excessive. Using the 6 dB rule, 31-bit-accurate emulation still maintains a dynamic range of around 186 dB, which is plenty of headroom even with rounding and truncation errors.

From the multiplication diagram shown in Figure 8.12, it is apparent that the multiplication of the least significant half-word $R1.L \times R0.L$ does not contribute much to the final result. In fact, if the result is truncated to 1.31, then this multiplication can only have an effect on the least significant bit of the 1.31 result. For many applications, the loss of accuracy due to this bit is balanced by the speeding up of the 32-bit multiplication through eliminating one 16-bit multiplication, one shift, and one addition.

The expression for 31-bit accurate multiplication is:

$$((R1.L \times R0.H) + (R1.H \times R0.L)) >> 16 + (R1.H \times R0.H)$$

On the Blackfin architecture, these instructions can be issued in parallel to yield an effective rate of two cycles for each 32-bit multiplication.

8.5   Audio Processing Methods

8.5.1   Getting Data to the Processor's Core

There are a number of ways to get audio data into the processor's core. For example, a foreground program can poll a serial port for new data, but this type of transfer is uncommon in embedded media processors, because it makes inefficient use of the core.

Instead, a processor connected to an audio codec usually uses a DMA engine to transfer the data from the codec link (like a serial port) to some memory space available to the processor. This transfer of data occurs in the background without the core's intervention. The only overhead is in setting up the DMA sequence and handling the interrupts once the data buffer has been received or transmitted.

8.5.2 Block Processing versus Sample Processing

Sample processing and block processing are two approaches for dealing with digital audio data. In the sample-based method, the processor crunches the data as soon as it's available. Here, the processing function incurs overhead during each sample period. Many filters (like FIR and IIR, described later) are implemented this way, because the effective latency is low.

Block processing, on the other hand, is based on filling a buffer of a specific length before passing the data to the processing function. Some filters are implemented using block processing because it is more efficient than sample processing. For one, the block method sharply reduces the overhead of calling a processing function for each sample. Also, many embedded processors contain multiple ALUs that can parallelize the computation of a block of data. What's more, some algorithms are, by nature, meant to be processed in blocks. For example, the Fourier transform (and its practical counterpart, the fast Fourier transform, or FFT) accepts blocks of temporal or spatial data and converts them into frequency domain representations.

8.5.3 Double-Buffering

In a block-based processing system that uses DMA to transfer data to and from the processor core, a "double buffer" must exist to handle the DMA transfers and the core. This is done so that the processor core and the core-independent DMA engine do not access the same data at the same time, causing a data coherency problem. To facilitate the processing of a buffer of length N, simply create a buffer of length $2 \times N$. For a bidirectional system, two buffers of length $2 \times N$ must be created. As shown in Figure 8.13a, the core processes the in1 buffer and stores the result in the out1 buffer, while the DMA engine is filling in0 and transmitting the data from out0. Figure 8.13b depicts that once the DMA engine is done with the left half of the double buffers, it starts transferring data into in1 and out of out1, while the core processes data from

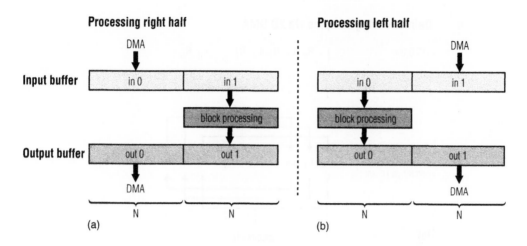

Processing right half   **Processing left half**

Figure 8.13: Double-buffering scheme for stream processing

in0 and into out0. This configuration is sometimes called "ping-pong buffering," because the core alternates between processing the left and right halves of the double buffers.

Note that, in real-time systems, the serial port DMA (or another peripheral's DMA tied to the audio sampling rate) dictates the timing budget. For this reason, the block processing algorithm must be optimized in such a way that its execution time is less than or equal to the time it takes the DMA to transfer data to/from one half of a double-buffer.

8.5.4 2D DMA

When data is transferred across a digital link like I²S, it may contain several channels. These may all be multiplexed on one data line going into the same serial port. In such a case, 2D DMA can be used to de-interleave the data so that each channel is linearly arranged in memory. Take a look at Figure 8.14 for a graphical depiction of this arrangement, where samples from the left and right channels are de-multiplexed into two separate blocks. This automatic data arrangement is extremely valuable for those systems that employ block processing.

8.5.5 Basic Operations

There are three fundamental building blocks in audio processing. They are the summing operation, multiplication, and time delay. Many more complicated effects and

Deinterleaving samples via 2D DMA

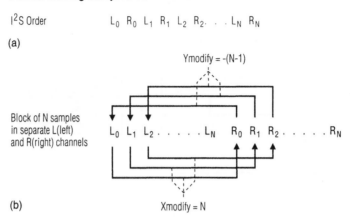

(a)

(b)

Figure 8.14: A 2D DMA engine used to de-interleave (a) I²S stereo data into (b) separate left and right buffers

algorithms can be implemented using these three elements. A summer has the obvious duty of adding two signals together. A multiplication can be used to boost or attenuate an audio signal. On most media processors, multiple summer and/or multiplier blocks can execute in a single cycle. A time delay is a bit more complicated. In many audio algorithms, the current output depends on a combination of previous inputs and/or outputs. The implementation of this delay effect is accomplished with a delay line, which is really nothing more than an array in memory that holds previous data. For example, an echo algorithm might hold 500 ms of input samples. The current output value can be computed by adding the current input value to a slightly attenuated prior sample. If the audio system is sample-based, then the programmer can simply keep track of an input pointer and an output pointer (spaced at 500 ms worth of samples apart), and increment them after each sampling period.

Since delay lines are meant to be reused for subsequent sets of data, the input and output pointers will need to wrap around from the end of the delay line buffer back to the beginning. In C/C++, this is usually done by appending the modulus operator (%) to the pointer increment.

This wrap-around may incur no extra processing cycles if you use a processor that supports circular buffering (see Figure 8.15). In this case, the beginning address and length of a circular buffer must be provided only once. During processing, the software

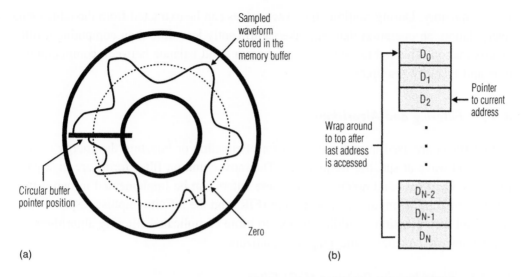

Figure 8.15: (a) Graphical representation of a delay line using a circular buffer (b) Layout of a circular buffer in memory

increments or decrements the current pointer within the buffer, but the hardware takes care of wrapping around to the beginning of the buffer if the current pointer falls outside of the buffer's boundaries. Without this automated address generation, the programmer would have to manually keep track of the buffer, thus wasting valuable processing cycles.

An echo effect derives from an important audio building block called the comb filter, which is essentially a delay with a feedback element. When multiple comb filters are used simultaneously, they can create the effect of reverberation.

8.5.6 Signal Generation

In some audio systems, a signal (for example, a sine wave) might need to be synthesized. Taylor Series function approximations can emulate trigonometric functions. Moreover, uniform random number generators are handy for creating white noise.

However, synthesis might not fit into a given system's processing budget. On fixed-point systems with ample memory, you can use a table lookup instead of generating a signal. This has the side effect of taking up precious memory resources, so hybrid methods can be used as a compromise. For example, you can store a coarse lookup table to save memory. During runtime, the exact values can be extracted from the table using interpolation, an operation that can take significantly less time than computing a full approximation. This hybrid approach provides a good balance between computation time and memory resources.

8.5.7   *Filtering and Algorithms*

Digital filters are used in audio systems for attenuating or boosting the energy content of a sound wave at specific frequencies. The most common filter forms are high-pass, low-pass, band-pass and notch. Any of these filters can be implemented in two ways. These are the finite impulse response filter (FIR) and the infinite impulse response filter (IIR), and they constitute building blocks to more complicated filtering algorithms like parametric equalizers and graphic equalizers.

8.5.7.1   *Finite Impulse Response (FIR) Filter*

The FIR filter's output is determined by the sum of the current and past inputs, each of which is first multiplied by a filter coefficient. The FIR summation equation, shown in Figure 8.16a, is also known as *convolution*, one of the most important operations in signal processing. In this syntax, x is the input vector, y is the output vector, and h holds the filter coefficients. Figure 8.16a also shows a graphical representation of the FIR implementation.

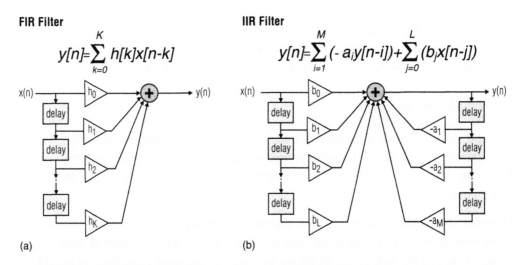

FIR Filter

$$y[n] = \sum_{k=0}^{K} h[k]x[n-k]$$

IIR Filter

$$y[n] = \sum_{i=1}^{M} (-a_i y[n-i]) + \sum_{j=0}^{L} (b_j x[n-j])$$

(a)   (b)

Figure 8.16: (a) FIR filter equation and structure (b) IIR filter equation and structure

Convolution is such a common operation in media processing that many processors are designed to execute a multiply-accumulate (MAC) instruction along with multiple data accesses (reads and writes) and pointer increments in one cycle.

8.5.7.2   Infinite Impulse Response (IIR) Filter

Unlike the FIR, whose output depends only on inputs, the IIR filter relies on both inputs and past outputs. The basic equation for an IIR filter is a difference equation, as shown in Figure 8.16b. Because of the current output's dependence on past outputs, IIR filters are often referred to as *recursive filters*. Figure 8.16b also gives a graphical perspective on the structure of the IIR filter.

8.5.7.3   Fast Fourier Transform

Quite often, we can do a better job describing an audio signal by characterizing its frequency composition. A Fourier transform takes a time-domain signal and translates it into the frequency domain; the inverse Fourier transform achieves the opposite, converting a frequency-domain representation back into the time domain. Mathematically, there are some nice property relationships between operations in the time domain and those in the frequency domain. Specifically, a time-domain convolution (or an FIR filter) is equivalent to a multiplication in the frequency domain. This tidbit would not be too practical if it weren't for a special optimized implementation of the Fourier transform called the fast Fourier transform (FFT). In fact, it is often more efficient to implement an FIR filter by transforming the input signal and coefficients into the frequency domain with an FFT, multiplying the transforms, and finally transforming the result back into the time domain with an inverse FFT.

There are other transforms that are used often in audio processing. Among them, one of the most common is the modified discrete cosine transform (MDCT), which is the basis for many audio compression algorithms.

8.5.8   Sample Rate Conversion

There are times when you will need to convert a signal sampled at one frequency to a different sampling rate. One situation where this is useful is when you want to decode an audio signal sampled at, say 8 kHz, but the DAC you're using does not support that sampling frequency. Another scenario is when a signal is oversampled, and converting to a lower sampling frequency can lead to a reduction in computation time.

The process of converting the sampling rate of a signal from one rate to another is called sampling rate conversion (or SRC).

Increasing the sampling rate is called interpolation, and decreasing it is called decimation. Decimating a signal by a factor of M is achieved by keeping only every Mth sample and discarding the rest. Interpolating a signal by a factor of L is accomplished by padding the original signal with L–1 zeros between each sample.

Even though interpolation and decimation factors are integers, you can apply them in series to an input signal to achieve a rational conversion factor. When you upsample by 5 and then downsample by 3, then the resulting resampling factor is 5/3 = 1.67.

To be honest, we oversimplified the SRC process a bit too much. In order to prevent artifacts due to zero-padding a signal (which creates images in the frequency domain), an interpolated signal must be low-pass-filtered before being used as an output or as an input into a decimator. This anti-imaging low-pass filter can operate at the input sample rate, rather than at the faster output sample rate, by using a special FIR filter structure that recognizes that the inputs associated with the L–1 inserted samples have zero values.

Similarly, before they're decimated, all input signals must be low-pass-filtered to prevent aliasing. The anti-aliasing low-pass filter may be designed to operate at the decimated sample rate, rather than at the faster input sample rate, by using a FIR filter structure that realizes the output samples associated with the discarded samples need not be computed. Figure 8.17 shows a flow diagram of a sample rate converter. Note that it is possible to combine the anti-imaging and anti-aliasing filter into one component for computational savings.

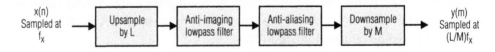

Figure 8.17: Sample-rate conversion through upsampling and downsampling

8.5.9 Audio Compression

Even though raw audio requires a lower bit rate than raw video, the amount of data is still substantial. The bit rate required to code a single CD-quality audio channel

(44.1 kHz at 16 bits) using the standard PCM method is 705.6 kbps—one minute of stereo sound requires over 10 Mbytes of storage! Sending this much data over a network or a serial connection is inefficient, and sometimes impossible. The solution comes in the form of compression algorithms called audio codecs. These software codecs, not to be confused with hardware ADCs and DACs discussed already, compress raw data either for low-bandwidth transfer or for storage, and decompress for the reverse effect.

There are lossless codecs and lossy codecs available for audio. Lossless codecs are constructed in such a way that a compressed signal can be reconstructed to contain the exact data as the original input signal. Lossless codecs are computationally intensive, and they can reduce audio bit rate by up to about ½. Lossy codecs can compress audio much more (10× or more, depending on desired quality), and the audio decoded from a lossy stream sounds very close to the original, even though information is lost forever in the encoding process. Lossy codecs can throw out data and still preserve audio integrity, because they are based on a psycho-acoustical model that takes advantage of our ears' physiology. In essence, a lossy codec can cheat by dropping data that will not affect how we'll ultimately perceive the signal. This technique is often referred to as "perceptual encoding."

Earlier, we mentioned frequency and temporal masking. Another useful feature for perceptual encoding—called *joint stereo encoding*—deals with multiple channels. The basic premise is that data in two or more channels is correlated. By decoupling unique features of each channel from the shared features, we can drastically reduce the data needed to encode the content. If one channel takes 196 kbps, then an encoder that recognizes the redundancy will allocate much less data than 2 × 196 kbps for a stereo stream, while still retaining the same perceived sound. The general rule of thumb is that multichannel audio can be efficiently encoded with an amount of data proportional to the square root of the number of channels (see Reference 1). In practice, audio encoders use two techniques: sub-band coding and transform coding. Sub-band coding splits the input audio signal into a number of sub-bands, using band-pass filters. A psycho-acoustical model is applied on the sub-bands to define the number of bits necessary to maintain a specified sound quality. Transform coding uses a transform like the FFT or an MDCT on a block of audio data. Then, a psycho-acoustical model is used to determine the proper quantization of the frequency components based on a masking threshold to ensure that the output sounds like the input signal.

Let's take a look at some of the currently available audio codecs. Some of the algorithms are proprietary and require a license before they can be used in a system. Table 8.6 lists common audio coding standards and the organizations responsible for them.

Table 8.6: Various audio codecs

Audio Coding Standard	(Licensing/Standardization Organization)
MP3	ISO/IEC
AAC	ISO/IEC
AC-3	Dolby Labs
Windows Media Audio	Microsoft
RealAudio	RealNetworks
Vorbis	Xiph.org
FLAC	Xiph.org

8.5.9.1 MP3

MP3 is probably the most popular lossy audio compression codec available today. The format, officially known as MPEG-1 Audio Layer 3, was released in 1992 as a complement to the MPEG-1 video standard from the Moving Pictures Experts Group. MPEG is a group of ISO/IEC, an information center jointly operated by the International Organization for Standardization and the International Electrotechnical Commission. MP3 was developed by the German Fraunhofer Institut Integrierte Schaltungen (Fraunhofer IIS), which holds a number of patents for MP3 encoding and decoding. Therefore, you must obtain a license before incorporating the MP3 algorithm into your embedded systems.

MP3 uses polyphase filters to separate the original signal into sub-bands. Then, the MDCT transform converts the signal into the frequency domain, where a psychoacoustical model quantizes the frequency coefficients. A CD-quality track can be MP3-encoded at a 128–196 kbps rate, thus achieving up to a 12:1 compression ratio.

8.5.9.2 AAC

Advanced Audio Coding (AAC) is a second-generation codec also developed by Fraunhofer IIS. It was designed to complement the MPEG-2 video format. Its main improvement over MP3 is the ability to achieve lower bit rates at equivalent sound quality.

8.5.9.3 AC-3

The AC-3 format was developed by Dolby Laboratories to efficiently handle multi-channel audio such as 5.1, a capability that was not implemented in the MP3 standard. The nominal stereo bit rate is 192 kbps, whereas it's 384 kbps for 5.1 surround sound.

A 5.1 surround-sound system contains five full-range speakers, including front left and right, rear left and right, and front center channels, along with a low-frequency (10 Hz–120 Hz) subwoofer.

8.5.9.4 WMA

Windows Media Audio (WMA) is a proprietary codec developed by Microsoft to challenge the popularity of MP3. Microsoft developed WMA with paid music distribution in mind, so they incorporated Digital Rights Management (DRM) into the codec. Besides the more popular lossy codec, WMA also supports lossless encoding.

8.5.9.5 RealAudio

RealAudio, developed by RealNetworks, is another proprietary format. It was conceived to allow the streaming of audio data over low bandwidth links. Many Internet radio stations use this format to stream their content. Recent updates to the codec have improved its quality to match that of other modern codecs.

8.5.9.6 Vorbis

Vorbis was created at the outset to be free of any patents. It is released by the Xiph.org Foundation as a completely royalty-free codec. A full Vorbis implementation for both floating-point and fixed-point processors is available under a free license from Xiph. org. Because it is free, Vorbis is finding its way into increasing numbers of embedded devices.

According to many subjective tests, Vorbis outperforms MP3, and it is therefore in the class of the newer codecs like WMA and AAC. Vorbis also fully supports multi-channel compression, thereby eliminating redundant information carried by the channels.

8.5.9.7 FLAC

FLAC is another open standard from the Xiph.org Foundation. It stands for Free Lossless Audio Codec, and as the name implies, it does not throw out any information from the original audio signal. This, of course, comes at the expense of much smaller achievable compression ratios. The typical compression range for FLAC is 30–70%.

8.5.10 Speech Compression

Speech compression is used widely in real-time communications systems like cell phones, and in packetized voice connections like Internet phones.

Since speech is more band-limited than full-range audio, it is possible to employ audio codecs, taking the smaller bandwidth into account. Almost all speech codecs do, indeed, sample voice data at 8 kHz. However, we can do better than just take advantage of the smaller frequency range. Since only a subset of the audible signals within the speech bandwidth is ever vocally generated, we can drive bit rates even lower. The major goal in speech encoding is a highly compressed stream with good intelligibility and short delays to make full-duplex communication possible.

The most traditional speech coding approach is code-excited linear prediction (CELP). CELP is based on linear prediction coding (LPC) models of the vocal tract and a supplementary residue codebook.

The idea behind using LPC for speech coding is founded on the observation that the human vocal tract can be roughly modeled with linear filters. We can make two basic kinds of sounds: voiced and unvoiced. Voiced sounds are produced when our vocal cords vibrate, and unvoiced sounds are created when air is constricted by the mouth, tongue, lips and teeth. Voiced sounds can be modeled as linear filters driven by a fundamental frequency, whereas unvoiced ones are modeled with random noise sources. Through these models, we can describe human utterances with just a few parameters. This allows LPC to predict signal output based on previous inputs and outputs. To complete the model, LPC systems supplement the idealized filters with residue (i.e., error) tables. The codebook part of CELP is basically a table of typical residues.

In real-time duplex communications systems, one person speaks while the other one listens. Since the person speaking is not contributing anything to the signal, some codecs implement features like voice activity detection (VAD) to recognize silence, and comfort noise generation (CNG) to simulate the natural level of noise without actually encoding it at the transmitting end.

Table 8.7: Various speech codecs

Speech Coding Standard	Bit Rate	Governing Body
GSM-FR	13 kbps	ETSI
GSM-EFR	12.2 kbps	ETSI
GSM-AMR	4.75, 5.15, 5.90, 6.70, 7.40, 7.95, 10.2, 12.2 kbps	3GPP
G.711	64 kbps	ITU-T
G.723.1	5.3, 6.3 kbps	ITU-T
G.729	6.4, 8, 11.8 kbps	ITU-T
Speex	2 – 44 kbps	Xiph.org

8.5.10.1 GSM

The GSM speech codecs find use in cell phone systems around the world. The governing body of these standards is the European Telecommunications Standards Institute (ETSI). There is actually an evolution of standards in this domain. The first one was GSM Full Rate (GSM-FR). This standard uses a CELP variant called Regular Pulse Excited Linear Predictive Coder (RPELPC). The input speech signal is divided into 20-ms frames. Each of those frames is encoded as 260 bits, thereby producing a total bit rate of 13 kbps. Free GSM-FR implementations are available for use under certain restrictions.

GSM Enhanced Full Rate (GSM-EFR) was developed to improve the quality of speech encoded with GSM-FR. It operates on 20-ms frames at a bit rate of 12.2 kbps, and it works in noise-free and noisy environments. GSM-EFR is based on the patented Algebraic Code Excited Linear Prediction (ACELP) technology, so you must purchase a license before using it in end products.

The 3rd Generation Partnership Project (3GPP), a group of standards bodies, introduced the GSM Adaptive Multi-Rate (GSM-AMR) codec to deliver even higher quality speech over lower-bit-rate data links by using an ACELP algorithm.

It uses 20-ms data chunks, and it allows for multiple bit rates at eight discrete levels between 4.75 kbps and 12.2 kbps. GSM-AMR supports VAD and CNG for reduced bit rates.

8.5.10.2 The "G-Dot" Standards

The International Telecommunication Union (ITU) was created to coordinate the standards in the communications industry, and the ITU Telecommunication Standardization Sector (ITU-T) is responsible for the recommendations of many speech codecs, known as the G.x standards.

8.5.10.3 G.711

G.711, introduced in 1988, is a simple standard when compared with the other options presented here. The only compression used in G.711 is companding (using either the μ-law or A-law standards), which compresses each data sample to 8 bits, yielding an output bit rate of 64 kbps.

8.5.10.4 G.723.1

G.723.1 is an ACELP-based dual-bit-rate codec, released in 1996, that targets Voice-Over-IP (VoIP) applications like teleconferencing. The encoding frame for G.723.1 is 30 ms. Each frame can be encoded in 20 or 24 bytes, thus translating to 5.3 kbps and 6.3 kbps streams, respectively. The bit rates can be effectively reduced through VAD and CNG. The codec offers good immunity against network imperfections like lost frames and bit errors. This speech codec is part of video conferencing applications described by the H.324 family of standards.

8.5.10.5 G.729

Another speech codec released in 1996 is G.729, which partitions speech into 10-ms frames, making it a low-latency codec. It uses an algorithm called Conjugate Structure ACELP (CS-ACELP). G.729 compresses 16-bit signals sampled at 8 kHz via 10-ms frames into a standard bit rate of 8 kbps, but it also supports 6.4 kbps and 11.8 kbps rates. VAD and CNG are also supported.

8.5.10.6 Speex

Speex is another codec released by Xiph.org, with the goal of being a totally patent-free speech solution. Like many other speech codecs, Speex is based on CELP with residue coding. The codec can take 8 kHz, 16 kHz, and 32 kHz linear PCM signals and code them into bit rates ranging from 2 to 44 kbps. Speex is resilient to network errors, and it supports voice activity detection. Besides allowing variable bit rates, another unique feature of Speex is stereo encoding. Source code is available from Xiph.org in both a floating-point reference implementation and a fixed-point version.

References

1. K.C. Pohlmann, *Principles of Digital Audio*, McGraw-Hill, 2000.
2. E.C. Ifeachor and B.W. Jervis, *Digital Signal Processing: A Practical Approach*, Addison-Wesley, 1999.
3. Steven Smith, *Digital Signal Processing: A Practical Guide for Engineers and Scientists*, Elsevier (Newnes), 2002.
4. B. Gold and N. Morgan, *Speech and Audio Signal Processing: Processing and Perception of Speech and Music*, Wiley & Sons, 2000.
5. Analog Devices, Inc., *Digital Signal Processing Applications Using the ADSP2100 Family*, Prentice Hall, 1992.
6. J. Tomarakos and D. Ledger, "Using the Low-Cost, High Performance ADSP1065L Digital Signal Processor for Digital Audio Applications," Analog Devices.
7. J. Sondermeyer, "EE-183: Rational Sample Rate Conversion with Blackfin Processors," http://www.analog.com/processors/resources/technicalLibrary/appNotes.html.
8. Analog Devices, Inc., "EE-186: Extended-Precision Fixed-Point Arithmetic on the Blackfin Processor Platform," http://www.analog.com/processors/resources/technicalLibrary/appNotes.html.
9. "Digital source coding of speech signals," http://www.ind.rwthaachen.de/research/speech_coding.html.
10. L. Dumond, "All About Decibels, Part I: What's your dB IQ?," ProRec.com.
11. Don Morgan, "It's All About Class," *Embedded Systems Programming*, August 31, 2001.

CHAPTER 9

Basics of Embedded Video
and Image Processing

David Katz
Rick Gentile

This (and the preceding) chapter comes from the book Embedded Media Processing *by David Katz and Rick Gentile, ISBN: 9780750679121. As we noted in the previous chapter, when Intel Corporation and Analog Devices collaborated to introduce the micro signal architecture (MSA) in 2000, the result was a processor uniting microcontroller (MCU) and DSP functionality into a single high-performance "convergent processing" device—a rather unique entry into the marketplace.*

What transpired with the MSA-based Blackfin® family was that a new class of application spaces opened up, from video-enabled handheld gadgets, to control-intensive industrial applications, to operating system/number-crunching combos running on a single processor core.

This chapter, "Basics of Embedded Video and Image Processing," deals with a basic differentiator between today's media processors and their recent forebears: that is, the ability to process high-quality video streams in real time. The authors try to put the "need to know" video basics in a single place without getting bogged down by format variations, geographic differences, legacy formats, etc.

Their aim is to inaugurate the user into video from a media processing standpoint, providing only the details most relevant to accomplishing the tasks at hand. After an overview of human visual perception as it relates to video, the text reviews video standards, color spaces, and pixel formats.

The authors then proceed to a discussion of video sources and displays that also includes an outline of the basic processing steps in an image pipeline. Next, they turn their attention to the media processor itself and walk the reader through an example video application, from raw input through processed display output. The chapter finishes with a high-level discussion of image and video compression.

—Clive "Max" Maxfield

9.1   Introduction

As consumers, we're intimately familiar with video systems in many embodiments. However, from the embedded developer's viewpoint, video represents a tangled web of different resolutions, formats, standards, sources and displays. Many references exist that delve into the details of all things video; outstanding among them are References 1 and 2.

In this chapter, our goal will not be to duplicate such excellent works. Rather, we will strive to untangle some of this intricate web, focusing on the most common circumstances you're likely to face in today's media processing systems. After reviewing the basics of video, we will discuss some common scenarios you may encounter in embedded multimedia design and provide some tips and tricks for dealing with challenging video design issues.

9.1.1   Human Visual Perception

Let's start by discussing a little physiology. Understanding how our eyes work has paved an important path in the evolution of video and imaging. As we'll see, video formats and compression algorithms both rely on the eye's responses to different types of stimuli.

Our eyes contain two types of vision cells: rods and cones. Rods are primarily sensitive to light intensity as opposed to color, and they give us night vision capability. Cones, on the other hand, are not tuned to intensity, but instead are sensitive to wavelengths of light between 400 nm (violet) and 770 nm (red). Thus, the cones provide the foundation for our color perception.

There are three types of cones, each with a different pigment that's either most sensitive to red, green or blue wavelengths, although there's a lot of overlap between the three responses. Taken together, the response of our cones peaks in the green region, at around 555 nm. This is why, as we'll see, we can make compromises in LCD displays by assigning the green channel more bits of resolution than the red or blue channels.

The discovery of the red, green and blue cones ties into the development of the *trichromatic color theory*, which states that almost any color of light can be conveyed by combining proportions of monochromatic red, green and blue wavelengths.

Because our eyes have lots more rods than cones, they are more sensitive to intensity than actual color. This allows us to save bandwidth in video and image representations by *subsampling* the color information.

Our perception of brightness is logarithmic, not linear. In other words, the actual intensity required to produce a 50% gray image (exactly between total black and total white) is only around 18% of the intensity we need to produce total white. This characteristic is extremely important in camera sensor and display technology, as we'll see in our discussion of *gamma correction*. Also, this effect leads to a reduced sensitivity to quantization distortion at high intensities, a trait that many media-encoding algorithms use to their advantage.

Another visual novelty is that our eyes adjust to the viewing environment, always creating their own reference for white, even in low-lighting or artificial-lighting situations. Because camera sensors don't innately act the same way, this gives rise to a *white balance* control in which the camera picks its reference point for absolute white.

The eye is less sensitive to high-frequency information than low-frequency information. What's more, although it can detect fine details and color resolution in still images, it cannot do so for rapidly moving images. As a result, *transform coding* (DCT, FFT, etc.) and *low-pass filtering* can be used to reduce total bandwidth needed to represent an image or video sequence.

Our eyes can notice a "flicker" effect at image update rates less than 50 to 60 times per second (50 to 60 Hz) in bright light. Under dim lighting conditions, this rate drops to about 24 Hz. Additionally, we tend to notice flicker in large uniform regions more so than in localized areas. These traits have important implications for *interlaced* video and display technologies.

9.1.2 What's a Video Signal?

At its root, a video signal is basically just a two-dimensional array of intensity and color data that is updated at a regular frame rate, conveying the perception of motion. On conventional cathode-ray tube (CRT) TVs and monitors, an electron beam modulated by an analog video signal such as that shown in Figure 9.1 illuminates phosphors on the

Breakdown of Luma Signal

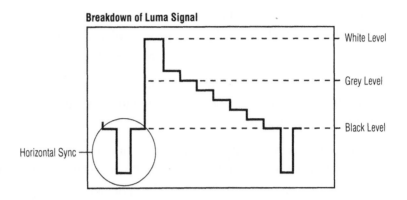

Figure 9.1: Example composition of Luma signal

screen in a top-to-bottom, left-to-right fashion. Synchronization signals embedded in the analog signal define when the beam is actively "painting" phosphors and when it is inactive, so that the electron beam can retrace from right to left to start on the next row, or from bottom to top to begin the next video field or frame. These synchronization signals are represented in Figure 9.2.

HSYNC is the horizontal synchronization signal. It demarcates the start of active video on each row (left to right) of a video frame. *Horizontal blanking* is the interval in which the electron gun retraces from the right side of the screen back over to the next row on the left side.

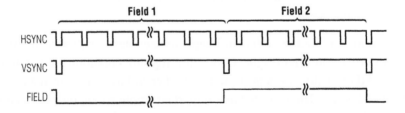

Figure 9.2: Typical timing relationships between HSYNC, VSYNC, FIELD

VSYNC is the vertical synchronization signal. It defines the start (top to bottom) of a new video image. *Vertical blanking* is the interval in which the electron gun retraces from the bottom right corner of the screen image back up to the top left corner.

FIELD distinguishes, for interlaced video, which field is currently being displayed. This signal is not applicable for progressive-scan video systems.

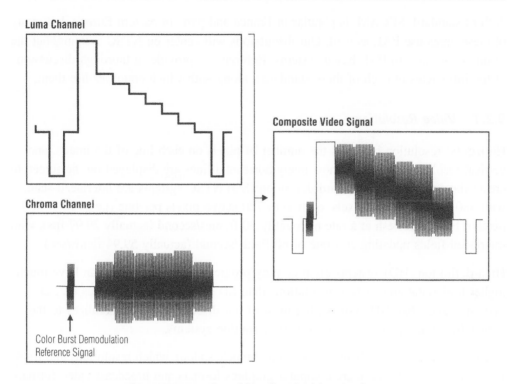

Luma Channel

Chroma Channel

Color Burst Demodulation
Reference Signal

Composite Video Signal

Figure 9.3: Analog video signal with color burst

The transmission of video information originated as a display of relative luminance from black to white—thus was born the black-and-white television system. The voltage level at a given point in space correlated to the brightness level of the image at that point.

When color TV became available, it had to be backward-compatible with black-and-white systems, so the color burst information was added on top of the existing luminance signal, as shown in Figure 9.3. Color information is also called chrominance. We'll talk more about it in our discussion on color spaces.

9.2 Broadcast TV—NTSC and PAL

Analog video standards differ in the ways they encode brightness and color information. Two standards dominate the broadcast television realm—NTSC and PAL. NTSC, devised by the National Television System Committee, is prevalent in Asia and North America, whereas PAL (Phase Alternation Line) dominates Europe and South America. PAL developed as an offshoot of NTSC, improving on its color distortion performance.

A third standard, SECAM, is popular in France and parts of eastern Europe, but many of these areas use PAL as well. Our discussions will center on NTSC systems, but the results relate also to PAL-based systems. Reference 2 provides a thorough discussion of the intricacies of each of these standards, along with which countries use them.

9.2.1   Video Resolution

Horizontal resolution indicates the number of pixels on each line of the image, and vertical resolution designates how many horizontal lines are displayed on the screen to create the entire frame. Standard definition (SD) NTSC systems are interlaced-scan, with 480 lines of active pixels, each with 720 active pixels per line (i.e., 720×480 pixels). Frames refresh at a rate of roughly 30 frames/second (actually 29.97 fps), with interlaced fields updating at a rate of 60 fields/second (actually 59.94 fields/sec).

High-definition (HD) systems often employ progressive scanning and can have much higher horizontal and vertical resolutions than SD systems. We will focus on SD systems rather than HD systems, but most of our discussion also generalizes to the higher frame and pixel rates of the high-definition systems.

When discussing video, there are two main branches along which resolutions and frame rates have evolved. These are computer graphics formats and broadcast video formats. Table 9.1 shows some common screen resolutions and frame rates belonging to each category. Even though these two branches emerged from separate domains with different requirements (for instance, computer graphics uses RGB progressive-scan schemes, while broadcast video uses YCbCr interlaced schemes), today they are used almost interchangeably in the embedded world. That is, VGA compares closely with the NTSC "D-1" broadcast format, and QVGA parallels CIF. It should be noted that, although D-1 is 720 pixels $\times$ 486 rows, it's commonly referred to as being 720×480 pixels (which is really the arrangement of the NTSC "DV" format used for DVDs and other digital video).

9.2.2   Interlaced versus Progressive Scanning

Interlaced scanning originates from early analog television broadcasts, where the image needed to be updated rapidly in order to minimize visual flicker, but the technology available did not allow for refreshing the entire screen this quickly. Therefore, each frame was "interlaced," or split into two fields, one consisting of odd-numbered scan lines, and the other composed of even-numbered scan lines, as depicted in Figure 9.4.

Table 9.1: Graphics vs. broadcast standards

Origin	Video Standard	Horizontal Resolution (Pixels)	Vertical Resolution (Pixels)	Total Pixels
Broadcast	QCIF	176	144	25,344
Graphics	QVGA	320	240	76,800
Broadcast	CIF	352	288	101,376
Graphics	VGA	640	480	307,200
Broadcast	NTSC	720	480	345,600
Broadcast	PAL	720	576	414,720
Graphics	SVGA	800	600	480,000
Graphics	XGA	1024	768	786,432
Broadcast	HDTV (720p)	1280	720	921,600
Graphics	SXGA	1280	1024	1,310,720
Graphics	UXGA	1600	1200	1,920,000
Broadcast	HDTV (1080i)	1920	1080	2,073,600

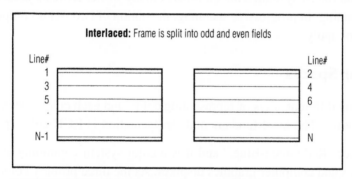

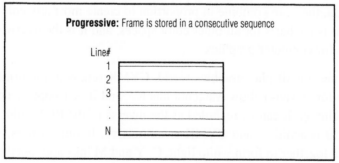

Figure 9.4: Interlaced scan vs. progressive scan illustration

The frame refresh rate for NTSC (PAL) was set at approximately 30 (25) frames/sec. Thus, large areas flicker at 60 (50) Hz, while localized regions flicker at 30 (25) Hz. This was a compromise to conserve bandwidth while accounting for the eye's greater sensitivity to flicker in large uniform regions.

Not only does some flickering persist, but interlacing also causes other artifacts. For one, the scan lines themselves are often visible. Because each NTSC field is a snapshot of activity occurring at 1/60 second intervals, a video frame consists of two temporally different fields. This isn't a problem when you're watching the display, because it presents the video in a temporally appropriate manner. However, converting interlaced fields into progressive frames (a process known as *deinterlacing*) can cause jagged edges when there's motion in an image. Deinterlacing is important because it's often more efficient to process video frames as a series of adjacent lines.

With the advent of digital television, progressive (that is, noninterlaced) scan has become a very popular input and output video format for improved image quality. Here, the entire image updates sequentially from top to bottom, at twice the scan rate of a comparable interlaced system. This eliminates many of the artifacts associated with interlaced scanning. In progressive scanning, the notion of two fields composing a video frame does not apply.

9.3   Color Spaces

There are many different ways of representing color, and each color system is suited for different purposes. The most fundamental representation is RGB color space.

RGB stands for "Red-Green-Blue," and it is a color system commonly employed in camera sensors and computer graphics displays. As the three primary colors that sum to form white light, they can combine in proportion to create most any color in the visible spectrum. RGB is the basis for all other color spaces, and it is the overwhelming choice of color space for computer graphics.

Just as RGB rules the display graphics world, CYMK reigns in the printing realm. CYMK stands for "Cyan-Yellow-Magenta-blacK," and it is a popular color space for printing and painting. It can be regarded as the inverse of the RGB color system, to the extent that RGB is additive, and CYMK is subtractive. In other words, whereas R, G and B light add together to form white light, C, Y and M inks add together to absorb all white. In other words, they create black. However, because it's difficult to create pure black, given the physical properties of printing media, true black is also added to the color system (hence, the fourth letter, "K").

9.3.1 Gamma Correction

"Gamma" is a crucial phenomenon to understand when dealing with color spaces. This term describes the nonlinear nature of luminance perception and display. Note that this is a twofold manifestation: the human eye perceives brightness in a nonlinear manner, and physical output devices (such as CRTs and LCDs) display brightness nonlinearly. It turns out, by way of coincidence, that human perception of luminance sensitivity is almost exactly the inverse of a CRT's output characteristics.

Stated another way, luminance on a display is roughly proportional to the input analog signal voltage raised to the power of gamma. On a CRT or LCD display, this value is ordinarily between 2.2 and 2.5. A camera's precompensation, then, scales the RGB values to the power of (1/gamma).

The upshot of this effect is that video cameras and computer graphics routines, through a process called "gamma correction," prewarp their RGB output stream both to compensate for the target display's nonlinearity and to create a realistic model of how the eye actually views the scene. Figure 9.5 illustrates this process.

Gamma-corrected RGB coordinates are referred to as $R'G'B'$ space, and the luma value Y' is derived from these coordinates. Strictly speaking, the term "luma" should only refer to this gamma-corrected luminance value, whereas the true "luminance" Y is a color science term formed from a weighted sum of R, G, and B (with no gamma correction applied). Reference 1 provides an in-depth discussion of proper terminology related to gamma, luma, and chroma in color systems.

Often when we talk about YCbCr and RGB color spaces in this text, we are referring to gamma-corrected components—in other words, $Y'CbCr$ or $R'G'B'$. However, because this notation can be distracting and doesn't affect the substance of our discussion, and since it's clear that gamma correction is essential (whether performed by the processor or by dedicated hardware) at sensor and display interfaces, we will confine ourselves to the YCbCr/RGB nomenclature even in cases where gamma adjustment has been applied. The exception to this convention is when we discuss actual color space conversion equations.

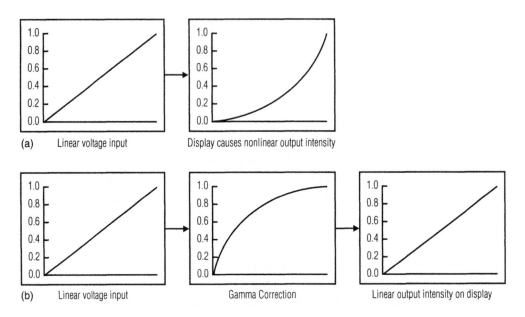

Figure 9.5: Gamma correction linearizes the intensity produced for a given input amplitude (a) Linear input voltage creates a nonlinear brightness characteristic on the display (b) By precorrecting for this distortion, display intensity follows input voltage linearly

While RGB channel format is a natural scheme for representing real-world color, each of the three channels is highly correlated with the other two. You can see this by independently viewing the R, G, and B channels of a given image—you'll be able to perceive the entire image in each channel. Also, RGB is not a preferred choice for image processing because changes to one channel must be performed in the other two channels as well, and each channel has equivalent bandwidth.

To reduce required transmission bandwidths and increase video compression ratios, other color spaces were devised that are highly uncorrelated, thus providing better compression characteristics than RGB does. The most popular ones—YPbPr, YCbCr, and YUV—all separate a luminance component from two chrominance components. This separation is performed via scaled color difference factors $(B'-Y')$ and $(R'-Y')$.

The Pb/Cb/U term corresponds to the $(B'-Y')$ factor, and the Pr/Cr/V term corresponds to the $(R'-Y')$ parameter. YPbPr is used in component analog video, YUV applies to composite NTSC and PAL systems, and YCbCr relates to component digital video.

Separating luminance and chrominance information saves image processing bandwidth. Also, as we'll see shortly, we can reduce chrominance bandwidth considerably via subsampling, without much loss in visual perception. This is a welcome feature for video-intensive systems.

As an example of how to convert between color spaces, the following equations illustrate translation between 8-bit representations of Y'CbCr and R'G'B' color spaces, where Y', R', G' and B' normally range from 16–235, and Cr and Cb range from 16–240.

$$Y' = (0.299)R' + (0.587)G' + (0.114)B'$$

$$Cb = -(0.168)R' - (0.330)G' + (0.498)B' + 128$$

$$Cr = (0.498)R' - (0.417)G' - (0.081)B' + 128$$

$$R' = Y' + 1.397(Cr - 128)$$

$$G' = Y' - 0.711(Cr - 128) - 0.343(Cb - 128)$$

$$B' = Y' + 1.765(Cb - 128)$$

9.3.2 Chroma Subsampling

With many more rods than cones, the human eye is more attuned to brightness and less to color differences. As luck (or really, design) would have it, the YCbCr color system allows us to pay more attention to Y, and less to Cb and Cr. As a result, by subsampling these chroma values, video standards and compression algorithms can achieve large savings in video bandwidth.

Before discussing this further, let's get some nomenclature straight. Before subsampling, let's assume we have a full-bandwidth YCbCr stream. That is, a video source generates a stream of pixel components in the form of Figure 9.6a. This is called "4:4:4 YCbCr." This notation looks rather odd, but the simple explanation is this: the first number is always 4, corresponding historically to the ratio between the luma sampling frequency and the NTSC color subcarrier frequency. The second number corresponds to the ratio between luma and chroma within a given line (horizontally): if there's no downsampling of chroma with respect to luma, this number is 4. The third number, if it's the same as the second digit, implies no vertical subsampling of chroma.

On the other hand, if it's a 0, there is 2:1 chroma subsampling between lines. Therefore, 4:4:4 implies that each pixel on every line has its own unique Y, Cr and Cb components.

Now, if we filter a 4:4:4 YCbCr signal by subsampling the chroma by a factor of two horizontally, we end up with 4:2:2 YCbCr. '4:2:2' implies that there are four luma values for every two chroma values on a given video line. Each (Y,Cb) or (Y,Cr) pair represents one pixel value. Another way to say this is that a chroma pair coincides spatially with every other luma value, as shown in Figure 9.6b. Believe it or not, 4:2:2 YCbCr qualitatively shows little loss in image quality compared with its 4:4:4 YCbCr source, even though it represents a savings of 33% in bandwidth over 4:4:4 YCbCr. As we'll discuss soon, 4:2:2 YCbCr is a foundation for the ITU-R BT.601 video recommendation, and it is the most common format for transferring digital video between subsystem components.

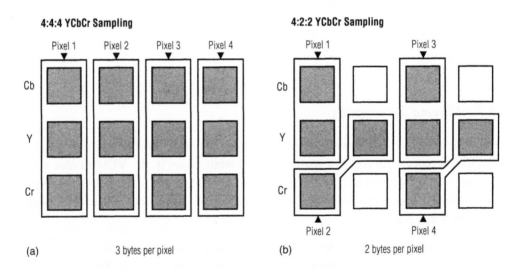

Figure 9.6: (a) 4:4:4 vs. (b) 4:2:2 YCbCr pixel sampling

Note that 4:2:2 is not the only chroma subsampling scheme. Figure 9.7 shows others in popular use. For instance, we could subsample the chroma of a 4:4:4 YCbCr stream by a factor of four horizontally, as shown in Figure 9.7c, to end up with a 4:1:1 YCbCr stream. Here, the chroma pairs are spatially coincident with every fourth luma value. This chroma filtering scheme results in a 50% bandwidth savings; 4:1:1 YCbCr is a

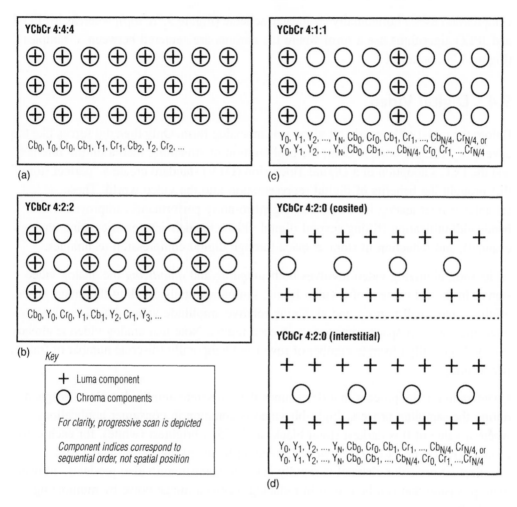

Figure 9.7: (a) YCbCr 4:4:4 stream and its chroma-subsampled derivatives (b) 4:2:2 (c) 4:1:1 (d) 4:2:0

popular format for inputs to video compression algorithms and outputs from video decompression algorithms.

Another format popular in video compression/decompression is 4:2:0 YCbCr; it's more complex than the others we've described for a couple of reasons. For one, the Cb and Cr components are each subsampled by 2, horizontally *and* vertically. This means we have to store multiple video lines in order to generate this subsampled stream. What's more, there are two popular formats for 4:2:0 YCbCr. MPEG-2 compression uses a horizontally co-located scheme (Figure 9.7d, top), whereas MPEG-1 and JPEG algorithms use a form where the chroma are centered between Y samples (Figure 9.7d, bottom).

9.4   Digital Video

Before the mid-1990s, nearly all video was in analog form. Only then did forces like the advent of MPEG-2 compression, the proliferation of streaming media on the Internet, and the FCC's adoption of a Digital Television (DTV) standard create a "perfect storm" that brought the benefits of digital representation into the video world. These advantages over analog include better signal-to-noise performance, improved bandwidth utilization (fitting several digital video channels into each existing analog channel), and reduction in storage space through digital compression techniques.

At its root, digitizing video involves both *sampling* and *quantizing* the analog video signal. In the 2D context of a video frame, sampling entails dividing the image space, gridlike, into small regions and assigning relative amplitude values based on the intensities of color space components in each region. Note that analog video is already sampled vertically (discrete number of rows) and temporally (discrete number of frames per second).

Quantization is the process that determines these discrete amplitude values assigned during the sampling process. Eight-bit video is common in consumer applications, where a value of 0 is darkest (total black) and 255 is brightest (white), for each color channel (R,G,B or YCbCr). However, it should be noted that 10-bit and 12-bit quantization per color channel is rapidly entering mainstream video products, allowing extra precision that can be useful in reducing received image noise by minimizing round-off error.

The advent of digital video provided an excellent opportunity to standardize, to a large degree, the interfaces to NTSC and PAL systems. When the International Telecommunication Union (ITU) met to define recommendations for digital video standards, it focused on achieving a large degree of commonality between NTSC and PAL standards, such that the two could share the same coding formats.

They defined two separate recommendations—ITU-R BT.601 and ITU-R BT.656. Together, these two define a structure that enables different digital video system components to interoperate. Whereas BT.601 defines the parameters for digital video transfer, BT.656 defines the interface itself.

9.4.1   ITU-R BT.601 (formerly CCIR-601)

BT.601 specifies methods for digitally coding video signals, employing the YCbCr color space for efficient use of channel bandwidth. It proposes 4:2:2 YCbCr as a preferred format for broadcast video. Synchronization signals (HSYNC, VSYNC, FIELD) and a clock are also provided to delineate the boundaries of active video regions.

Each BT.601 pixel component (Y, Cr, or Cb) is quantized to either 8 or 10 bits, and both NTSC and PAL have 720 pixels of active video per line. However, they differ in their vertical resolution. While 30 frames/sec NTSC has 525 lines (including vertical blanking, or retrace, regions), the 25 frame/sec rate of PAL is accommodated by adding 100 extra lines, or 625 total, to the PAL frame.

BT.601 specifies Y with a nominal range from 16 (total black) to 235 (total white). The color components Cb and Cr span from 16 to 240, but a value of 128 corresponds to no color. Sometimes, due to noise or rounding errors, a value might dip outside the nominal boundaries, but never all the way to 0 or 255.

9.4.2   ITU-R BT.656 (formerly CCIR-656)

Whereas BT.601 outlines how to digitally encode video, BT.656 actually defines the physical interfaces and data streams necessary to implement BT.601. It defines both bit-parallel and bit-serial modes. The bit-parallel mode requires only a 27 MHz clock (for NTSC, 30 frames/sec) and 8 or 10 data lines (depending on the resolution of pixel components). All synchronization signals are embedded in the data stream, so no extra hardware lines are required. Figure 9.8 shows some common timing relationships using embedded frame syncs, including the standard BT.656 4:2:2 YCbCr format, as well as "BT.656-style" 4:4:4 YCbCr and RGB formats.

The bit-serial mode requires only a multiplexed 10-bit-per-pixel serial data stream over a single channel, but it involves complex synchronization, spectral shaping and clock recovery conditioning. Furthermore, the bit clock rate runs close to 300 MHz, so it can be challenging to implement bit-serial BT.656 in many systems. In this chapter, we'll focus our attention on the bit-parallel mode only.

The frame partitioning and data stream characteristics of ITU-R BT.656 are shown in Figures 9.9 and 9.10, respectively, for 525/60 (NTSC) and 625/50 (PAL) systems.

Common Digital Video Formats

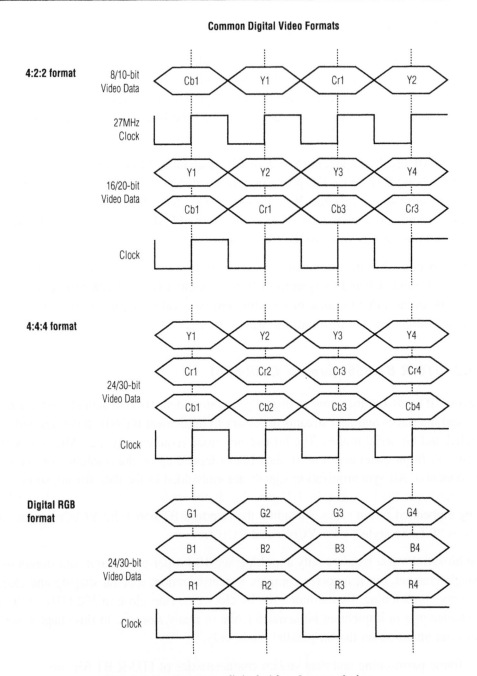

Figure 9.8: Common digital video format timing

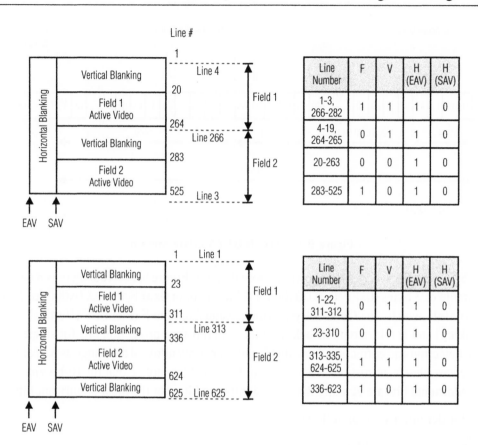

Figure 9.9: ITU-R BT.656 frame partitioning

In BT.656, the Horizontal (H), Vertical (V), and Field (F) signals are sent as an embedded part of the video data stream in a series of bytes that form a control word. The Start of Active Video (SAV) and End of Active Video (EAV) signals indicate the beginning and end of data elements to read in on each line. SAV occurs on a 1-to-0 transition of H, and EAV occurs on a 0-to-1 transition of H. An entire field of video is composed of Active Video, Horizontal Blanking (the space between an EAV and SAV code) and Vertical Blanking (the space where V = 1).

A field of video commences on a transition of the F bit. The "odd field" is denoted by a value of F = 0, whereas F = 1 denotes an even field. Progressive video makes no distinction between Field 1 and Field 2, whereas interlaced video requires each field to be handled uniquely, because alternate rows of each field combine to create the actual video frame.

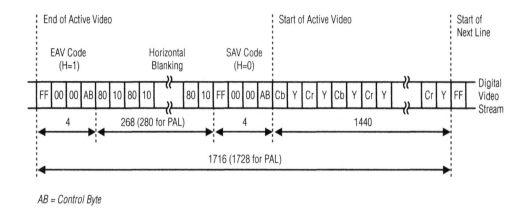

AB = Control Byte

Figure 9.10: ITU-R BT.656 data stream

The SAV and EAV codes are shown in more detail in Figure 9.11. Note there is a defined preamble of 3 bytes (0xFF, 0x00, 0x00 for 8-bit video, or 0x3FF, 0x000, 0x000 for 10-bit video), followed by the Control Byte, which, aside from the F (Field), V (Vertical Blanking) and H (Horizontal Blanking) bits, contains four protection bits for single-bit error detection and correction. Note that F and V are only allowed to change as part of EAV sequences (that is, transitions from H = 0 to H = 1). Also, notice that for 10-bit video, the two additional bits are actually the least-significant bits, not the most-significant bits.

The bit definitions are as follows:

- F = 0 for Field 1

- F = 1 for Field 2

	8-bit Data								10-bit Data	
	D9 (MSB)	D8	D7	D6	D5	D4	D3	D2	D1	D0
Preamble	1	1	1	1	1	1	1	1	1	1
	0	0	0	0	0	0	0	0	0	0
	0	0	0	0	0	0	0	0	0	0
Control Byte	1	F	V	H	P3	P2	P1	P0	0	0

P3:P0 = Error detection/correction bits

Figure 9.11: SAV/EAV preamble codes

- V = 1 during Vertical Blanking

- V = 0 when not in Vertical Blanking

- H = 0 at SAV

- H = 1 at EAV

- P3 = V XOR H

- P2 = F XOR H

- P1 = F XOR V

- P0 = F XOR V XOR H

The vertical blanking interval (the time during which V = 1) can be used to send nonvideo information, like audio, teletext, closed-captioning, or even data for interactive television applications. BT.656 accommodates this functionality through the use of ancillary data packets. Instead of the "0xFF, 0x00, 0x00" preamble that normally precedes control bytes, the ancillary data packets all begin with a "0x00, 0xFF, 0xFF" preamble. For further information about ancillary data types and formats, refer to Reference 7.

Assuming that ancillary data is not being sent, during horizontal and vertical blanking intervals the (Cb, Y, Cr, Y, Cb, Y, ...) stream is (0x80, 0x10, 0x80, 0x10, 0x80, 0x10...). Also, note that because the values 0x00 and 0xFF hold special value as control preamble demarcators, they are not allowed as part of the active video stream. In 10-bit systems, the values (0x000 through 0x003) and (0x3FC through 0x3FF) are also reserved, so as not to cause problems in 8-bit implementations.

9.5 A Systems View of Video

Figure 9.12 shows a typical end-to-end embedded digital video system. In one case, a video source feeds into a media processor (after being digitized by a video decoder, if necessary). There, it might be compressed via a software encoder before being stored locally or sent over the network.

In an opposite flow, a compressed stream is retrieved from a network or from mass storage. It is then decompressed via a software decoder and sent directly to a digital output display (like a TFT-LCD panel), or perhaps it is instead converted to analog form by a video encoder for display on a conventional CRT.

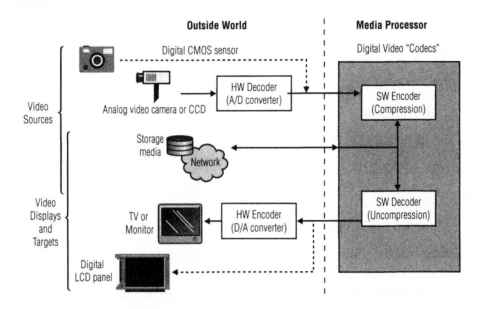

Figure 9.12: System video flow for analog/digital sources and displays

Keep in mind that compression and decompression represent only a subset of possible video processing algorithms that might run on the media processor. Still, for our purposes, they set a convenient template for discussion. Let's examine in more detail the video-specific portions of these data flows.

9.5.1   Video Sources

9.5.1.1 Analog Video Sources

A video decoder chip converts an analog video signal (e.g., NTSC, PAL, CVBS, S-Video) into a digital form (usually of the ITU-R BT.601/656 YCbCr or RGB variety). This is a complex, multi-stage process. It involves extracting timing information from the input, separating luma from chroma, separating chroma into Cr and Cb components, sampling the output data, and arranging it into the appropriate format. A serial interface such as SPI or I^{2}C configures the decoder's operating parameters. Figure 9.13 shows a block diagram of a representative video decoder.

9.5.1.2   Digital Video Sources

CCDs and CMOS Sensors   Camera sources today are overwhelmingly based on either charge-coupled device (CCD) or CMOS technology. Both of these technologies convert light into electrical signals, but they differ in how this conversion occurs.

In CCD devices, an array of millions of light-sensitive picture elements, or pixels, spans the surface of the sensor. After exposure to light, the accumulated charge over the entire CCD pixel array is read out at one end of the device and then digitized via an Analog Front End (AFE) chip or CCD processor. On the other hand, CMOS sensors directly digitize the exposure level at each pixel site.

In general, CCDs have higher quality and better noise performance, but they are not power-efficient. CMOS sensors are easy to manufacture and have low power dissipation, but at reduced quality. Part of the reason for this is because the transistors at each pixel site tend to occlude light from reaching part of the pixel. However, CMOS has started giving CCD a run for its money in the quality arena, and increasing numbers of mid-tier camera sensors are now CMOS-based.

Regardless of their underlying technology, all pixels in the sensor array are sensitive to grayscale intensity—from total darkness (black) to total brightness (white). The extent to which they're sensitive is known as their "bit depth." Therefore, 8-bit pixels can distinguish between 2^8, or 256, shades of gray, whereas 12-bit pixel values differentiate between 4096 shades. Layered over the entire pixel array is a color filter that segments each pixel into several color-sensitive "subpixels." This arrangement allows a measure of different color intensities at each pixel site. Thus, the color at each pixel location can be viewed as the sum of its red, green and blue channel light content, superimposed in an additive manner. The higher the bit depth, the more colors that can be generated in the RGB space. For example, 24-bit color (8 bits each of R,G,B) results in 2^{24}, or 16.7 million, discrete colors.

Bayer Pattern   In order to properly represent a color image, a sensor needs three color samples—most commonly, red, green and blue—for every pixel location. However, putting three separate sensors in every camera is not a financially tenable solution (although lately such technology is becoming more practical). What's more, as sensor resolutions increase into the 5–10 megapixel range, it becomes apparent that some form of image compression is necessary to prevent the need to output 3 bytes (or worse yet, three 12-bit words for higher-resolution sensors) for each pixel location.

Not to worry, because camera manufacturers have developed clever ways of reducing the number of color samples necessary. The most common approach is to use a Color Filter Array (CFA), which measures only a single color at any given pixel location.

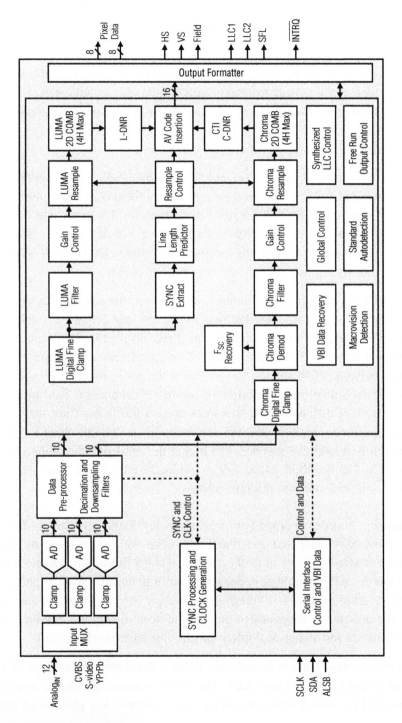

Figure 9.13: Block diagram of ADV7183B video decoder

Then, the results can be interpolated by the image processor to appear as if three colors were measured at every location.

The most popular CFA in use today is the Bayer pattern, shown in Figure 9.14. This scheme, invented by Kodak, takes advantage of the fact that the human eye discerns differences in green-channel intensities more than red or blue changes. Therefore, in the Bayer color filter array, the green subfilter occurs twice as often as either the blue or red subfilter. This results in an output format informally known as "4:2:2 RGB," where four green values are sent for every two red and blue values.

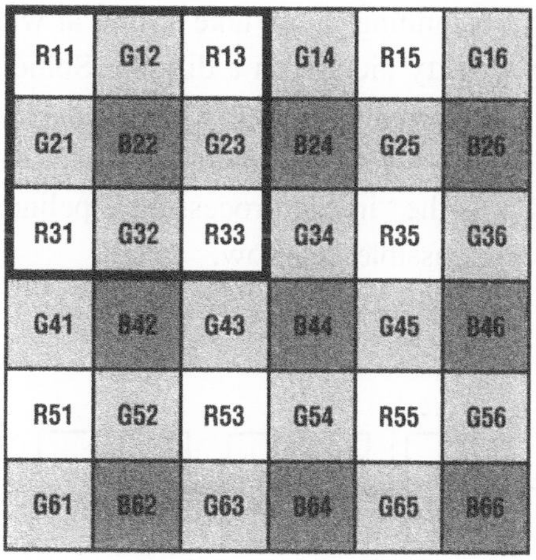

R11	G12	R13	G14	R15	G16
G21	B22	G23	B24	G25	B26
R31	G32	R33	G34	R35	G36
G41	B42	G43	B44	G45	B46
R51	G52	R53	G54	R55	G56
G61	B62	G63	B64	G65	B66

$G_{22}=(G_{12}+G_{21}+G_{23}+G_{32})/4$

$R_{22}=(R_{11}+R_{13}+R_{31}+R_{33})/4$

$B_{22}=B_{22}$, actual sampled value

Figure 9.14: Bayer pattern image sensor arrangement

Connecting to Image Sensors   CMOS sensors ordinarily output a parallel digital stream of pixel components in either YCbCr or RGB format, along with horizontal and vertical synchronization signals and a pixel clock. Sometimes, they allow for an external clock and sync signals to control the transfer of image frames out from the sensor.

CCDs, on the other hand, usually hook up to an Analog Front End (AFE) chip, such as the AD9948, that processes the analog output signal, digitizes it, and generates appropriate timing to scan the CCD array. A processor supplies synchronization signals to the AFE, which needs this control to manage the CCD array. The digitized parallel output stream from the AFE might be in 10-bit, or even 12-bit, resolution per pixel component.

Recently, LVDS (low-voltage differential signaling) has become an important alternative to the parallel data bus approach. LVDS is a low-cost, low-pin-count, high-speed serial interconnect that has better noise immunity and lower power consumption than the standard parallel approach. This is important as sensor resolutions and color depths increase, and as portable multimedia applications become more widespread.

Image Pipe   Of course, the picture-taking process doesn't end at the sensor; on the contrary, its journey is just beginning. Let's take a look at what a raw image has to go through before becoming a pretty picture on a display. Sometimes this is done within the sensor electronics module (especially with CMOS sensors), and other times these steps must be performed by the media processor. In digital cameras, this sequence of processing stages is known as the "image processing pipeline," or just "image pipe." Refer to Figure 9.15 for one possible dataflow.

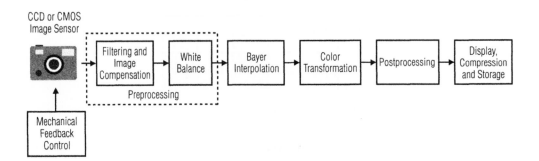

Figure 9.15: Example image processing pipeline

Mechanical Feedback Control   Before the shutter button is even released, the focus and exposure systems work with the mechanical camera components to control lens position based on scene characteristics. *Auto-exposure* algorithms measure brightness over discrete scene regions to compensate for overexposed or underexposed areas by manipulating shutter speed and/or aperture size. The net goals here are to maintain relative contrast between different regions in the image and to achieve a target average luminance.

Auto-focus algorithms divide into two categories. Active methods use infrared or ultrasonic emitters/receivers to estimate the distance between the camera and the object being photographed. Passive methods, on the other hand, make focusing decisions based on the received image in the camera.

In both of these subsystems, the media processor manipulates the various lens and shutter motors via PWM output signals. For auto-exposure control, it also adjusts the Automatic Gain Control (AGC) circuit of the sensor.

Preprocessing As we discussed earlier, a sensor's output needs to be *gamma-corrected* to account for eventual display, as well as to compensate for nonlinearities in the sensor's capture response.

Since sensors usually have a few inactive or defective pixels, a common preprocessing technique is to eliminate these via median filtering, relying on the fact that sharp changes from pixel to pixel are abnormal, since the optical process blurs the image somewhat.

Filtering and Image Compensation This set of algorithms accounts for the physical properties of lenses that warp the output image compared to the actual scene the user is viewing. Different lenses can cause different distortions; for instance, wide-angle lenses create a "barreling" or "bulging" effect, while telephoto lenses create a "pincushion" or "pinching" effect. Lens *shading distortion* reduces image brightness in the area around the lens. *Chromatic aberration* causes color fringes around an image. The media processor needs to mathematically transform the image in order to correct for these distortions.

Another area of preprocessing is *image stability compensation*, or *hand-shaking correction*. Here, the processor adjusts for the translational motion of the received image, often with the help of external transducers that relate the real-time motion profile of the sensor.

White Balance Another stage of preprocessing is known as *white balance*. When we look at a scene, regardless of lighting conditions, our eyes tend to normalize everything to the same set of natural colors. For instance, an apple looks deep red to us whether we're indoors under fluorescent lighting or outside in sunny weather. However, an image sensor's "perception" of color depends largely on lighting conditions, so it needs to map its acquired image to appear "lighting-agnostic" in its final output. This mapping can be done either manually or automatically.

In manual systems, you point your camera at an object you determine to be "white," and the camera will then shift the "color temperature" of all images it takes to accommodate this mapping. Automatic White Balance (AWB), on the other hand, uses inputs from the image sensor and an extra white balance sensor to determine what should be regarded as "true white" in an image. It tweaks the relative gains between the R, G and B channels of the image. Naturally, AWB requires more image processing than manual methods, and it's another target of proprietary vendor algorithms.

Bayer Interpolation Demosaicking, or interpolation of the Bayer data, is perhaps the most crucial and numerically intensive operation in the image pipeline. Each camera manufacturer typically has its own "secret recipe," but in general, the approaches fall into a few main algorithm categories.

Nonadaptive algorithms like bilinear interpolation or bicubic interpolation are among the simplest to implement, and they work well in smooth areas of an image. However, edges and texture-rich regions present a challenge to these straightforward implementations. Adaptive algorithms, those that change behavior based on localized image traits, can provide better results.

One example of an adaptive approach is *edge-directed reconstruction*. Here, the algorithm analyzes the region surrounding a pixel and determines in which direction to perform interpolation. If it finds an edge nearby, it interpolates along the edge, rather than across it. Another adaptive scheme assumes a *constant hue* for an entire object, and this prevents abrupt changes in color gradients within individual objects.

Many other demosaicking approaches exist, some involving frequency-domain analysis, Bayesian probabilistic estimation, and even neural networks. For an excellent survey of the various methods available, refer to Reference 11.

Color Transformation In this stage, the interpolated RGB image is transformed to the targeted output color space (if not already in the right space). For compression or display to a television, this will usually involve an RGB $\rightarrow$ YCbCr matrix transformation, often with another gamma correction stage to accommodate the target display. The YCbCr outputs may also be chroma subsampled at this stage to the standard 4:2:2 format for color bandwidth reduction with little visual impact.

Postprocessing In this phase, the image is perfected via a variety of filtering operations before being sent to the display and/or storage media. For instance, edge enhancement, pixel thresholding for noise reduction, and color-artifact removal are all common at this stage.

Display/Compression/Storage Once the image itself is ready for viewing, the image pipe branches off in two different directions. In the first, the postprocessed image is output to the target display, usually an integrated LCD screen (but sometimes an NTSC/PAL television monitor, in certain camera modes). In the second, the image is sent to the media processor's compression algorithm, where industry-standard compression techniques (JPEG, for instance) are applied before the picture is stored locally in some storage medium (usually a nonvolatile flash memory card).

9.5.2 Video Displays

9.5.2.1 Analog Video Displays

Video Encoder A video encoder converts a digital video stream into an analog video signal. It typically accepts a YCbCr or RGB video stream in either ITU-R BT.656 or BT.601 format and converts it to a signal compliant with one of several different output standards (e.g., NTSC, PAL, SECAM). A host processor controls the encoder via a serial interface like SPI or I^2C, programming such settings as pixel timing, input/output formats, and luma/chroma filtering. Figure 9.16 shows a block diagram of a representative encoder. Video encoders commonly output in one or more of the following analog formats:

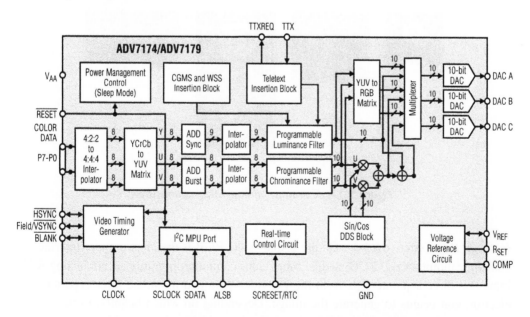

Figure 9.16: Block diagram of ADV7179 video encoder

CVBS—This acronym stands for Composite Video Baseband Signal (or Composite Video Blanking and Syncs.) Composite video connects through the ubiquitous yellow RCA jack shown in Figure 9.17a. It contains luma, chroma, sync and color burst information all on the same wire.

S Video—Using the jack shown in Figure 9.17b sends the luma and chroma content separately. Separating the brightness information from the color difference signals dramatically improves image quality, which accounts for the popularity of S Video connections on today's home theater equipment.

Component Video—Also known as YPbPr, this is the analog version of YCbCr digital video. Here, the luma and each chroma channel are brought out separately, each with its own timing. This offers maximum image quality for analog transmission. Component connections are very popular on higher-end home theater system components like DVD players and A/V receivers (Figure 9.17c).

Analog RGB has separate channels for red, green and blue signals. This offers similar image quality to component video, but it's normally used in the computer graphics realm (see Figure 9.17d), whereas component video is primarily employed in the consumer electronics arena.

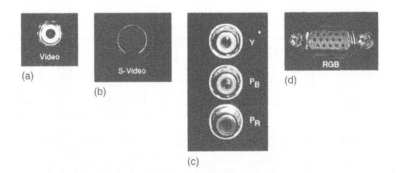

Figure 9.17: Common analog video connectors

Cathode Ray Tubes (CRTs) On the display side, RGB is the most popular interface to computer monitors and LCD panels. Most older computer monitors accept analog RGB inputs on three separate pins from the PC video card and modulate three separate electron gun beams to generate the image. Depending on which beam(s) excite a phosphor point on the screen, that point will glow either red, green, blue, or some combination of these colors. This is different from analog television, where a composite signal, one that includes all color information superimposed on a single input, modulates a single electron beam. Newer computer monitors use DVI, or Digital Visual Interface, to accept RGB information in both digital and analog formats.

The main advantages of CRTs are that they're very inexpensive and can produce more colors than a comparably sized LCD panel. Also, unlike LCDs, they can be viewed from any angle. On the downside, CRTs are very bulky, emit considerable electromagnetic radiation, and can cause eyestrain due to their refresh-induced flicker.

9.5.2.2   Digital Video Displays

Liquid Crystal Display (LCD) Panels   There are two main categories of LCD technology: passive matrix and active matrix. In the former (whose common family members include STN, or Super Twisted Nematic, derivatives), a glass substrate imprinted with rows forms a "liquid crystal sandwich" with a substrate imprinted with columns. Pixels are constructed as row-column intersections. Therefore, to activate a given pixel, a timing circuit energizes the pixel's column while grounding its row. The resultant voltage differential untwists the liquid crystal at that pixel location, which causes it to become opaque and block light from coming through.

Straightforward as it is, passive matrix technology does have some shortcomings. For one, screen refresh times are relatively slow (which can result in "ghosting" for fastmoving images). Also, there is a tendency for the voltage at a row-column intersection to bleed over into neighboring pixels, partly untwisting the liquid crystals and blocking some light from passing through the surrounding pixel area. To the observer, this blurs the image and reduces contrast. Moreover, viewing angle is relatively narrow.

Active matrix LCD technology improves greatly upon passive technology in these respects. Basically, each pixel consists of a capacitor and transistor switch. This arrangement gives rise to the more popular term, "Thin-Film Transistor Liquid Crystal Display (TFT-LCD)." To address a particular pixel, its row is enabled, and then a voltage is applied to its column. This has the effect of isolating only the pixel of interest, so others in the vicinity don't turn on. Also, since the current to control a given pixel is reduced, pixels can be switched at a faster rate, which leads to faster refresh rates for TFT-LCDs over passive displays. What's more, modulating the voltage level applied to the pixel allows many discrete levels of brightness. Today, it is common to have 256 levels, corresponding to 8 bits of intensity.

Connecting to a TFT-LCD panel can be a confusing endeavor due to all of the different components involved. First, there's the *panel* itself, which houses an array of pixels arranged for strobing by row and column, referenced to the pixel clock frequency.

The *Backlight* is often a CCFL (Cold Cathode Fluorescent Lamp), which excites gas molecules to emit bright light while generating very little heat. CCFLs exhibit durability, long life, and straightforward drive requirements. LEDs are also a popular backlight method, mainly for small- to mid-sized panels. They have the advantages of low cost, low operating voltage, long life, and good intensity control. However, for larger panel sizes, LED backlights can draw a lot of power compared with comparable CCFL solutions.

An *LCD Controller* contains most of the circuitry needed to convert an input video signal into the proper format for display on the LCD panel. It usually includes a timing generator that controls the synchronization and pixel clock timing to the individual pixels on the panel. However, in order to meet LCD panel size and cost requirements, sometimes timing generation circuitry needs to be supplied externally. In addition to the standard synchronization and data lines, timing signals are needed to drive the individual rows and columns of the LCD panel. Sometimes, spare general-purpose PWM (pulse-width modulation) timers on a media processor can substitute for this separate chip, saving system cost.

Additional features of LCD controller chips include on-screen display support, graphics overlay blending, color lookup tables, dithering and image rotation. The more elaborate chips can be very expensive, often surpassing the cost of the processor to which they're connected.

An *LCD Driver* chip is necessary to generate the proper voltage levels to the LCD panel. It serves as the "translator" between the output of the LCD Controller and the LCD Panel. The rows and columns are usually driven separately, with timing controlled by the timing generator. Liquid crystals must be driven with periodic polarity inversions, because a dc current will stress the crystal structure and ultimately deteriorate it. Therefore, the voltage polarity applied to each pixel varies either on a per-frame, per-line, or per-pixel basis, depending on the implementation.

With the trend toward smaller, cheaper multimedia devices, there has been a push to integrate these various components of the LCD system. Today, integrated TFT-LCD modules exist that include timing generation and drive circuitry, requiring only a data bus connection, clocking/synchronization lines, and power supplies. Some panels are also available with a composite analog video input, instead of parallel digital inputs.

OLED (Organic Light-Emitting Diode) Displays The "Organic" in OLED refers to the material that's encased between two electrodes. When charge is applied through this organic substance, it emits light. This display technology is still very new, but it holds promise by improving upon several deficiencies in LCD displays. For one, it's a self-emissive technology and does not require a backlight. This has huge implications for saving panel power, cost and weight—an OLED panel can be extremely thin. Additionally, it can support a wider range of colors than comparable LCD panels can, and its display of moving images is also superior to that of LCDs. What's more, it supports a wide viewing angle and provides high contrast. OLEDs have an electrical signaling and data interface similar to that of TFT-LCD panels.

For all its advantages, so far the most restrictive aspect of the OLED display is its limited lifetime. The organic material breaks down after a few thousand hours of use, although this number has now improved in some displays to over 10,000 hours—quite suitable for many portable multimedia applications. It is here where OLEDs have their brightest future—in cellphones, digital cameras, and the like. However, it is also quite possible that we'll ultimately see televisions or computer monitors based on OLED technology.

9.6   Embedded Video Processing Considerations

9.6.1   Video Port Features

To handle video streams, processors must have a suitable interface that can maintain a high data transfer rate into and out of the part. Some processors accomplish this through an FPGA and/or FIFO connected to the processor's external memory interface. Typically, this device will negotiate between the constant, relatively slow stream of video into and out of the processor, and the sporadic but speedy bursting nature of the external memory controller.

However, there are problems with this arrangement. For example, FPGAs and FIFOs are expensive, often costing as much as the video processor itself. Additionally, using the external memory interface for video transfer steals bandwidth from its other prime use in these systems—moving video buffers back and forth between the processor core and external memory.

Therefore, a dedicated video interface is highly preferable for media processing systems. On Blackfin processors, this is the Parallel Peripheral Interface (PPI). The PPI is a multifunction parallel interface that can be configured between 8 and 16 bits in width. It supports bidirectional data flow and includes three synchronization lines and a clock pin for connection to an externally supplied clock. The PPI can gluelessly decode ITU-R BT.656 data and can also interface to ITU-R BT.601 video streams. It is flexible and fast enough to serve as a conduit for high-speed analog-to-digital converters (ADCs) and digital-to-analog converters (DACs). It can also emulate a host interface for an external processor. It can even act as a glueless TFT-LCD controller.

The PPI has some built-in features that can reduce system costs and improve data flow. For instance, in BT.656 mode the PPI can decode an input video stream and automatically ignore everything except active video, effectively reducing an NTSC input video stream rate from 27 Mbytes/s to 20 Mbytes/s, and markedly reducing the amount of off-chip memory needed to handle the video. Alternately, it can ignore active video regions and only read in ancillary data that's embedded in vertical blanking intervals. These modes are shown pictorially in Figure 9.18.

Likewise, the PPI can ignore every other field of an interlaced stream; in other words, it will not forward this data to the DMA controller. While this instantly decimates input bandwidth requirements by 50%, it also eliminates 50% of the source content, so sometimes this tradeoff might not be acceptable. Nevertheless, this can be a useful feature when the input video resolution is much greater than the required output resolution.

On a similar note, the PPI allows "skipping" of odd- or even-numbered elements, again saving DMA bandwidth for the skipped pixel elements. For example, in a 4:2:2 YCbCr stream, this feature allows only luma or chroma elements to be read in, providing convenient partitioning of an algorithm between different processors; one can read in the luma, and the other can read the chroma. Also, it provides a simple way to convert an image or video stream to grayscale (luma-only). Finally, in highspeed converter

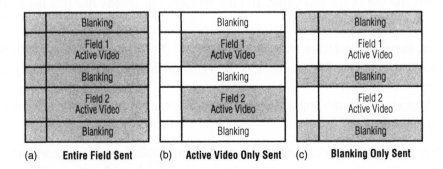

Figure 9.18: Selective masking of BT.656 regions in PPI

applications with interleaved I/Q data, this feature allows partitioning between these in-phase and quadrature components.

Importantly, the PPI is format-agnostic, in that it is not hardwired to a specific video standard. It allows for programmable row lengths and frame sizes. This aids applications that need, say, CIF or QCIF video instead of standard NTSC/PAL formats. In general, as long as the incoming video has the proper EAV/SAV codes (for BT.656 video) or hardware synchronization signals (for BT.601 video), the PPI can process it.

9.6.1.1 Packing

Although the BT.656 and BT.601 recommendations allow for 10-bit pixel elements, this is not a very friendly word length for processing. The problem is, most processors are very efficient at handling data in 8-bit, 16-bit or 32-bit chunks, but anything in-between results in data movement inefficiencies. For example, even though a 10-bit pixel value is only 2 bits wider than an 8-bit value, most processors will treat it as a 16-bit entity with the 6 most significant bits (MSBs) set to 0. Not only does this consume bandwidth on the internal data transfer (DMA) buses, but it also consumes a lot of memory—a disadvantage in video applications, where several entire frame buffers are usually stored in external memory.

A related inefficiency associated with data sizes larger than 8 bits is nonoptimal packing. Usually, a high-performance media processor will imbue its peripherals with a data packing mechanism that sits between the outside world and the internal data movement buses of the processor, and its goal is to minimize the overall bandwidth burden that the data entering or exiting the peripheral places on these buses.

Therefore, an 8-bit video stream clocking into a peripheral at 27 Mbytes/s might be packed onto a 32-bit internal data movement bus, thereby requesting service from this bus at a rate of only 27/4, or 6.75 MHz. Note that the overall data transfer rate remains the same (6.75 MHz × 32 bits = 27 Mbytes/s). In contrast, a 10-bit video stream running at 27 Mbytes/s would only be packed onto the 32-bit internal bus in two 16-bit chunks, reducing the overall transfer rate to 27/2, or 13.5 MHz. In this case, since only 10 data bits out of every 16 are relevant, 37.5% of the internal bus bandwidth is not used efficiently.

9.6.1.2 Possible Data Flows

It is instructive to examine some ways in which a video port connects in multimedia systems, to show how the system as a whole is interdependent on each component. In Figure 9.19a, an image source sends data to the PPI. The DMA engine then dispositions it to L1 memory, where the data is processed to its final form before being sent out through a high-speed serial port. This model works very well for low-resolution video processing and for image compression algorithms like JPEG, where small blocks of video (several lines worth) can be processed and are subsequently never needed again. This flow also can work well for some data converter applications.

In Figure 9.19b, the video data is not routed to L1 memory, but instead is directed to L3 memory. This configuration supports algorithms such as MPEG-2 and MPEG-4, which require storage of intermediate video frames in memory in order to perform temporal compression. In such a scenario, a bidirectional DMA stream between L1 and L3 memories allows for transfers of pixel macroblocks and other intermediate data.

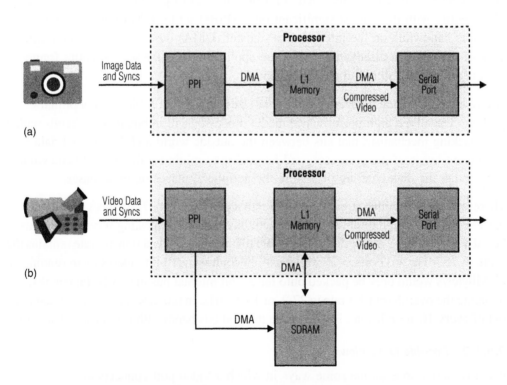

Figure 9.19: Possible video port data transfer scenarios

9.6.2   Video ALUs

Most video applications need to deal with 8-bit data, since individual pixel components (whether RGB or YCbCr) are usually byte quantities. Therefore, 8-bit video ALUs and byte-based address generation can make a huge difference in pixel manipulation. This is a nontrivial point, because embedded media processors typically operate on 16-bit or 32-bit boundaries.

The Blackfin processor has some instructions that are geared to processing 8-bit video data efficiently. Table 9.2 shows a summary of the specific instructions that can be used together to handle a variety of video operations.

Let's look at a few examples of how these instructions can be used.

The Quad 8-bit Subtract/Absolute/Accumulate (SAA) instruction is well-suited for block-based video motion estimation. The instruction subtracts 4 pairs of bytes, takes the absolute value of each difference, and accumulates the results. This all happens within a single cycle. The actual formula is shown below:

$$SAA = \sum_{i=0}^{N-1} \sum_{j=0}^{N-1} |a(i,j) - b(i,j)|$$

Consider the macroblocks shown in Figure 9.20a. The reference frame of 16 pixels × 16 pixels can be further divided into four groups. A very reasonable assumption is that neighboring video frames are correlated to each other. That is, if there is motion, then pieces of each frame will move in relation to macroblocks in previous frames. It takes less information to encode the movement of macroblocks than it does to encode each video frame as a separate entity—MPEG compression uses this technique.

This motion detection of macroblocks decomposes into two basic steps. Given a *reference macroblock* in one frame, we can search all surrounding macroblocks (*target macroblocks*) in a subsequent frame to determine the closest match. The offset in location between the reference macroblock (in Frame n) and the best-matching target macroblock (in Frame n + 1) is the motion vector.

Figure 9.20b shows how this can be visualized in a system.

- Circle = some object in a video frame
- Solid square = reference macroblock

Table 9.2: Native Blackfin video instructions

Instruction	Description	Algorithm Use
Byte Align	Copies a contiguous four-byte unaligned word from a combination of two data registers.	Useful for aligning data bytes for subsequent SIMD instructions.
Dual 16-Bit Add/Clip	Adds two 8-bit unsigned values to two 16-bit signed values, then limits to an 8-bit unsigned range.	Used primarily for video motion compensation algorithms.
Dual 16-Bit Accumulator Extraction w/ Addition	Adds together the upper half words and lower half words of each accumulator and loads into a destination register.	Used in conjunction with the Quad SAA instruction for motion estimation.
Quad 8-Bit Add	Adds two unsigned quad byte number sets.	Useful for providing packed data arithmetic typical in video processing applications.
Quad 8-Bit Average – Byte	Computes the arithmetic average of two quad byte number sets byte-wise.	Supports binary interpolation used in fractional motion search and motion estimation.
Quad 8-Bit Average – Half Word	Computes the arithmetic average of two quad byte number sets byte-wise and places results on half word boundaries.	Supports binary interpolation used in fractional motion search and motion estimation.
Quad 8-Bit Pack	Packs four 8-bit values into one 32-bit register.	Prepares data for ALU operation.
Quad 8-Bit Subtract	Subtracts two quad byte number sets, byte-wise.	Provides packed data arithmetic in video applications.
Quad 8-Bit Subtract – Absolute - Accumulate	Subtracts four pairs of values, takes that absolute value, and accumulates.	Useful for block-based video motion estimation.
Quad 8-Bit Unpack	Copies four contiguous bytes from a pair of source registers.	Unpacks data after ALU operation.

- Dashed square = search area for possible macroblocks

- Gray square = best-matching target macroblock (i.e., the one representing the motion vector of the circle object)

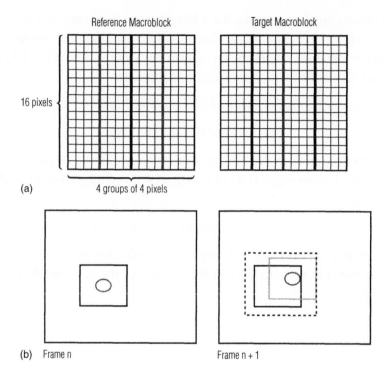

Figure 9.20: Illustration of Subtract-Absolute-Accumulate (SAA) instruction

The SAA instruction on a Blackfin processor is single-cycle because it utilizes four 8-bit ALUs in each clock cycle. We can implement the following loop to iterate over each of the four entities shown in Figure 9.20.

```
/* used in a loop that iterates over an image block */

/* I0 and I1 point to the pixels being operated on */

SAA (R1:0,R3:2) || R1 = [I0++] || R2 = [I1++]; /* compute absolute
    difference and accumulate */

SAA (R1:0,R3:2) (R) || R0 = [I0++] || R3 = [I1++];

SAA (R1:0,R3:2) || R1 = [I0 ++ M3] || R2 = [I1++M1]; /* after
    fetch of 4th word of target block, pointer is made to point to
    the next row */

SAA (R1:0,R3:2) (R) || R0 = [I0++] || R2 = [I1++];
```

Let's now consider another example, the 4-Neighborhood Average computation whose basic kernel is shown in Figure 9.21a. Normally, four additions and one division (or multiplication or shift) are necessary to compute the average. The BYTEOP2P instruction can accelerate the implementation of this filter.

The value of the center pixel of Figure 9.21b is defined as:

$$x = \text{Average}(xN, xS, xE, xW)$$

The BYTEOP2P can perform this kind of average on two pixels (Figures 9.21c,d) in one cycle. So, if $x1 = \text{Average}(x1N, x1S, x1E, x1W)$, and $x2 = \text{Average}(x2N, x2S, x2E, x2W)$, then

```
R3 = BYTEOP2P(R1:0, R3:2);
```

will compute both pixel averages in a single cycle, assuming the x1 (N, S, E, W) information is stored in registers R1 and R0, and the x2 (N, S, E, W) data is sourced from R3 and R2.

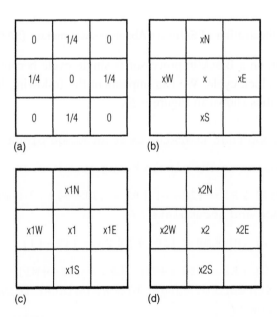

Figure 9.21: Neighborhood Average computation

9.6.3   DMA Considerations

Two-dimensional DMA capability offers several system-level benefits. For starters, two-dimensional DMA can facilitate transfers of macroblocks to and from external memory, allowing data manipulation as part of the actual transfer. This eliminates the overhead typically associated with transferring noncontiguous data. It can also allow the system to minimize data bandwidth by selectively transferring, say, only the desired region of an image, instead of the entire image.

As another example, 2D DMA allows data to be placed into memory in a sequence more natural to processing. For example, as shown in Figure 9.22, RGB data may enter a processor's L2 memory from a CCD sensor in interleaved RGB888 format, but using 2D DMA, it can be transferred to L3 memory in separate R, G and B planes. Interleaving/deinterleaving color space components for video and image data saves additional data moves prior to processing.

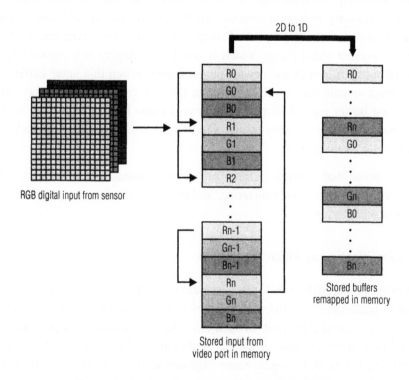

Figure 9.22: Deinterleaving data with 2D DMA

9.6.3.1   *Planar versus Interleaved Buffer Formats*

How do you decide whether to structure your memory buffers as interleaved or planar? The advantage to interleaved data is that it's the natural output format of image sensors, and the natural input format for video encoders. However, planar buffers (that is, separate memory regions for each pixel component) are often more effective structures for video algorithms, since many of them (JPEG and MPEG included) work on luma and chroma channels separately. What's more, accessing planar buffers in L3 is more efficient than striding through interleaved data, because the latency penalty for SDRAM page misses is spread out over a much larger sample size when the buffers are structured in a planar manner.

9.6.3.2   *Double-Buffering*

We have previously discussed the need for double-buffering as a means of ensuring that current data is not overwritten by new data until you're ready for this to happen. Managing a video display buffer serves as a perfect example of this scheme. Normally, in systems involving different rates between source video and the final displayed content, it's necessary to have a smooth switchover between the old content and the new video frame. This is accomplished using a double-buffer arrangement. One buffer points to the present video frame, which is sent to the display at a certain refresh rate. The second buffer fills with the newest output frame. When this latter buffer is full, a DMA interrupt signals that it's time to output the new frame to the display. At this point, the first buffer starts filling with processed video for display, while the second buffer outputs the current display frame. The two buffers keep switching back and forth in a "ping-pong" arrangement.

It should be noted that multiple buffers can be used, instead of just two, in order to provide more margin for synchronization, and to reduce the frequency of interrupts and their associated latencies.

9.6.4   *Classification of Video Algorithms*

Video processing algorithms fall into two main categories: spatial and temporal. It is important to understand the different types of video processing in order to discern how the data is treated efficiently in each case. Many algorithms use a combination of spatial and temporal techniques (dubbed "spatiotemporal processing") to achieve their desired results. For now, let's gain some exposure to the different classifications of image and video processing algorithms.

Spatial processing can either be pixel-wise or block-based. Pixel-by-pixel processing is spatially local. Every pixel is touched in turn to create a single output, and no result accumulation is necessary from previous (or future) pixels to generate the output. An example of pixel-wise processing is color space conversion from RGB to YCbCr. Often, this type of computation can be performed on the fly as the data arrives, without the need to store source image data to external memory.

Block-based algorithms, while spatially localized, typically need a group of pixel data in order to generate a result. A simple example of this is an averaging engine, where a given pixel is low-pass filtered by averaging its value with those of its neighbors. A more complex example is a 5×5 2D convolution kernel used to detect edge information in an image frame, drawing on the information from a wide neighborhood of surrounding pixels. In these block-based algorithms, there needs to be some source image storage in memory, but the entire video frame doesn't necessarily need to be saved, since it might be acceptable to operate on a few lines of video at a time. This could make the difference between needing external SDRAM for frame storage versus using faster, smaller on-chip L1 or L2 memory as a line buffer.

Thus far, we've considered only spatially *local* processing. However, another class of algorithms involves spatially *global* processing. These routines operate on an entire image or video frame at a time, and usually need to traverse often between an L3 frame buffer and L1 storage of source data and intermediate results. Examples of spatially global algorithms include the Discrete Cosine Transform (DCT), the Hough Transform that searches for line segments within an image, and histogram equalization algorithms that depend on analysis of the entire image before modifying the image based on these results.

Temporal processing algorithms strive to exploit changes or similarities in specific pixels or pixel regions between frames. One example of temporal processing is conversion of an interlaced field to a progressive scan format via line doubling and/or filtering. Here, the assumption is that two interlaced fields are temporally similar enough to allow one field's content alone to be used to generate the "missing" field. When this is not the case—for instance, on motion boundaries or scene changes—there will be an aberration in the video output, but the preprocessing serves to soften this effect.

Another example where temporal processing is important is in video surveillance systems. Here, each frame is compared to the preceding one(s) to determine if there are enough differences to warrant an "alarm condition." In other words, an algorithm determines if something moved, vanished or appeared unexpectedly across a sequence of frames.

The majority of widely used video compression algorithms combine spatial and temporal elements into a category called spatiotemporal processing. Spatially, each image frame is divided into macroblocks of, say, 16 × 16 pixels. Then, these macroblocks are tracked and compared frame-to-frame to derive motion estimation and compensation approximations.

9.6.5 Bandwidth Calculations

Let's spend a few minutes going through some rough bandwidth calculations for video, in order to bring to light some important concepts about estimating required throughput.

First, we'll consider a progressive-scan VGA (640 × 480) CMOS sensor that connects to a processor's video port, sending 8-bit 4:2:2 YCbCr data at 30 frames/sec.

(640 × 480 pixels/frame)(2 bytes/pixel)(30 frames/sec) ≈ 18.4 Mbytes/second

This represents the raw data throughput into the processor. Often, there will also be blanking information transmitted on each row, such that the actual pixel clock here would be somewhere around 24 MHz. By using a sensor mode that outputs 16 bits at a time (luma and chroma on one clock cycle), the required clock rate is halved, but the total throughput will remain unchanged, since the video port takes in twice as much data on each clock cycle.

Now, let's switch over to the display side. Let's consider a VGA LCD display with a "RGB565" characteristic. That is, each RGB pixel value is packed into 2 bytes for display on an LCD screen (as we'll discuss soon). The nuance here is that LCDs usually require a refresh rate somewhere in the neighborhood of 50 to 80 Hz. Therefore, unless we use a separate LCD controller chip with its own frame memory, we need to update the display at this refresh rate, even though our sensor input may only be changing at a 30 frames/sec rate. So we have, for example,

(640 × 480 pixels/frame)(2 bytes/pixel)(75 frames/sec refresh) = 46.08 Mbytes/second

Since we typically transfer a parallel RGB565 word on every clock cycle, our pixel clock would be somewhere in the neighborhood of 25 MHz (accounting for blanking regions).

9.6.5.1 Sample Video Application Walk-Through

Let's walk through the sample system of Figure 9.23 to illustrate some fundamental video processing steps present in various combinations in an embedded video application. In the diagram, an interlaced-scan CMOS sensor sends a 4:2:2 YCbCr video stream through the processor's video port, at which point it is deinterlaced and scan-rate converted. Then it passes through some computational algorithm(s) and is prepared for output to an LCD panel. This preparation involves chroma resampling, gamma correction, color conversion, scaling, blending with graphics, and packing into the appropriate output format for display on the LCD panel. Note that this system is only provided as an example, and not all of these components are necessary in a given system. Additionally, these steps may occur in different order than shown here.

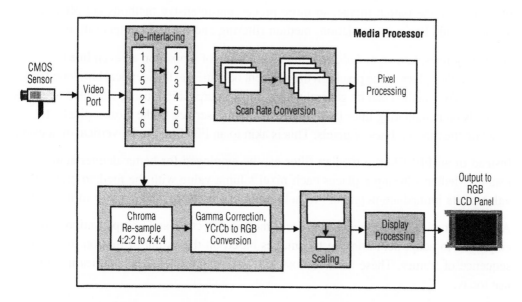

Figure 9.23: Example flow of Video In to LCD Out, with processing stages in-between

9.6.6 Deinterlacing

When taking video source data from a camera that outputs interlaced NTSC data, it's often necessary to deinterlace it so that the odd and even lines are interleaved in memory, instead of being located in two separate field buffers. Deinterlacing is not only needed for efficient block-based video processing, but also for displaying interlaced video in progressive format (for instance, on an LCD panel). There are many ways to deinterlace, each with its own benefits and drawbacks. We'll review some methods here, but for a good discussion of deinterlacing approaches, refer to Reference 13.

Perhaps the most obvious solution would be to simply read in Field 1 (odd scan lines) and Field 2 (even scan lines) of the source data and interleave these two buffers to create a sequentially scanned output buffer. While this technique, called *weave*, is commonly used, it has the drawback of causing artifacts related to the fact that the two fields are temporally separated by 60 Hz, or about 16.7 ms. Therefore, while this scheme may suit relatively static image frames, it doesn't hold up well in scenes with high motion content.

To mitigate these artifacts and conserve system bandwidth, sometimes it's acceptable to just read in Field 1 and duplicate it into Field 2's row locations in the output buffer. This "line doubling" approach understandably reduces image resolution and causes blockiness in the output image, so more processing-intensive methods are often used. These include linear interpolation, median filtering and motion compensation.

Averaging lines, known as *bob*, determines the value of a pixel on an even line by averaging the values of the pixels on the adjacent odd lines, at the identical horizontal index location. This method can be generalized to linear interpolation, using weighted averages of pixels on neighboring lines (with highest weights ascribed to the closest pixels) to generate the missing lines of pixels. This is akin to an FIR filter in the vertical dimension.

Instead of an FIR filter, a median filter can be employed for better deinterlacing results. Median filtering replaces each pixel's luma value with the median gray-scale value of its immediate neighbors to help eliminate high-frequency noise in the image.

As an alternative deinterlacing approach, motion detection and compensation can be employed to adapt deinterlacing techniques based on the extent of motion in a given sequence of frames. These are the most advanced deinterlacing techniques in general use today.

9.6.7 Scan Rate Conversion

Once the video has been deinterlaced, a scan-rate conversion process may be necessary, in order to insure that the input frame rate matches the output display refresh rate. In order to equalize the two, fields may need to be dropped or duplicated. Of course, as with deinterlacing, some sort of filtering is preferable in order to smooth out high-frequency artifacts caused by creating abrupt frame transitions.

A special case of frame rate conversion that's employed to convert a 24 frame/sec stream (common for 35-mm and 70-mm movie recordings) to the 30 frame/sec required by NTSC video is *3:2 pulldown*. For instance, motion pictures recorded at 24 fps would run 25% faster (= 30/24) if each film frame is used only once in an NTSC video system. Therefore, 3:2 pulldown was conceived to adapt the 24 fps stream into a 30 fps video sequence. It does so by repeating frames in a certain periodic pattern, shown in Figure 9.24.

3:2 Pulldown and Frame Rate Conversion
(24 fps progressive movie → 30 fps interlaced TV)

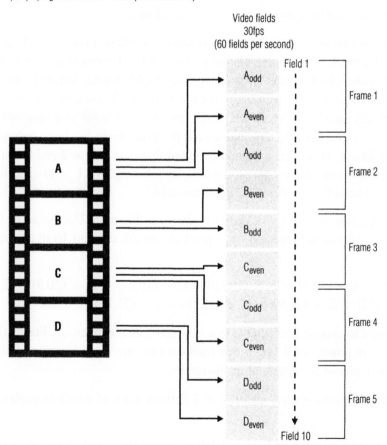

Figure 9.24: 3:2 pulldown frame repetition pattern

9.6.8 Pixel Processing

As we discussed above, there are lots of video algorithms in common use, and they're stratified into spatial and temporal classifications. One particularly common video operator is the two-dimensional (2D) convolution kernel, which is used for many different forms of image filtering. We'll discuss the basics of this kernel here.

9.6.8.1 2D Convolution

Since a video stream is really an image sequence moving at a specified rate, image filters need to operate fast enough to keep up with the succession of input images. Thus, it is imperative that image filter kernels be optimized for execution in the lowest possible number of processor cycles. This can be illustrated by examining a simple image filter set based on two-dimensional convolution.

Convolution is one of the fundamental operations in image processing. In a two-dimensional convolution, the calculation performed for a given pixel is a weighted sum of intensity values from pixels in its immediate neighborhood. Since the neighborhood of a mask is centered on a given pixel, the mask usually has odd dimensions. The mask size is typically small relative to the image, and a 3 × 3 mask is a common choice, because it is computationally reasonable on a per-pixel basis, but large enough to detect edges in an image. However, it should be noted that 5 × 5, 7 × 7 and beyond are also widely used. Camera image pipes, for example, can employ 11 × 11 (and larger!) kernels for extremely complex filtering operations.

The basic structure of the 3 × 3 kernel is shown in Figure 9.25a. As an example, the output of the convolution process for a pixel at row 20, column 10 in an image would be:

$Out(20,10) = A \times (19,9) + B \times (19,10) + C \times (19,11) + D \times (20,9) + E \times (20,10) + F \times (20,11) + G \times (21,9) + H \times (21,10) + I \times (21,11)$

It is important to choose coefficients in a way that aids computation. For instance, scale factors that are powers of 2 (including fractions) are preferred because multiplications can then be replaced by simple shift operations.

Figure 9.25b–e shows several useful 3 × 3 kernels, each of which is explained briefly below.

The Delta Function shown in Figure 9.25b is among the simplest image manipulations, passing the current pixel through without modification.

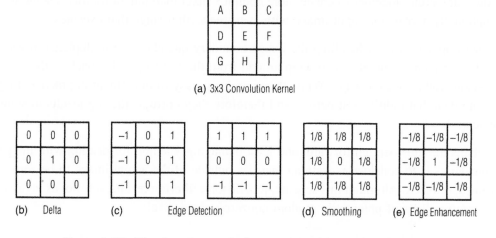

(a) 3x3 Convolution Kernel

(b) Delta (c) Edge Detection (d) Smoothing (e) Edge Enhancement

Figure 9.25: The 3 × 3 convolution mask and how it can be used

Figure 9.25c shows two popular forms of an edge detection mask. The first one detects vertical edges, while the second one detects horizontal edges. High output values correspond to higher degrees of edge presence.

The kernel in Figure 9.25d is a smoothing filter. It performs an average of the 8 surrounding pixels and places the result at the current pixel location. This has the result of "smoothing," or low-pass filtering, the image.

The filter in Figure 9.25e is known as an "unsharp masking" operator. It can be considered as producing an edge-enhanced image by subtracting from the current pixel a smoothed version of itself (constructed by averaging the 8 surrounding pixels).

9.6.9 Dealing with Image Boundaries

What happens when a function like 2D convolution operates on pixels near an image's border regions? To properly perform pixel filtering requires pixel information "outside" these boundaries. There are a couple of remedies for this situation. The simplest is just to ignore these edge regions. That is, consider that a 5 × 5 convolution kernel needs 2 pixels to the left, top, bottom and right of the current pixel in order to perform properly. Therefore, why not just shave two rows off of the image in each direction, so as to guarantee that the kernel will always act on real data? Of course, this isn't always an ideal approach, since it throws out real image data. Also, in cases where filters are strung together to create more complex pixel manipulations, this scheme will continually narrow the input image with every new filter stage that executes.

Other popular ways of handling the image boundary quandary are to duplicate rows and/or columns of pixels, or to wrap around from the left (top) edge back to the previous right (bottom) edge. While these might be easy to implement in practice, they create data that didn't exist before, and therefore they corrupt filtering results to some extent.

Perhaps the most straightforward, and least damaging, method for dealing with image boundaries is to consider everything that lies outside of the actual image to be zero-valued, or black. Although this scheme, too, distorts filtering results, it is not as invasive as creating lines of potentially random nonzero-valued pixels.

9.6.10 *Chroma Resampling, Gamma Correction and Color Conversion*

Ultimately, the data stream in our example needs to be converted to RGB space. We already discussed earlier how to convert between 4:4:4 YCbCr and RGB spaces, via a 3×3 matrix multiplication. However, up to this point, our pixel values are still in 4:2:2 YCbCr space. Therefore, we need to resample the chroma values to achieve a 4:4:4 format. Then the transformation to RGB will be straightforward, as we've already seen.

Resampling from 4:2:2 to 4:4:4 involves interpolating Cb and Cr values for those Y samples that are missing one of these components. A clear-cut way to resample is to interpolate the missing chroma values from their nearest neighbors by simple averaging. That is, a missing Cb value at a pixel site would be replaced by the average of the nearest two Cb values. Higher-order filtering might be necessary for some applications, but this simplified approach is often sufficient. Another approach is to replicate the relevant chroma values of neighboring pixels for those values that are missing in the current pixel's representation.

In general, conversions from 4:1:1 space to 4:2:2 or 4:4:4 formats involves only a one-dimensional filter (with tap values and quantities consistent with the level of filtering desired). However, resampling from 4:2:0 format into 4:2:2 or 4:4:4 format involves vertical sampling as well, necessitating a two-dimensional convolution kernel.

Because chroma resampling and YCbCr $\rightarrow$ RGB conversion are both linear operations, it is possible to combine the steps into a single mathematical process, thus achieving 4:2:2 YCbCr $\rightarrow$ RGB conversion efficiently.

At this stage, gamma correction is often necessary as well. Because of its nonlinear nature, gamma correction is most efficiently performed via a table lookup, prior to color space conversion. Then the result of the conversion process will yield gamma-corrected RGB components (commonly known as R′G′B′), the appropriate format for output display.

9.6.11   Scaling and Cropping

Video scaling allows the generation of an output stream whose resolution is different from that of the input format. Ideally, the fixed scaling requirements (input data resolution, output panel resolution) are known ahead of time, in order to avoid the computational load of arbitrary scaling between input and output streams.

Depending on the application, scaling can be done either upwards or downwards. It is important to understand the content of the image to be scaled (e.g., the presence of text and thin lines). Improper scaling can make text unreadable or cause some horizontal lines to disappear in the scaled image.

The easiest way to adjust an input frame size to an output frame that's smaller is simply to crop the image. For instance, if the input frame size is 720 × 480 pixels, and the output is a VGA frame (640 × 480 pixels), you can drop the first 40 and the last 40 pixels on each line. The advantage here is that there are no artifacts associated with dropping pixels or duplicating them. Of course, the disadvantage is that you'd lose 80 pixels (about 11%) of frame content. Sometimes this isn't too much of an issue, because the leftmost and rightmost extremities of the screen (as well as the top and bottom regions) are often obscured from view by the mechanical enclosure of the display.

If cropping isn't an option, there are several ways to downsample (reduce pixel and/or line count) or upsample (increase pixel and/or line count) an image that allow tradeoffs between processing complexity and resultant image quality.

9.6.11.1   Increasing or Decreasing Pixels per Row

One straightforward method of scaling involves either dropping pixels (for downsampling) or duplicating existing pixels (for upsampling). That is, when scaling down to a lower resolution, some number of pixels on each line (and/or some number of lines per frame) can be thrown away. While this certainly reduces processing load, the results will yield aliasing and visual artifacts.

A small step up in complexity uses linear interpolation to improve the image quality. For example, when scaling down an image, filtering in either the horizontal or vertical direction obtains a new output pixel, which then replaces the pixels used during the interpolation process. As with the previous technique, information is still thrown away, and again artifacts and aliasing will be present.

If the image quality is paramount, there are other ways to perform scaling without reducing the quality. These methods strive to maintain the high frequency content of the image consistent with the horizontal and vertical scaling, while reducing the effects of aliasing. For example, assume an image is to be scaled by a factor of Y:X. To accomplish this scaling, the image could be upsampled ("interpolated") by a factor of Y, filtered to eliminate aliasing, and then downsampled ("decimated") by a factor of X, in a manner analogous to Figure 8.17 for audio. In practice, these two sampling processes can be combined into a single multirate filter.

9.6.11.2 Increasing or Reducing Lines per Frame

The guidelines for increasing or reducing the number of pixels per row generally extend to modifying the number of lines per frame of an image. For example, throwing out every other line (or one entire interlaced field) provides a quick method of reducing vertical resolution. However, as we've mentioned above, some sort of vertical filtering is necessary whenever removing or duplicating lines, because these processes introduce artifacts into the image. The same filter strategies apply here: simple vertical averaging, higher-order FIR filters, or multirate filters to scale vertically to an exact ratio.

9.6.12 Display Processing

9.6.12.1 Alpha Blending

Often it is necessary to combine two image and/or video buffers prior to display. A practical example of this is overlaying of icons like signal strength and battery level indicators onto a cellular phone's graphics display. An example involving two video streams is picture-in-picture functionality.

When combining two streams, you need to decide which stream "wins" in places where content overlaps. This is where alpha blending comes in. It defines a variable alpha (α) that indicates a "transparency factor" between an overlay stream and a background stream as follows:

Output value = $\alpha \times$ (foreground pixel value) + $(1 - \alpha) \times$ (background pixel value)

As the equation shows, an α value of 0 results in a completely transparent overlay, whereas a value of 1 results in a completely opaque overlay that disregards the background image entirely.

Sometimes α is sent as a separate channel along with the pixel-wise luma and chroma information. This results in the notation "4:2:2:4," where the last digit indicates an alpha key that accompanies each 4:2:2 pixel entity. Alpha is coded in the same way as the luma component, but often only a few discrete levels of transparency are needed for most applications. Sometimes a video overlay buffer is premultiplied by alpha or premapped via a lookup table, in which case it's referred to as a "shaped" video buffer.

9.6.12.2 Compositing

The act of compositing involves positioning an overlay buffer inside a larger image buffer. Common examples are a "picture-in-picture" mode on a video display, and placement of graphics icons over the background image or video. In general, the composition function can take several iterations before the output image is complete. In other words, there may be many "layers" of graphics and video that combine to generate a composite image.

Two-dimensional DMA capability is very useful for compositing, because it allows the positioning of arbitrarily sized rectangular buffers inside a larger buffer. One thing to keep in mind is that any image cropping should take place after the composition process, because the positioned overlay might violate any newly cropped boundaries. Of course, an alternative is to ensure that the overlay won't violate the boundaries in the first place, but this is sometimes asking too much!

9.6.12.3 Chroma Keying

The term "chroma keying" refers to a process by which a particular color (usually blue or green) in a background image is replaced by the content in an overlay image when the two are composited together. This provides a convenient way to combine two video images by purposefully tailoring parts of the first image to be replaced by the appropriate sections of the second image. Chroma keying can be performed in either software or hardware on a media processor. As a related concept, sometimes a "transparent color" can be used in an overlay image to create a pixelwise "$\alpha = 0$" condition when a separate alpha channel is not available.

9.6.12.4 Rotation

Many times it is also necessary to rotate the contents of an output buffer before display. Almost always, this rotation is in some multiple of 90 degrees. Here's another instance where 2D DMA is extremely useful.

9.6.12.5 Output Formatting

Most color LCD displays targeted for consumer applications (TFT-LCDs) have a digital RGB interface. Each pixel in the display actually has 3 subpixels—one each with red, green and blue filters—that the human eye resolves as a single color pixel. For example, a 320 × 240 pixel display actually has 960 × 240 pixel components, accounting for the R, G and B subpixels. Each subpixel has 8 bits of intensity, thus forming the basis of the common 24-bit color LCD display.

The three most common configurations use either 8 bits per channel for RGB (RGB888 format), 6 bits per channel (RGB666 format), or 5 bits per channel for R and B, and 6 bits for G (RGB565 format).

RGB888 provides the greatest color clarity of the three. With a total of 24 bits of resolution, this format provides over 16 million shades of color. It offers the high resolution and precision needed in high-performance applications like LCD TVs.

The RGB666 format is popular in portable electronics. Providing over 262,000 shades of color, this format has a total of 18 bits of resolution. However, because the 18-pin (6 + 6 + 6) data bus doesn't conform nicely to 16-bit processor data paths, a popular industry compromise is to use 5 bits each of R and B, and 6 bits of G (5 + 6 + 5 = a 16-bit data bus) to connect to a RGB666 panel. This scenario works well because green is the most visually important color of the three. The least-significant bits of both red and blue are tied at the panel to their respective most-significant bits. This ensures a full dynamic range for each color channel (full intensity down to total black).

9.7 Compression/Decompression

Over the past decade or so, image, video and audio standards all have undergone revolutionary changes. As Figure 9.26 conveys, whereas historical progress was primarily based on storage media evolution, the dawn of digital media has changed this path into one of compression format development. That is, once in the digital realm, the main challenge becomes "How much can I condense this content while maximizing

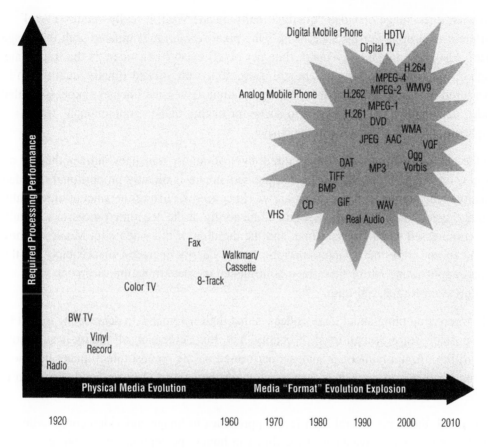

Figure 9.26: Historical progression of media

attainable quality?" Naturally, consumers view this format proliferation as a distraction that obscures the real prize—access to a plethora of multimedia content.

9.7.1   Lossless versus Lossy Compression

"Lossless" compression implies that the original content can be reconstructed identically, whereas "lossy" compression allows some degradation in reconstruction quality in order to achieve much higher compression ratios. In JPEG, for instance, lossless compression typically achieves ratios of perhaps 2:1, whereas lossy schemes can compress a source image by 10:1 or more.

So what does image or video "quality" really mean? Well, it really entails several different measures. For instance, what's the pixel *resolution* compared with the image size? Also, how much *color depth* (bits per pixel) exists? For video, is the frame rate high or low? Too low a frame rate will cause noticeable flicker. Finally, quality can be measured by what is sufficient for the application. Given the intensive processing that video demands, it's advantageous to strive for quality that's "good enough" for an application, but not excessively expensive.

In the compression domain, the required level of quality translates into another set of issues. For one, the bit rate of the compressed stream is directly proportional to video quality. This bit rate is bounded by real-world concerns, such as the amount of available storage space for the compressed source, the ability of the decoding processor to handle the compressed stream in real time, and the duration of the video clip. Moreover, how high can you raise the compression ratio before quality degrades unacceptably for the target application? All of these factors influence the determination of proper compression format and ratio.

The interesting thing about lossy codecs is that their net goal is to achieve the lowest bit rate possible for a desired level of quality. Yet, this perception of quality is subjective and differs from a numerical analysis performed on the reconstituted image. In other words, all that matters is that we humans think the end result looks good—it doesn't matter if the metrics tell us otherwise.

In general, there are several high-level approaches to image and video compression, most of which take advantage of the limits in human perception. For instance, as with 4:2:2 YCbCr broadcast video, we can reduce the color bandwidth in an image to account for the fact that our eyes are much more sensitive to intensity than to color. Spatial techniques reduce compressed image size by accounting for visual similarities within image regions, such that a small amount of information can be made to represent a large object or scene portion.

For video, temporal compression takes advantage of similarities between neighboring frames to encode only the differences between them, instead of considering each frame as its own separate image to be compressed.

9.7.2 Image Compression

Let's start by surveying some widespread image compression schemes. We'll then extend our discussion to popular video compression formats.

9.7.2.1 BMP

The BMP bitmapped file representation is known to all Microsoft Windows® users as its native graphics format. A BMP file can represent images at color depths up to 24 bits (8 bits each for R, G and B). Unfortunately, though, this format is normally an uncompressed representation, and file sizes can be quite unwieldy at 24 bits per pixel. As a result, the BMP format is not Internet-friendly. Luckily, other file formats have been devised that provide a range of "quality versus file size" trade-offs to facilitate image transfer across a network.

9.7.2.2 GIF

GIF, or the Graphics Interchange Format, was developed by CompuServe in the 1980s for their own network infrastructure. This is still a popular Internet format today, especially for compressing computer graphics images with limited color palettes, since GIF itself is palette-based, allowing only 256 colors in a single image. GIF compresses better than JPEG does for these types of images, but JPEG excels for photographs and other images having many more than just 256 colors.

9.7.2.3 PNG

The development of the royalty-free "portable network graphics (PNG)" format is a direct product of some licensing fees that beset the GIF format. Although much of the GIF-relevant patents have expired, PNG is still seen as an improvement over GIF in color depth, alpha channel support (for transparency), and other areas. It is considered a successor to GIF.

9.7.2.4 JPEG

JPEG is perhaps the most popular compressed image format on the Internet today. As noted above, it's mainly intended for photographs, but not for line drawings and other graphics with very limited color palettes. The 10x–20x compression ratios of JPEG really got people excited about digital photos and sharing their pictures over bandwidth-limited networks. Of course, the higher the compression rate, the larger the loss of information. Nevertheless, even at relatively high rates, JPEG results in a much smaller file size but retains a comparable visual quality to the original bitmapped image.

9.7.2.5 JPEG2000

JPEG2000, also known as J2K, is a successor to JPEG, addressing some of its fundamental limitations while remaining backward-compatible with it. It achieves better compression ratios (roughly 30% better for excellent quality), and it performs well on binary (black and white) images, computer graphics, and photographs alike. Like JPEG, it has both lossy and lossless modes. J2K also supports "region of interest" compression, where selective parts of an image can be encoded in higher quality than other regions are.

J2K isn't DCT-based like JPEG. Instead, it's wavelet-based. Essentially, wavelets involve a frequency-domain transform like the DCT, but they also include information pertaining to spatial location of the frequency elements. Reference 3 offers a good overview of wavelet technology.

9.7.3   Video Compression

Video data is enormously space-consuming, and its most daunting aspect is that it just keeps coming! One second of uncompressed 4:2:2 NTSC video requires 27 Mbytes of storage, and a single minute requires 1.6 Gbytes! Because raw video requires such dedicated high bandwidth, the industry takes great pains to avoid transferring it in uncompressed form whenever possible. Compression algorithms have been devised that reduce the bandwidth requirements for NTSC/PAL video from tens of megabytes per second down to just a few megabits per second, with adjustable tradeoffs in video quality for bandwidth efficiency.

Compression ratio itself is highly dependent on the content of the video being compressed, as well as on the compression algorithm being used. Some encoders, like MPEG-4 and H.264, have the ability to encode foreground and background information separately, leading to higher compression than those that deal with the entire image as a whole. Images that have very little spatial variation (e.g., big blocks of uniform color), as well as video that remains fairly static from frame to frame (e.g., slow-changing backgrounds) can be highly compressed. On the other hand, detailed images and rapidly varying content reduce compression ratios.

Most embedded media processing developers tend to treat video codecs as "drop-in" code modules that the processor vendor or a third-party source has optimized specifically for that platform. In other words, most embedded multimedia designers today expect to acquire the video codec, as opposed to design it from scratch. Thus, their chief concern is which codec(s) to pick for their application. Let's review a little about some popular ones below. For detailed operation about how a particular algorithm works, refer to References 2 and 3.

There is no right answer in choosing a video codec—it really depends on what your end application needs are. If the content source, distribution and display are all well controlled, it may be possible to use a proprietary codec for increased performance that's not encumbered by the dictates of the standards bodies. Or, if the idea of paying royalties doesn't appeal to you, open-source codecs are very attractive alternatives. Content protection, network streaming capability, and bit rate characteristics (constant versus variable, encoding options, etc.) are other important features that differentiate one codec from another. And don't forget to consider what codecs are already ported to your chosen media processor, how many implementations exist, how well-supported they are, and what it will cost to use them (up-front and per-unit).

Table 9.3 lists many of the popular video coding standards, as well as their originating bodies. Let's review a little about each here.

Table 9.3: Popular video coding standards

Video Coding Standards	Year Introduced	Originating Body
M-JPEG	1980s	ISO
H.261	1990	ITU-T
MPEG-1	1993	ISO/IEC
MPEG-2 (H.262)	1994/1995	ITU-T and ISO/IEC
H.263, H.263+, H.263++	1995–2000	ITU-T
MPEG-4	1998–2001	ISO/IEC
DV	1996	IEC
QuickTime	1990s	Apple Computer
RealVideo	1997	RealNetworks
Windows Media Video	1990s	Microsoft
Ogg Theora	2002 (alpha)	Xiph.org Foundation
H.264 (MPEG-4 Part 10)	2003	ITU-T and ISO/IEC

9.7.3.1   Motion JPEG (M-JPEG)

Although not specifically encompassed by the JPEG standard, M-JPEG offered a convenient way to compress video frames before MPEG-1 formalized a method. Essentially, each video frame is individually JPEG-encoded, and the resulting compressed frames are stored (and later decoded) in sequence.

There also exists a Motion JPEG2000 codec that uses J2K instead of JPEG to encode individual frames. It is likely that this codec will replace M-JPEG at a rate similar to the adoption of J2K over JPEG.

9.7.3.2   H.261

This standard, developed in 1990, was the first widespread video codec. It introduced the idea of segmenting a frame into 16×16 "macroblocks" that are tracked between frames to determine motion compensation vectors. It is mainly targeted at video-conferencing applications over ISDN lines ($p \times 64$ kbps, where p ranges from 1 to 30). Input frames are typically CIF at 30 fps, and output compressed frames occupy 64–128 kbps for 10 fps resolution. Although still used today, it has mostly been superseded by its successor, H.263.

9.7.3.3   MPEG-1

When MPEG-1 entered the scene in the early 1990s, it provided a way to digitally store audio and video and retrieve it at roughly VHS quality. The main focus of this codec was storage on CD-ROM media. Specifically, the prime intent was to allow storage and playback of VHS-quality video on a 650–750 Mbyte CD, allowing creation of a so-called "Video CD (VCD)." The combined video/audio bit stream could fit within a bandwidth of 1.5 Mbit/sec, corresponding to the data retrieval speed from CD-ROM and digital audio tape systems at that time.

At high bit rates, it surpasses H.261 in quality (allowing over 1 Mbps for CIF input frames). Although CIF is used for some source streams, another format called SIF is perhaps more popular. It's 352×240 pixels per frame, which turns out to be about ½ of a full 720×480 NTSC frame. MPEG-1 was intended for compressing SIF video at 30 frames/sec, progressive-scan. Compared to H.261, MPEG-1 adds bidirectional motion prediction and half-pixel motion estimation. Although still used by diehards for VCD creation, MPEG-1 pales in popularity today compared with MPEG-2.

9.7.3.4   MPEG-2

Driven by the need for scalability across various end markets, MPEG-2 soon improved upon MPEG-1, with the capability to scale in encoded bit rate from 1 Mbit/sec up to 30 Mbit/sec. This opened the door to higher-performance applications, including DVD videos and standard-and high-definition TV. Even at the lower end of MPEG-2 bit rates, the quality of the resulting stream is superior to that of an MPEG-1 clip.

This complex standard is composed of 10 parts. The "Visual" component is also called H.262. Whereas MPEG-1 focused on CDs and VHS-quality video, MPEG-2 achieved DVD-quality video with an input conforming to BT.601 (NTSC 720 × 480 at 30 fps), and an output in the range of 4–30 Mbps, depending on which "performance profile" is chosen. MPEG-2 supports interlaced and progressive scanning.

9.7.3.5   H.263

This codec is ubiquitous in videoconferencing, outperforming H.261 at all bit rates. Input sources are usually QCIF or CIF at 30 fps, and output bit rates can be less than 28.8 kbps at 10 fps, for the same performance as H.261. Therefore, whereas H.261 needs an ISDN line, H.263 can use ordinary phone lines. H.263 finds use in end markets like video telephony and networked surveillance (including Internet-based applications).

9.7.3.6   MPEG-4

MPEG-4 starts from a baseline of H.263 and adds several improvements. Its prime focus is streaming multimedia over a network. Because the network usually has somewhat limited bandwidth, typical input sources for MPEG-4 codecs are CIF resolution and below. MPEG-4 allows different types of coding to be applied to different types of objects. For instance, static background textures and moving foreground shapes are treated differently to maximize the overall compression ratio. MPEG-4 uses several different performance profiles, the most popular among them "Simple" (similar to MPEG-1) and "Advanced Simple" (field-based like MPEG-2). Simple Profile is suitable for low video resolutions and low bit rates, like streaming video to cellphones. Advanced Simple Profile, on the other hand, is intended for higher video resolutions and higher bit rates.

9.7.3.7   DV

DV is designed expressly for consumer (and subsequently professional) video devices. Its compression scheme is similar in nature to Motion JPEGs, and it accepts the BT.601

sampling formats for luma and chroma. Its use is quite popular in camcorders, and it allows several different "playability" modes that correspond to different bit rates and/or chroma subsampling schemes. DV is commonly sent over a IEEE 1394 ("FireWire") interface, and its bit rate capability scales from the 25 Mbit/sec commonly used for standard-definition consumer-grade devices, up beyond 100 Mbit/sec for high-definition video.

9.7.3.8 QuickTime

Developed by Apple Computer, QuickTime comprises a collection of multimedia codecs and algorithms that handle digital video, audio, animation, images and text. QuickTime 7.0 complies with MPEG-4 and H.264. In fact, the QuickTime file format served as the basis for the ISO-based MPEG-4 standard, partly because it provided a full end-to-end solution, from video capture, editing and storage, to content playback and distribution.

9.7.3.9 RealVideo

RealVideo is a proprietary video codec developed by RealNetworks. RealVideo started as a low-bit-rate streaming format to PCs, but now it extends to the portable device market as well, for streaming over broadband and cellular infrastructures. It can be used for live streaming, as well as for video-on-demand viewing of a previously downloaded file. RealNetworks bundles RealVideo with RealAudio, its proprietary audio codec, to create a RealMedia container file that can be played with the PC application RealPlayer.

9.7.3.10 Windows Media Video (WMV)/VC-1

This codec is a Microsoft-developed variant on MPEG-4 (starting from WMV7). It also features digital rights management (DRM) for control of how content can be viewed, copied, modified or replayed. In an effort to standardize this proprietary codec, Microsoft submitted WMV9 to the Society of Motion Picture and Television Engineers (SMPTE) organization, and it's currently a draft standard under the name "VC-1."

9.7.3.11 Theora

Theora is an open-source, royalty-free video codec developed by the Xiph.org Foundation, which also has developed several open-source audio codecs. Theora is based on the VP3 codec that On2 Technologies released into the public domain. It mainly competes with low-bit-rate codecs like MPEG-4 and Windows Media Video.

9.7.3.12 H.264

H.264 is also known as MPEG-4 Part 10, H.26L, or MPEG-4 AVC ("Advanced Video Coding") profile. It actually represents a hallmark of cooperation, in the form of a joint definition between the ITU-T and ISO/IEC committees. H.264's goal is to reduce bit rates by 50% or more for comparable video quality, compared with its predecessors. It works across a wide range of bit rates and video resolutions. H.264's dramatic bit rate reductions come at the cost of significant implementation complexity. This complexity is what limits H.264 encoding at D-1 format today to only higher-end media processors, often requiring two discrete devices that split the processing load.

9.7.4 Encoding/Decoding on an EMP

Today, many of the video compression algorithms mentioned above (e.g., MPEG-4, WMV9 and H.264) are all competing for industry and consumer acceptance. Their emergence reflects the shift from dedicated media stored and played on a single device to a "streaming media" concept, where content flows between "media nodes" connected by wireless or wireline networks. These transport networks might be either low-bandwidth (e.g., Bluetooth) or high-bandwidth (e.g., 100BaseT Ethernet), and it's essential to match the streaming content with the transport medium. Otherwise, the user experience is lackluster—slow-moving video, dropped frames and huge processing delays.

As Figure 9.27 shows, these new algorithms all strive to provide higher-resolution video at lower bit rates than their predecessors (e.g., MPEG-1 and MPEG-2) and at comparable or better quality. But that's not all; they also extend to many more applications than the previous generation of standards by offering features like increased scalability (grabbing only the subset of the encoded bit stream needed for the application), better immunity to errors, and digital rights management capabilities (to protect content from unauthorized viewing or distribution).

However, the downside of these newer algorithms is that they generally require even more processing power than their predecessors in order to achieve their remarkable results.

Decoding is usually a much lighter processing load than encoding, by about half. Decoders have an operational flow that's usually specified by the standards body that designed the compression algorithm. This same body, however, does not dictate how an encoder must work. Therefore, encoders tend to use a lot more processing power in

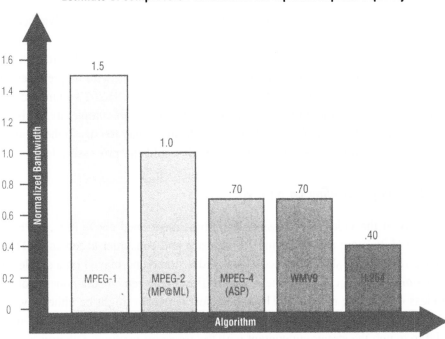

Figure 9.27: Progress in video encode algorithms leads to lower bit rates and higher storage density for comparable video quality

order to squeeze the input stream into the lowest possible output bit rate. In fact, the model of "Encode on the PC, Decode on the Device" is still common today for devices such as PDAs, MP3 players, and the like.

With the proliferation of video and audio standards, interoperability between systems has become increasingly challenging. Along with the burgeoning number of available formats, the numbers and types of networks and media devices have also grown significantly. As discussed above, these different network service levels result in multiple compression standards being used to fulfill the multimedia requirements of a system.

9.7.4.1 *Transcoding*

In the end, the consumer demands a seamless pathway through which to share, transfer and experience media. Today, "transcoding" of multimedia content is helping to remove barriers between different encoding/decoding platforms. Transcoding allows data to move transparently between and within wired and wireless networks. At the highest level, it is simply a way to convert from one encoding scheme to another. Importantly, it allows bit stream reduction with a manageable loss in content quality, appropriate for the available data channel. In other words, transcoding allows compression of a data stream to meet a target storage size or a specific transmission path. For example, MPEG-2 content from a DVD server might be transcoded into MPEG-4 before transmission across a wireless network to a PDA, thus resulting in a significant bit stream reduction (which translates into dollars saved). On the PDA, the data can then be transcoded into an appropriate format before being played.

The type of operation described above is known as "format transcoding." Another related process is "sample-rate-reduction transcoding," which lowers the streaming media bit rate by reducing the displayed resolution or frame rate. In either case, transcoding is performed in order to match the needs of the target device, and thus it can place a large burden on the system processor.

As far as transcoding is concerned, totally programmable solutions like convergent processors have an advantage over fixed-function ASICs, because they allow completely flexible conversion between arbitrary media formats. Additionally, the fast clock rates of convergent processors can enable real-time transcoding, where the processor can decode the input stream while simultaneously encoding it into the target format. This speeds up the process and saves interim storage space. That is, the entire source stream doesn't have to be downloaded to memory before starting to transcode it into the desired format. In this scenario, the processor performs what would otherwise be an effort by multiple ASICs that are each geared toward a limited set of video encoding or decoding functions.

References

1. Charles Poynton, *Digital Video and HDTV: Algorithms and Interfaces*, Boston: Morgan Kaufman, 2003.
2. Keith Jack, *Video Demystified*, Fourth edition, Elsevier (Newnes), 2004.
3. Peter Symes, *Video Compression Demystified*, New York: McGraw-Hill, 2001.
4. ITU-R BT.601-5 — Studio Encoding Parameters of Digital Television for Standard 4:3 and Wide-Screen 16:9 Aspect Ratios, 1995.
5. ITU-R BT.656-4—Interfaces for Digital Component Video Signals in 525-Line and 625-Line Television Systems Operating at the 4:2:2 Level of Recommendation ITU-R BT.601 (Part A), 1998.
6. Analog Devices, Inc., ADSP-BF533 Blackfin Processor Hardware Reference, Rev 3.0, September 2004.
7. Keith Jack, "BT.656 Video Interface for ICs," Intersil, Application Note AN9728.2, July 2002.

8. "Digital CCD Camera Design Guide," Cirrus Logic, Application Note AN80, December 1998.

9. Andrew Wilson, "CMOS Sensors Contend for Camera Designs," Vision Systems Design, September 2001.

10. Rajeev Ramanath, Wesley E. Snyder, Youngjun Yoo, and Mark S. Drew, "Color Image Processing Pipeline," IEEE Signal Processing Magazine (22:1), 2005.

11. Bahadir K. Gunturk, John Glotzbach Yucel Altunbasak, Ronald W. Schafer, and Russl M. Mersereau, "Demosaicking: Color Filter Array Interpolation," IEEE Signal Processing Magazine (22:1), 2005.

12. Gary Sullivan and Stephen Estrop, "Video Rendering with 8-Bit YUV Formats," MSDN Library, August 2003, http://Msdn.microsoft.com/library/default.asp?url=/library/en-us/dnwmt/html/yuvformats.asp.

13. G. de Haan and EB Bellers, "Deinterlacing—An overview," Proceedings of the IEEE, Vol. 86, No. 9, Sep. 1998.

14. Matthias Braun, Martin Hahn, Jens-Rainer Ohm, and Maati Talmi, "Motion Compensating Real-Time Format Converter for Video on Multimedia Displays," Heinrich-Hertz-Institut and MAZ Brandenburg, Germany, http://www.ient.rwthaachen.de/engl/team/ohm/publi/icip971.pdf.

15. Dan Ramer, "What the Heck is 3:2 Pulldown?," www.dvdfile.com/news/special_report/production_a_z /3_2_pulldown.htm).

16. LeCroy Applications Brief, "Video Basics," Number LAB 1010. Revision 4, 2001,http://www.lecroy.com/tm/library/LABs/PDF/LAB1010.pdf.

17. Rajeev Ramanath, "Interpolation Methods for the Bayer Color Array," Master of Science Thesis, Department of Electrical and Computer Engineering, North Carolina State University, 2000, http://www.lib.ncsu.edu/theses/available/etd-20001120-131644/unrestricted/etd.pdf.

18. Martyn Williams, "Benefits of Future Displays Debated," PC World Online, April 9, 2003, http://www.pcworld.com/news/article/0,aid,110209,00.asp.

19. National Instruments, "Anatomy of a Video Signal," http://zone.ni.com/devzone/conceptd.nsf/webmain/0C19487AA97D229C86256 85E00803830?OpenDocument.

20. Kelin J. Kuhn, "EE 498: Conventional Analog Television — An Introduction," University of Washington, http://www.ee.washington.edu/conselec/CE/kuhn/ntsc/95x4.htm.

21. Wave Report, "OLED Tutorial," 2005, http://www.wave-report.com/tutorials/oled.htm.

22. Gary A. Thom and Alan R. Deutermann,"A Comparison of Video Compression Algorithms," Delta Information Systems, 2001, http://www.delta-info.com/DDV/downloads/Comparison%20White%20Paper.pdf.

23. Jan Rychter, "Video Codecs Explained," 2003, http://www.sentivision.com/technology/video-codecs/codecs-explained-en.html.

24. Robert Currier, "Digital Video Codec Choices, Part 1," 1995,http://www.synthetic-ap.com/qt/codec1.html.

25. Robert Kremens, Nitin Sampat, Shyam Venkataraman, and Thomas Yeh, "System Implications of Implementing Auto-Exposure on Consumer Digital Cameras," New York, http://www.cis.rit.edu/~rlkpci/ei99_Final.pdf.

26. Scott Wilkinson, "An Eye for Color, Part 1," *Ultimate AV Magazine*, February 2004, http://ultima-teavmag.com/features/204eye/index1.html.

27. Keith Jack, "YCbCr to RGB Considerations," Intersil, Application Note AN9717, March 1997.

CHAPTER 10

Programming Streaming FPGA Applications Using Block Diagrams in Simulink

Brian C. Richards
Chen Chang
John Wawrzynek
Robert W. Brodersen

This chapter—which was written by Brian C. Richards, Chen Chang, John Wawrzynek, Robert W. Brodersen—comes from the book Reconfigurable Computing: The Theory and Practice of FPGA-Based Computation *edited by Scott Hauck and André DeHon, ISBN: 9780123705228. (Phew, try saying this ten times quickly.)*

The book itself is divided into six major parts—hardware, programming, compilation/ mapping, application development, case studies, and future trends. The chapter presented here comes from "Part II: Programming Reconfigurable Systems."

"Although a system designer can use hardware description languages, such as VHDL and Verilog to program field-programmable gate arrays (FPGAs), the algorithm developer typically uses higher-level descriptions to refine an algorithm. As a result, an algorithm described in a language such as C or MATLAB® from The MathWorks is frequently re-entered by hand by the system designer, after which the two descriptions must be verified and refined manually. This can be extremely time-consuming.

"To avoid reentering a design when translating from a high-level simulation language to HDL, the algorithm developer can describe a system from the beginning using block diagrams in Simulink® from The Mathworks.

"In this chapter, a high-performance image-processing system is described using Simulink and mapped to an FPGA-based platform using a design flow built around the Xilinx System Generator tools. The system implements edge detection in real time on a digitized video stream and produces a corresponding video stream labeling the edges. The edges

can then be viewed on a high-resolution monitor. This design demonstrates how to describe a high-performance parallel datapath, implement control subsystems, and interface to external devices, including embedded processors."

—Clive "Max" Maxfield

Although a system designer can use hardware description languages, such as VHDL and Verilog to program field-programmable gate arrays (FPGAs), the algorithm developer typically uses higher-level descriptions to refine an algorithm. As a result, an algorithm described in a language such as Matlab or C is frequently reentered by hand by the system designer, after which the two descriptions must be verified and refined manually. This can be time consuming.

To avoid reentering a design when translating from a high-level simulation language to HDL, the algorithm developer can describe a system from the beginning using block diagrams in Matlab Simulink [1]. Other block diagram environments can be used in a similar way, but the tight integration of Simulink with the widely used Matlab simulation environment allows developers to use familiar data analysis tools to study the resulting designs. With Simulink, a single design description can be prepared by the algorithm developer and refined jointly with the system architect using a common design environment.

The single design entry is enabled by a library of Simulink operator primitives that have a direct mapping to HDL, using matching Simulink and HDL models that are cycle accurate and bit accurate between both domains. Examples and compilation environments include System Generator from Xilinx [2], Synplify DSP from Synplicity [3], and the HDL Coder from The Mathworks [1]. Using such a library, nearly any synchronous multirate system can be described, with high confidence that the result can be mapped to an FPGA given adequate resources.

In this chapter, a high-performance image-processing system is described using Simulink and mapped to an FPGA-based platform using a design flow built around the Xilinx System Generator tools. The system implements edge detection in real time on a digitized video stream and produces a corresponding video stream labeling the edges. The edges can then be viewed on a high-resolution monitor. This design demonstrates how to describe a high-performance parallel datapath, implement control subsystems, and interface to external devices, including embedded processors.

10.1   Designing High-Performance Datapaths using Stream-Based Operators

Within Simulink we employ a Synchronous Dataflow computational model (SDF). Each operator is executed once per clock cycle, consuming input values and producing new output values once per clock tick. This discipline is well suited for streambased design, encouraging both the algorithm designer and the system architect to describe efficient datapaths with minimal idle operations.

Clock signals and corresponding clock enable signals do not appear in the Simulink block diagrams using the System Generator libraries, but are automatically generated when an FPGA design is compiled. To support multirate systems, the System Generator library includes up-sample and down-sample blocks to mark the boundaries of different clock domains. When compiled to an FPGA, clock enable signals for each clock domain are automatically generated.

All System Generator components offer compile time parameters, allowing the designer to control data types and refine the behavior of the block. Hierarchical blocks, or *subsystems* in Simulink, can also have user-defined parameters, called *mask parameters*. These can be included in block property expressions within that subsystem to provide a means of generating a variety of behaviors from a single Simulink description. Typical mask parameters include data type and precision specification and block latency to control pipeline stage insertion. For more advanced library development efforts, the mask parameters can be used by a Matlab program to create a custom schematic at compile time.

The System Generator library supports fixed-point or Boolean data types for mapping to FPGAs. Fixed-point data types include signed and unsigned values, with bit width and decimal point location as parameters. In most cases, the output data types are inferred automatically at compile time, although many blocks offer parameters to define them explicitly.

Pipeline operators are explicitly placed into a design either by inserting delay blocks or by defining a delay parameter in selected functional blocks. Although the designer is responsible for balancing pipeline operators, libraries of high-level components have been developed and reused to hide pipeline-balancing details from the algorithm developer.

The Simulink approach allows us to describe highly concurrent SDF systems where many operators—perhaps the entire dataflow path—can operate simultaneously. With modern FPGAs, it is possible to implement these systems with thousands of simultaneous operators running at the system clock rate with little or no control logic, allowing complex, high-performance algorithms to be implemented.

10.2 An Image-Processing Design Driver

The goal of the edge detection design driver is to generate a binary bit mask from a video source operating at up to a 200 MHz pixel rate, identifying where likely edges are in an image. The raw color video is read from a neighboring FPGA over a parallel link, and the image intensity is then calculated, after which two 3×3 convolutional Sobel operator filters identify horizontal and vertical edges; the sum of their absolute values indicates the relative strength of a feature edge in an image. A runtime programmable gain (variable multiplier) followed by an adjustable threshold maps the resulting pixel stream to binary levels to indicate if a given pixel is labeled as an edge of a visible feature. The resulting video mask is then optionally mixed with the original color image and displayed on a monitor.

Before designing the datapaths in the edge detection system, the data and control specification for the video stream sources and sinks must be defined. By convention, stream-based architectures are implemented by pairing data samples with corresponding control tags and maintaining this pairing through the architecture. For this example, the video datastreams may have varying data types as the signals are processed whereas the control tags are synchronization signals that track the pipeline delays in the video stream. The input video stream and output display stream represent color pixel data using 16 bits—5 bits for red, 6 bits for green, and 5 bits for blue unsigned pixel intensity values. Intermediate values might represent video data as 8-bit grayscale intensity values or as 1-bit threshold detection mask values.

As the datastreams flow through the signal-processing datapath, the operators execute at a constant 100 MHz sample rate, with varying pipeline delays through the system. The data, however, may arrive at less than 100 MHz, requiring a corresponding `enable` signal to tag valid data. Additionally, `hsync`, `vsync`, and `msync` signals are defined to be true for the first pixel of each row, frame, and movie sequence, respectively, allowing a large variety of video stream formats to be supported by the same design.

Once a streaming format has been specified, library components can be developed that forward a video stream through a variety of operators to create higher-level functions while maintaining valid, pipeline-delayed synchronization signals. For blocks with a pipeline latency that is determined by mask parameters, the synchronization signals must also be delayed based on the mask parameters so that the resulting synchronization signals match the processed datastream.

10.2.1 Converting RGB Video to Grayscale

The first step in this example is to generate a grayscale video stream from the RGB input data. The data is converted to intensity using the NTSC RGB-to-Y matrix:

$$Y = 0.3 \times \text{red} + 0.59 \times \text{green} + 0.11 \times \text{blue}$$

This formula is implemented explicitly as a block diagram, shown in Figure 10.1, using constant gain blocks followed by adders. The constant multiplication values are defined as floating-point values and are converted to fixed point according to mask parameters in the gain model. This allows the precision of the multiplication to be defined separately from the gain, leaving the synthesis tools to choose an implementation. The scaled results are then summed with an explicit adder tree.

Note that if the first adder introduces a latency of `adder_delay` clock cycles, the b input to the second adder, `add2`, must also be delayed by `adder_delay` cycles to maintain the cycle alignment of the RGB data. Both the `Delay1` block and the `add1` block have a subsystem mask parameter defining the delay that the block will introduce, provided by the mask parameter dialog as shown in Figure 10.2. Similarly, the synchronization signals must be delayed by three cycles corresponding to one cycle for the gain blocks, one cycle for the first adder, and one cycle for the second adder. By designing subsystems with configurable delays and data precision parameters, library components can be developed to encourage reuse of design elements.

10.2.2 Two-dimensional Video Filtering

The next major block following the RGB-to-grayscale conversion is the edge detection filter itself (Figure 10.3), consisting of two pixel row delay lines, two 3 × 3 kernels, and a simplified magnitude detector. The delay lines store the two rows of pixels preceding the current row of video data, providing three streams of vertically aligned pixels that are connected to the two 3 × 3 filters—the first one detecting horizontal edges and the second detecting vertical edges. These filters produce two signed fixed-point streams of pixel values, approximating the edge gradients in the source video image.

On every clock cycle, two 3 × 3 convolution kernels must be calculated, requiring several parallel operators. The operators implement the following convolution kernels:

Sobel *X* Gradient:

–1	0	+1
–2	0	+2
–1	0	+1

Sobel *Y* Gradient:

+1	+2	+1
0	0	0
–1	–2	–1

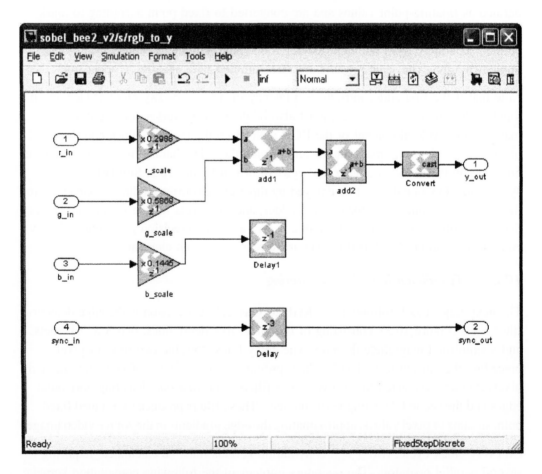

Figure 10.1: An RGB-to-Y (intensity) Simulink diagram

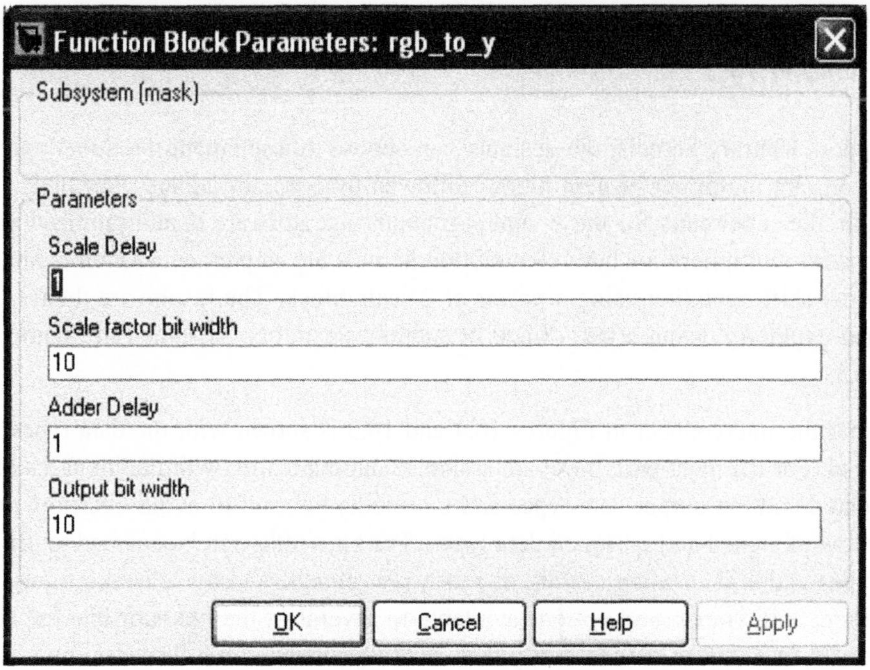

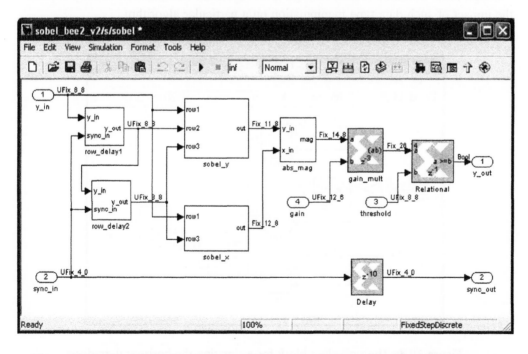

Figure 10.2: A dialog describing mask parameters for the `rgb_to_y` block

Figure 10.3: The Sobel edge detection filter, processing an 8-bit video datastream to produce a stream of Boolean values indicating edges in the image

To support arbitrary kernels, the designer can choose to implement the Sobel operators using constant multiplier or gain blocks followed by a tree of adders. For this example, the subcircuits for the *x*- and *y*-gradient operators are hand-optimized so that the nonzero multipliers for both convolution kernels are implemented with a single hardwired shift operation using a power-of-2 scale block. The results are then summed explicitly, using a tree of add or subtract operators, as shown in Figures 10.4 and 10.5.

Note that the interconnect in Figures 10.4 and 10.5 is shown with the data types displayed. For the most part, these are assigned automatically, with the input data types propagated and the output data types and bit widths inferred to avoid overflow or underflow of signed and unsigned data types. The bit widths can be coerced to different data types and widths using casting or reinterpret blocks, and by selecting saturation, truncation, and wraparound options available to several of the operator blocks. The designer must exercise care to verify that such adjustments to a design do not change the behavior of the algorithm.

Through these Simulink features a high-level algorithm designer can directly explore the impact of such data type manipulation on a particular algorithm.

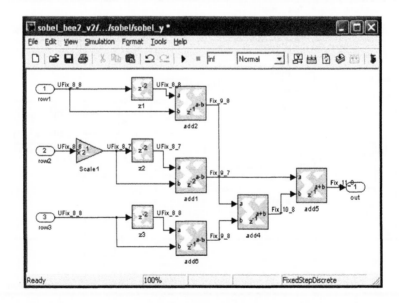

Figure 10.4: The `sobel_y` **block for estimating the horizontal gradient in the source image**

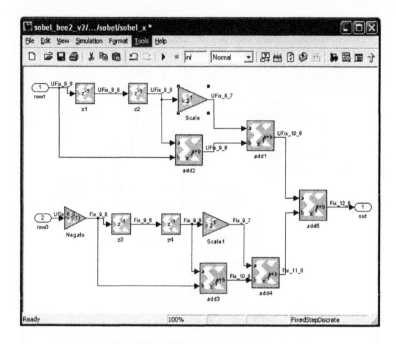

Figure 10.5: The `sobel_x` **block for estimating the vertical gradient in the source image**

Once the horizontal and vertical intensity gradients are calculated for the neighborhood around a given pixel, the likelihood that the pixel is near the boundary of a feature can be calculated. To label a pixel as a likely edge of a feature in the image, the magnitude of the gradients is approximated and the resulting nonnegative value is scaled and compared to a given threshold. The magnitude is approximated by summing the absolute values of the horizontal and vertical edge gradients, which, although simpler than the exact magnitude calculation, gives a result adequate for our applications.

A multiplier and a comparator follow the magnitude function to adjust the sensitivity to image noise and lighting changes, respectively, resulting in a 1-bit mask that is nonzero if the input pixel is determined to be near the edge of a feature. To allow the user to adjust the gain and threshold values interactively, the values are connected to gain and threshold input ports on the filter (see Figure 10.6).

To display the resulting edge mask, an overlay datapath follows the edge mask stream, allowing the mask to be recombined with the input RGB (red, green, blue) signal in a variety of ways to demonstrate the functionality of the system in real time.

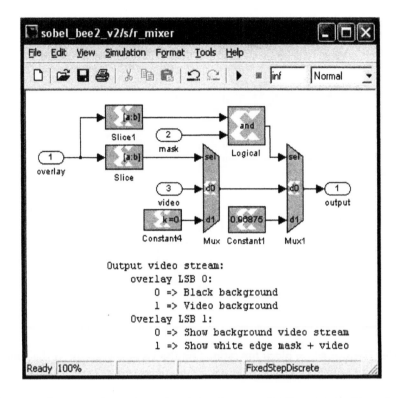

Figure 10.6: One of three video mixers for choosing displays of the filtered results

The overlay input is read as a 2-bit value, where the LSB 0 bit selects whether the background of the image is black or the original RGB, and the LSB 1 bit selects whether or not the mask is displayed as a white overlay on the background. Three of these mixer subsystems are used in the main video-filtering subsystem, one for each of the red, green, and blue video source components.

The three stream-based filtering subsystems are combined into a single subsystem, with color video in and color video out, as shown in Figure 10.7. Note that the color data fed straight through to the red, green, and blue mixers is delayed. The delay, 13 clock cycles in this case, corresponds to the pipeline delay through both the rgb_to_y block and the Sobel edge detection filter itself. This is to ensure that the background original image data is aligned with the corresponding pixel results from the filter. The sync signals are also delayed, but this is propagated through the filtering blocks and does not require additional delays.

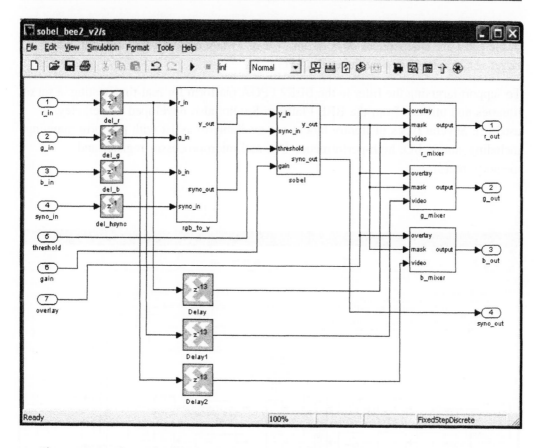

Figure 10.7: The main filtering subsystem, with RGB-to-Y, Sobel, and mixer blocks

10.2.3   Mapping the Video Filter to the BEE2 FPGA Platform

Our design, up to this point, is platform independent—any Xilinx component supported by the System Generator commercial design flow can be targeted. The next step is to map the design to the BEE2 platform—a multiple-FPGA design, developed at UC Berkeley [4], that contains memory to store a stream of video data and an HDMI interface to output that data to a high-resolution monitor.

For the Sobel edge detection design, some ports are for video datastreams and others are for control over runtime parameters. The three user-controllable inputs to the filtering subsystem, threshold, gain, and overlay are connected to external input ports, for connection to the top-level testbench. The filter, included as a subsystem of this testbench design, is shown in Figures 10.8 and 10.9. So far, the library primitives used in the filter are independent of both the type of FPGA that will be used and the target testing platform containing the FPGA.

To support targeting the filter to the BEE2 FPGA platform for real-time testing, a set of libraries and utilities from the BEE Platform Studio, also developed at Berkeley, is used [5]. Several types of library blocks are available to assist with platform mapping, including simple I/O, high-performance I/O, and microprocessor register and memory interfaces.

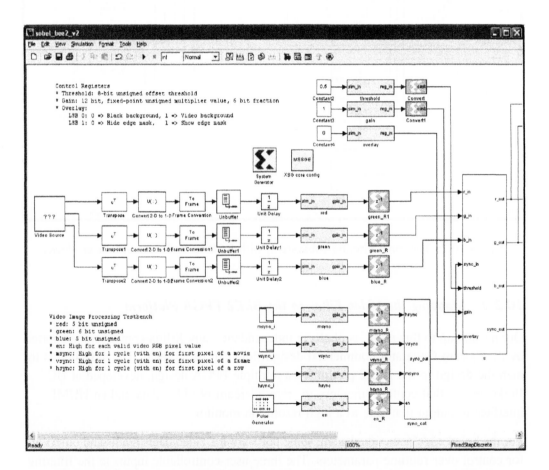

Figure 10.8: The top-level video testbench, with input, microprocessor register, and configuration blocks

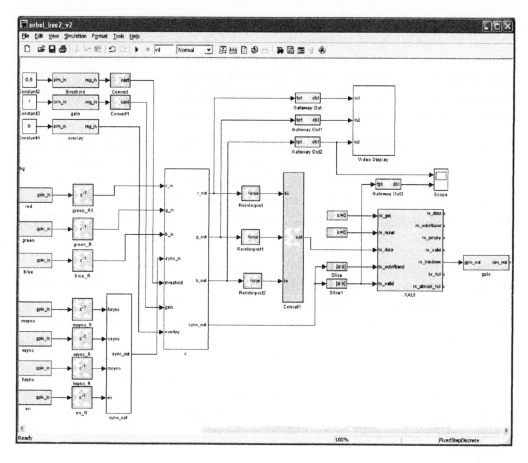

Figure 10.9: The output section of the top-level testbench, with a 10G XAUI interface block

The strategy for using the Simulink blocks to map a design to an FPGA assumes that a clear boundary is defined to determine which operators are mapped to the FPGA hardware and which are for simulation only. The commercial tools and design flows for generating FPGA bit files assume that there are input and output library blocks that appear to Simulink as, respectively, double-precision to fixed-point conversion and fixed-point to double type conversion blocks. For simulation purposes, these blocks allow the hardware description to be simulated with a software testbench to verify basic functionality before mapping the design to hardware. They also allow the designer to assign the FPGA pin locations for the final configuration files.

The BEE Platform Studio (BPS) [5] provides additional I/O blocks that allow the designer to select pin locations symbolically, choosing pins that are hardwired to other FPGAs, LEDs, and external connections on the platform. The designer is only required to select a platform by setting BPS block parameters, and does not need to keep track of I/O pin locations. This feature allows the designer to experiment with architectural tradeoffs without becoming a hardware expert.

In addition to the basic I/O abstractions, the BPS allows high-performance or analog I/O devices to be designed into a system using high-level abstractions. For the video-testing example, a 10 Gbit XAUI I/O block is used to output the color video stream to platform-specific external interfaces. The designer selects the port to be used on the actual platform from a pulldown menu of available names, hiding most implementation details.

A third category of platform-specific I/O enables communication with embedded microprocessors, such as the Xilinx MicroBlaze soft processor core or the embedded PowerPC available on several FPGAs. Rather than describe the details of the microprocessor subsystem, the designer simply selects which processor on a given platform will be used and a preconfigured platform-specific microprocessor subsystem is then generated and included in the FPGA configuration files. For the video filter example, three microprocessor registers are assigned and connected to the threshold, gain, and overlap inputs to the filter using general-purpose I/O (GPIO) blocks. When the BPS design flow is run, these CPU register blocks are mapped to GPIO registers on the selected platform, and C header files are created to define the memory addresses for the registers.

10.3 Specifying Control in Simulink

On the one hand, Simulink is well suited to describing highly pipelined streambased systems with minimal control overhead, such as the video with synchronization signals described in the earlier video filter example. These designs assume that each dataflow operator is essentially running in parallel, at the full clock rate. On the other hand, control tasks, such as state machines, tend to be inherently sequential and can be more challenging to describe efficiently in Simulink. Approaches to describing control include:

- Counters, registers, and logic to describe controllers
- Matlab M-code descriptions of control blocks

- VHDL or Verilog hand-coded or compiled descriptions

- Embedded microprocessors

To explore the design of control along with a stream-based datapath, consider the implementation of a synchronous delay line based on a single-port memory. The approach described here is to alternate between writing two data samples and reading two data samples on consecutive clock cycles. A simpler design could be implemented using dual-port memory on an FPGA, but the one we are using allows custom SOC designs to use higher-density single-port memory blocks.

10.3.1 Explicit Controller Design with Simulink Blocks

The complete synchronous delay line is shown in Figure 10.10. The control in this case is designed around a counter block, where the least significant bit selects between the two words read or written from the memory on a given cycle and the upper counter bits determine the memory address. In addition to the counter, control-related blocks include *slice* blocks to select bit fields and Boolean *logic* blocks. For this design,

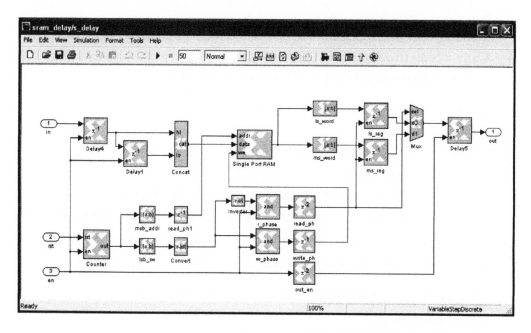

Figure 10.10: A simple datapath with associated explicit control

the block diagram is effective for describing control, but minor changes to the controller can require substantial redesign.

10.3.2 Controller Design Using the Matlab M Language

For a more symbolic description of the synchronous delay line controller, the designer can use the Matlab "M" language to define the behavior of a block, with the same controller described previously written as a Matlab function. Consider the code in Listing 10.1 that is saved in the file sram_delay_cntl.m.

Listing 10.1: The delay line controller described with the Matlab function
sram_delay_cntl.m

```
function [addr, we, sel] = sram_delay_cntl(rst, en, counter_bits,
counter_max)
% sram_delay_cntl - MCode implementation block.
% Author: Brian Richards, 11/16/2005, U. C. Berkeley
%
% The following Function Parameter Bindings should be declared in
% the MCode block Parameters (sample integer values are given):
    % {'counter_bits', 9, 'counter_max', 5}

% Define all registers as persistent variables.
persistent count,
    count = xl_state(0, {xlUnsigned, counter_bits, 0});
persistent addr_reg,
    addr_reg = xl_state(0, {xlUnsigned, counter_bits-1, 0});
persistent we_reg,    we_reg   = xl_state(0, {xlBoolean});
persistent sel_reg_1, sel_reg_1 = xl_state(0, {xlBoolean});
persistent sel_reg_2, sel_reg_2 = xl_state(0, {xlBoolean});

% Delay the counter output, and split the lsb from
% the upper bits.
addr = addr_reg;
```

```
addr_reg = xl_slice(count, counter_bits-1, 1);
count_lsb = xfix({xlBoolean}, xl_slice(count, 0, 0));

% Write-enable logic
we = we_reg;
we_reg = count_lsb & en;

% MSB–LSB select logic
sel = sel_reg_2;
sel_reg_2 = sel_reg_1;
sel_reg_1 = ~count_lsb & en;

% Update the address counter:
if (rst | (en & (count == counter_max)))
    count = 0;
elseif (en)
    count = count + 1;
else
    count = count;
end
```

To add the preceding controller to a design, the Xilinx M-code block can be dragged from the Simulink library browser and added to the subsystem. A dialog box then asks the designer to select the file containing the M source code, and the block sram_delay_cntl is automatically created and added to the system (see Figure 10.11).

There are several advantages to using the M-code description compared to its explicit block diagram equivalent. First, large, complex state machines can be described and documented efficiently using the sequential M language. Second, the resulting design will typically run faster in Simulink because many finegrained blocks are replaced by a single block. Third, the design is mapped to an FPGA by generating an equivalent VHDL RTL description and synthesizing the resulting controller; the synthesis tools can produce different results depending on power, area, and speed constraints, and can optimize for different FPGA families.

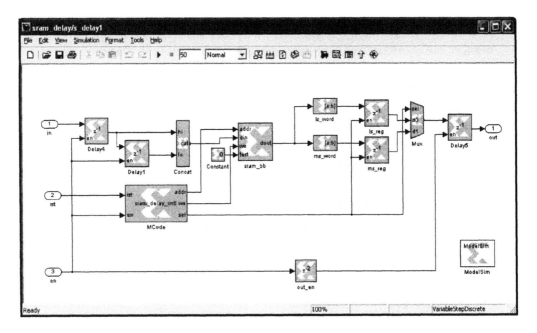

Figure 10.11: A simple datapath using a controller described in Matlab code

10.3.3   Controller Design Using VHDL or Verilog

As in the M language approach just described, a controller can also be described with a *black box* containing VHDL or Verilog source code. This approach can be used for both control and datapath subsystems and has the benefit of allowing IP to be included in a Simulink design.

The VHDL or Verilog subsystems must be written according to design conventions to ensure that the subsystem can be mapped to hardware. Clocks and enables, for example, do not appear on the generated Simulink block, but must be defined in pairs (e.g., clk_sg, ce_sg) for each implied data rate in the system. Simulink designs that use these VHDL or Verilog subsystems can be verified by cosimulation between Simulink and an external HDL simulator, such as Modelsim [6]. Ultimately, the same description can be mapped to hardware, assuming that the hardware description is synthesizable.

10.3.4   Controller Design Using Embedded Microprocessors

The most elaborate controller for an FPGA is the embedded microprocessor. In this case, control can be defined by running compiled or interpreted programs on the microprocessor. On the BEE2 platform, a tiny shell can be used interactively to control datapath settings, or a custom C-based program can be built using automatically generated header files to symbolically reference hardware devices.

A controller implemented using an embedded microprocessor is often much slower than the associated datapath hardware, perhaps taking several clock cycles to change control parameters. This is useful for adjusting parameters that do not change frequently, such as threshold, gain, and overlay in the Sobel filter. The BEE Platform Studio design flow uses the Xilinx Embedded Development Kit (EDK) to generate a controller running a command line shell, which allows the user to read and modify configuration registers and memory blocks within the FPGA design. Depending on the platform, this controller can be accessed via a serial port, a network connection, or another interface port.

The same embedded controller can also serve as a source or sink for low-bandwidth datastreams. An example of a user-friendly interface to such a source or sink is a set of Linux 4.2 kernel extensions developed as part of the BEE operating system, BORPH [7]. BORPH defines the notion of a hardware process, where a bit file and associated interface information is encapsulated in an executable .bof file. When launched from the Linux command line, a software process is started that programs and then communicates with the embedded processor on a selected FPGA. To the end user, hardware sources and sinks in Simulink are mapped to Linux files or pipes, including standard input and standard output. These file interfaces can then be accessed as software streams to read from or write to a stream-based FPGA design for debugging purposes or for applications with low-bandwidth continuous datastreams.

10.4   Component Reuse: Libraries of Simple and Complex Subsystems

In the previous sections, low-level primitives were described for implementing simple datapath and control subsystems and mapping them to FPGAs. To make this methodology attractive to the algorithm developer and system architect, all of these capabilities are combined to create reusable library components, which can be parameterized for a variety of applications; many of them have been tested in a variety of applications.

10.4.1   Signal-processing Primitives

One example of a rich library developed for the BPS is the Astronomy library, which was codeveloped by UC Berkeley and the Space Sciences Laboratory [8,9] for use in a variety of high-performance radio astronomy applications. In its simplest form, this library comprises a variety of complex-valued operators based on Xilinx System Generator real-valued primitives. These blocks are implemented as Simulink subsystems with optional parameters defining latency or data type constraints.

10.4.2   Tiled Subsystems

To enable the development of more sophisticated library components, Simulink supports the use of Matlab M language programs to create or modify the schematic within a subsystem based on parameters passed to the block. With the Simulink Mask Editor, initialization code can be added to a subsystem to place other Simulink blocks and to add interconnect to define a broad range of implementations for a single library component.

Figure 10.12 illustrates an example of a tiled cell, the `biplex_core` FFT block, which accepts several implementation parameters. The first parameters define the size and precision of the FFT operator, followed by the quantization behavior (truncation or rounding) and the overflow behavior of adders (saturation or wrapping). The pipeline latencies of addition and multiplication operators are also user selectable within the subsystem.

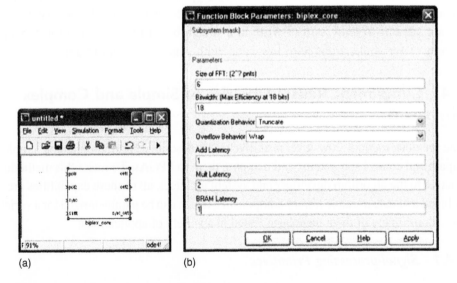

(a)   (b)

Figure 10.12: The biplex—core dual-channel FFT block (a), with the parameter dialog box (b)

Automatically tiled library components can conditionally use different subsystems, and can have multiple tiling dimensions. An alternative to the streambased `biplex_core` block shown in Figure 10.13, a parallel FFT implementation, is also available, where the number of I/O ports changes with the FFT size parameter. An 8-input, 8-output version is illustrated in Figure 10.14. The parallel FFT tiles butterfly subsystems in two dimensions and includes parameterized pipeline registers so that the designer can explore speed versus pipeline latency tradeoffs.

In addition to the FFT, other commonly used high-level components include a poly-phase filter bank (PFB), data delay and reordering blocks, adder trees, correlator functions, and FIR filter implementations. Combining these platform-independent subsystems with the BPS I/O and processor interface library described in Section 10.2.3, an algorithm designer can take an active role in the architectural development of high-performance stream-based signal-processing applications.

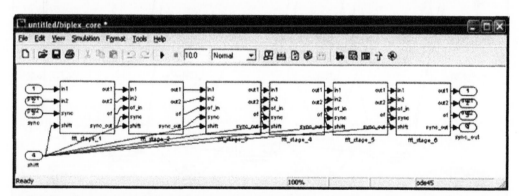

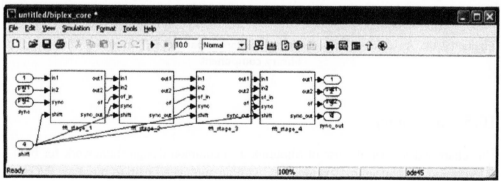

Figure 10.13: Two versions of the model schematic for the `biplex_core` library component, with the size of the FFT set to 6 (2^6) and 4 (2^4). The schematic changes dynamically as the parameter is adjusted

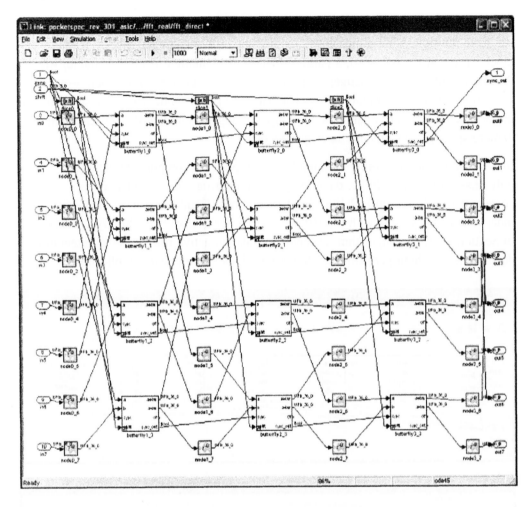

Figure 10.14: An automatically generated 8-channel parallel FFT from the `fft_direct` library component

10.5   Summary

This chapter described the use of Simulink as a common design framework for both algorithm and architecture development, with an automated path to program FPGA platforms. This capability, combined with a rich library of high-performance parameterized stream-based DSP components, allows new applications to be developed and tested quickly.

The real-time Sobel video edge detection described in this chapter runs on the BEE2 platform, shown in Figure 10.15, which has a dedicated LCD monitor connected to it. Two filtered video samples are shown, with edges displayed with and without the original source color video image.

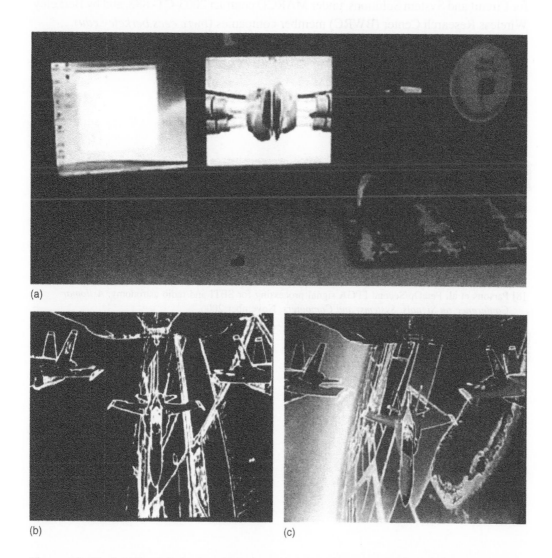

Figure 10.15: (a) The Sobel edge detection filter running on the BEE2, showing the BEE2 console and video output on two LCD displays, with (b, c) two examples of edge detection results based on interactive user configuration from the console

For more information on the BPS and related software, visit *http://bee2. eecs.berkeley. edu*, and for examples of high-performance stream-based library components, see the Casper Project [9].

Acknowledgments This work was funded in part by C2S2, the MARCO Focus Center for Circuit and System Solutions, under MARCO contract 2003-CT- 888, and by Berkeley Wireless Research Center (BWRC) member companies (*bwrc.eecs.berkeley.edu*). The BEE Platform Studio development was done jointly with the Casper group at the Space Sciences Laboratory (*ssl.berkeley.edu/casper*).

References

[1] *http://www.mathworks.com.*
[2] *http://www.xilinx.com.*
[3] *http://www.synplicity.com.*
[4] Chang, J. Wawrzynek, R.W. Brodersen. BEE2: A high-end reconfigurable computing system. *IEEE Design and Test of Computers* 22(2), March/April 2005.
[5] C. Chang. *Design and Applications of a Reconfigurable Computing System for High Performance Digital Signal Processing*, Ph.D. thesis, University of California, Berkeley, 2005.
[6] http://www.mentor.com.
[7] K. Camera, H. K.-H. So, R. W. Brodersen. An integrated debugging environment for reprogrammble hardware systems. *Sixth International Symposium on Automated and Analysis-Driven Debugging*, September, 2005.
[8] Parsons et al. PetaOp/Second FPGA signal processing for SETI and radio astronomy. *Asilomar Conference on Signals, Systems, and Computers*, November 2006.
[9] *http://casper.berkeley.edu/papers/asilomar—2006.pdf.*

Index